Introduction to the Senses

From Biology to Computer Science

An understanding of the senses – vision, hearing, touch, chemical and other non-human senses – is important not only for many fields of biology but also in applied areas such as human computer interaction, robotics and computer games. Using information theory as a unifying framework, this is a wide-ranging survey of sensory systems, covering all known senses.

The book draws on three unifying principles to examine senses: the Nyquist sampling theorem; Shannon's information theory; and the creation of different streams of information to subserve different tasks. This framework is used to discuss the fascinating role of sensory adaptation in the context of environment and lifestyle. Providing a fundamental grounding in sensory perception, the book then demonstrates how this knowledge can be applied to the design of human–computer interfaces and virtual environments. It is an ideal resource for both graduate and undergraduate students of biology, engineering (robotics) and computer science.

Terry R.J. Bossomaier is Professor in Computer Systems and Director of the Centre for Research in Complex Systems (CRiCS) at Charles Sturt University, Bathurst, Australia. He is particularly interested in sensory information in the context of vision and hearing, and was involved in the creation of the first degree course in computer games in Australia in 2000.

Introduction to the Senses

From Biology to Computer Science

TERRY R.J. BOSSOMAIER

Charles Sturt University, Bathurst, Australia

CAMBRIDGE UNIVERSITY PRESS
Cambridge, New York, Melbourne, Madrid, Cape Town,
Singapore, São Paulo, Delhi, Mexico City

Cambridge University Press
The Edinburgh Building, Cambridge CB2 8RU, UK

Published in the United States of America by Cambridge University Press, New York

www.cambridge.org
Information on this title: www.cambridge.org/9780521812665

First published 2012

Printed in the United Kingdom at the University Press, Cambridge

A catalogue record for this publication is available from the British Library

Library of Congress Cataloguing in Publication data
Bossomaier, Terry R. J. (Terry Richard John)
Introduction to the senses : from biology to computer science / Terry Bossomaier.
    p.   cm.
ISBN 978-0-521-81266-5 (hardback)
1. Senses and sensation.   2. Human-machine systems.   3. Perception.   I. Title.
QP431.B59   2012
612.8084′6 – dc23   2012006050

ISBN 978-0-521-81266-5 Hardback

Contents

Colour plates are to be found between pp. 268 and 269.

Foreword

What are the commonalities of information gathering and processing in all living creatures? This is the implicit question that underpins Terry Bossomaier's ambitious book *The Senses*. His is a Herculean task and one to be greatly applauded.

Bossomaier addresses the senses using the tools of contemporary information science, in an attempt to provide a unifying perspective, one that allows for quantitative comparison of senses between the species.

This fascinates me. It is now nearly 35 years since Simon Laughlin, Doekele Stavenga and I introduced information theory to understand the design of eyes, both compound eyes of insects as well as the simple eyes of humans. We recognised that the fundamental limitations to resolving the power of eyes are the wave (diffraction) and particle (photon noise) nature of light. By appreciating their interrelation we derived insight into the design and limitations of eyes, especially between the optical image quality and the visual photoreceptor mosaic. The capacity of the eye to perceive its spatial environment was quantified by determining the number of different pictures that can be reconstructed by its array of visual cells. We were then able to decide on the best compromise between an animal's capacity for fine detail and contrast sensitivity. In a series of papers, including those with Bossomaier and A. Hughes, we went on to use the tools of information theory to study various aspects of eye design. It was a rewarding and rich endeavour, but one at that time limited to vision.

Now, in this book you will read how Bossomaier significantly generalises these early applications of information theory and applies them to all the senses, not just to vision.

But, Bossomaier not only stresses the importance of information theory in sensory processing, he also stresses the importance of contextualising it to the animal's lifestyle. Indeed, as I have often emphasised, there is no single optimum design of any particular sense as physicists might like, but rather a variety of different solutions, each dictated by the animal's lifestyle within its environment. Two animals can share the same habitat, yet they can have radically different eyes. This realisation is made all the more transparent by comparative studies of many species besides humans. It is fascinating to learn how diverse animals adapt and make use of the physical properties of the environment, from the earth's magnetic field to the sky's polarisation, from heat, electricity and sound to the odours associated with chemicals and how these continue to inspire the implementation of sensory apparatus in man-made devices such as robots. Interestingly in this regard, it was the birefringent nature of the fly's visual photoreceptors that provided an early inspiration for single mode, single polarisation optical fibres for telecommunication and sensor applications.

Most importantly Bossomaier stresses that information is not passively collected by the senses, but instead is profoundly influenced by what the animal already knows and anticipates. So, animals perceive not only what they sense. Sensory detail can be modulated and even suppressed by higher level concepts. This remains a crucially important area of research and one which consumes my attention.

Indeed, over the last decade or so our thinking about the general strategy of sensory processing and the crucial role of top-down processing has been undergoing a paradigm shift. As Bossomaier says, 'Across many thousands of studies lies the assumption that information is processed sequentially, from sensors through to high level perception deep within the brain. Yet sensory systems abound in recurrent connections, which strongly suggest that the situation is far more complex.'

One of the most enjoyable aspects of the book is its comprehensive survey of the senses and the alluring accounts of the way they are adapted to remarkable advantage by animals. This no doubt is contributed to by the fact that Bossomaier approaches the subject from the left field, with both enthusiasm and receptivity to new ideas but admirably armed with the necessary scientific tools and panoramic knowledge of the senses to impart deep insight.

Professor Allan Snyder FRS
Director of the Centre for the Mind and
150th Anniversary Chair at the University of Sydney

Acknowledgements

The ideas for this book go back a long way. An interest in photography, images and the sounds of musical instruments goes right back to my father. My friends and colleagues at Ilford Limited put my understanding of image physics and perception on a firm footing, but John Lewis and Alan Kragh were especially motivating and influential.

The vision community at the Australian National University was a uniquely stimulating environment, with exciting discussions with many people, too numerous to list, but my grasp of sensory processing would not have been possible without Peter Bishop, Geof Henry, Abbie Hughes, Bill Levick, Ted Maddess, Daniel Osorio, Srini Srinivasan and Allan Snyder.

Thanks also to the many computer science games students who read the drafts of this book. Finally, as someone passionate about imagery but with no obvious drawing skills, some help with diagram production proved essential. Rebecca Wotzko fulfilled the role admirably with remarkable patience.

1 Introduction and overview

The human brain is the most complex phenomenon in the known universe.
John Eccles (Popper & Eccles, 1977)

A visitor from the mild climate of the UK to Rochester, New York State, in the middle of summer, receives a sensory shock. Apart from being much, much warmer, the visceral impact is huge. The light is brighter, the colouring of the birds is dramatic and the scent of the trees and plants is just so strong. At night the circadias are almost deafening. The information about the world gathered by sensory systems is the core idea this book will explore.

It so happens that Rochester is the home of Eastman Kodak and other major imaging companies such as Xerox and Bausch and Lomb, the place where a lot of important research on image physics, capture and storage took place. Images have played a part in human culture since the earliest cave paintings, but as computers have got faster, dynamic images, from mobile phones to giant plasma displays increasingly dominate our lives. Reproduced sound has gone beyond the radio to the ubiquitous MP3 player, seen on countless commuters, runners and diverse workers. But there are other senses, not yet so widespread in the artificial world. This book lies at the interface of the sensory biological world and man-made systems. It also projects forward to new computer interfaces and virtual environments not far down the track.

The senses of many animals, especially human beings, are very powerful general purpose information-seeking systems. A cat soon learns to recognise the sound of metal on metal of tin opener on tin or the sheepdog the distinctive whistle from which he receives his instructions. Yet neither sound would have been likely to occur in feline or canine evolution until the last century. But conversely, all sensory systems are finely tuned to lifestyle requirements. A cat can hear sounds of higher frequency and at lower intensity but lacks human trichromatic vision. A bloodhound has much higher odour sensitivity. An eagle has higher visual resolution, in fact the best measured resolution of the animal kingdom. General purpose does not imply the absence of specialisation.

But information gathering from the environment is not the only objective. Specialised tuning for animal communication is widespread throughout nature. In the case of man, the sound frequencies of speech occur in the region of greatest sensitivity of human hearing. But in many animals' sensory mechanisms the signalling systems are even more precise and monofunctional. For example:

- Chemical signals such as the pheromones used in mating (§9.4.1). Ants are the pheromone kings. They leave chemical trails which help with navigation on social organisation using about 20 chemicals, or pheromones, for indicating food sources and other things useful to the anthill. Here the olfactory sense is tightly coupled to particular molecules which are generated by the animal.
- Symbiotic development of colour in flowers and insect visual systems for optimum pollination (§8.6.3).
- Danger signals such as the bitterness of alkaloids (§9.5) in plants or the bright colours of the coral snake.
- The bright colours of ripe fruit or young leaves (§8.6).
- Fireflies, which synchronise their flashes to produce remarkably pulsating clouds (Strogatz, 2003), or the deafening sound of circadias at night. Sometimes these signals are symbiotic. Insects and plants coevolve spectral and pattern cues to control pollination and access to nectar (§8.6.3). The sound of the circadias and the flashing of fireflies at night are specific communication signals.
- Warning signals such as the tuning of moth to bat (predator) echo-location signals and cricket to beating of wings of a predatory wasp (§5.4.3.1).

The senses form the fabric of our awareness, our consciousness, of the world. Some people, such as Ray Jackendoff (1987), argue that sensory inputs are our primary form of consciousness, an issue to which we return in the last chapter. Thus it is perhaps hard to realise what we are missing or to grasp what it is like to be with an altered sense, except perhaps when we lose our glasses, get a thick head cold or ear infection. But some animals use cues for which we do not have any senses at all (or at any rate they have atrophied or are beyond conscious awareness). Platypus, for example, use electrical signals (§11.1) while birds and turtles use magnetic cues as if they were following a compass (§11.3).

Animals have several quite distinct senses[1], but the central idea of information gathering fuses with another one of *information streaming*, splitting sensory input into multiple quasi-independent channels.

Now imagine that we enter a time machine and fast forward a hundred years or so. What will the senses look like then? Unfortunately, many animals may be extinct and Jurassic Park DNA technology may not have been successful at bringing them back. Some sensory mechanisms may have died with them. But we can be fairly certain that these sensory modalities will still be around. Robots, already taking their first tentative steps, in labs around the worlds, and even vacuuming our floors, will possess specially engineered senses. Some may be unlike anything in the biological world (imagine a robot for cleaning up radiation). But insatiable demand for computer games will also bring us creatures which operate in entirely virtual worlds (*animats*). Such creatures may even exist in virtual worlds with different fundamental physics, in the manner of George Gamow's Dreams of Mr Tomkins (Gamow, 1944).

---

[1] One might argue that there are some transitional senses, say, somewhere between hearing and touch, and there are certainly some common mechanisms (hair cells in the ear, whiskers, tactile sensors in the skin).

In today's computer games, the behaviour of creatures, usually referred to as non-player characters, NPCs, is tightly scripted. What an NPC can see at any time and whatever actions it can take depend upon it being given very specific information about the game state. But as computers get ever faster, and as networks of processors get ever more effective at sharing computational load, there is a future where the game has a simulated virtual world where artificial life learns and evolves, a world of animats. Throughout the book the richness and diversity of the animal kingdom on Earth will appear time and time again, a huge resource for building sensors, robots or animats.

Thus our exploration of the senses will take us on a diverse journey, from fundamental physics to the nature of consciousness itself. But, although, according to Lao Tzu, '*A journey of a thousand miles starts with a single step*' (Tao Te Ching, 604 BC–531 BC) let us first take satellite view of the book's geography and features.

Animal senses differ in many ways, both in detailed biology and in functional properties. Thus this book explores two perspectives of sensory systems: the vacuum cleaner sucking in data about the world and the selective generation and reception of precise signals. But the understanding of biological senses is a huge undertaking, thus to make the book tractable, the perspective is one of *extracting and processing information from the environment*, trying to make this quantitative wherever possible. This perspective provides a theoretical and illustrative framework for building artificial sensors, understanding human–computer interaction and creating artificial creatures for real or virtual worlds.

Aside from describing and building theories of sensory information processing, this perspective subserves two agendas. One is the building of an animat in a game or virtual world, such as the fish in Terzopolous virtual aquarium (Terzopoulos *et al.*, 1994). It is the newest: very few attempts have been made in this direction so far. In the Will Wright game, Spore, players build their own life forms with control over mesh, texture and animation. Future developments where these life forms have realistic senses are not hard to imagine.

The other is that of providing enough information for the human senses to be satisfied with a synthetic representation. This is old technology for vision and hearing. The telephone system has a frequency and volume range just sufficient to make speech intelligible. Movies set the frame rate at the minimum possible to avoid jerkiness from frame to frame. In the very first movies, the frame rate wasn't quite high enough, while today, the resolution of image displays still needs to get better, while the problems of managing room acoustics are still from solved in the consumer domain.

But the computational applications of the other senses are not so well developed. Smell and taste systems are only just beginning to hit the market at the beginning of the millenium. Haptic sensors, data gloves and other devices have been around for longer, yet they are not commonplace and are still relatively expensive. Tactile feedback systems are appearing in computer games, and here we see an important trend. The games market is huge, upwards of US $20 billion aready, and rising rapidly. It now drives graphics hardware and is likely to drive pretty much all these other research tools into commercial use.

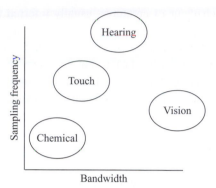

Figure 1.1 Illustration of how senses vary in sampling frequency and bandwidth. For example, vision has by far the highest data collection rate; hearing samples a much higher rate but takes in less per sample.

Like sensory systems, themselves, the book has three parallel themes:

1. What are the limits to animal senses: what is the sensitivity and dynamic range and where is it at its best and worst in the animal kingdom?
2. What are the bandwidth and computational requirements of the different senses – in other words what do we need to implement such a sense in an artificial life or robotic context; how much storage and bandwidth do we need? In some senses there are precise estimates, down to bit rates per neuron. In others estimates rely on secondary information, such as number of elementary detectors and numbers of neurons. Figure 1.1 illustrates how the senses vary in total information bandwidth and in the frequency with which they sample the environment.
3. How is information streamed into different categories or functional components? Different streams may subserve different functions, e.g. colour is not used extensively in human spatial navigation. Building a robot or animat can exploit such streaming by selecting only the processing streams necessary for its behaviour. Two predominant streams are the *where* and *what*, which anatomically become the *dorsal* and *ventral* streams (§2.3.2).

Intertwined within these themes are three other ideas: (i) signal representation; (ii) metric properties; and (iii) constancies and adaptation. An animal's lifestyle within an environment defines a useful bandwidth and some degree of specialisation (Snyder *et al.*, 1988; Zeki, 1999). So a rabbit's retina is quite different to a cat's, reflecting their herbivore/prey, carnivore/hunter lifestyles, even though they are of similar size and inhabit similar worlds. An example of this is the *scaling principle*, described by Snyder and Miller for *Falconiformes*, birds of prey such as eagles and hawks. The higher the bird flies (and usually the larger it is, with eagles at the top), the greater the visual resolution it needs to detect prey on the ground. This leads to an elegant scaling of focal length of lens in the eye with altitude of flight (Snyder & Miller, 1978).

Two mechanisms which occur repeatedly in biological senses are:

- **Inhibition,** the blocking of the activity of some cells by others, is frequently the mechanism by which cell populations achieve their collective goals (§2.4.2).
- **Adaptation,** the adjusting of sensitivity to match the input level just as we set the ISO sensitivity on a camera or the signal going to a microphone.

In part these arise from the characteristics of neural processing, such as the energy cost of neural activity (§2.4.3), but they have also found their way into artificial systems for computational speed and efficiency.

The senses have some built-in elements, but they almost always embody some rapid development in the very early stages after birth, as soon as they receive input from the external world. This period of rapid change occurs during a *critical period* after which further change ceases. There is a lesson for artificial senses here. There are some early transformations of the raw data which can be set up and left indefinitely. Whether this is done by learning in the environment or whether it may be possible to wire up a sensor in advance for the statistics of the environment in which it will work will be somewhat application dependent.

1.1   Structure of the book

The overarching structure of the book is to first examine some general concepts of brain and senses, then to develop the two key theoretical tools of representation and information theory, and finally to examine the senses in turn. The final chapter examines sensory integration. Although biology features strongly throughout, robots, or other man-made systems, also need sensors of the external world and many of the information collection and computation issues are very similar. But a robot will have quite different internal monitoring systems. Thus the main thrust of the book is on the processing of external data coming in from the world, with only a cursory look at the complex sensory networks which monitor the workings of the human body.

An emerging theme of the book is the similarities in organisation and computation within the senses, which will highlight the variations in our knowledge from one domain to another. To make this easier to understand, each sensory modality chapter covers several different dimensions: the physical stimulus, transduction to a neural signal, pathways to the brain, psychophysical properties and artificial systems. The focus is primarily human but other animals appear where theory or experimental data is more detailed.

The chapters that concentrate on a particular sense follow a fairly standard sequence. Sensory systems are complex, with many components, so the first thing to do is to get an overview, followed by four more or less separate dimensions.

The nature of the physical stimulus itself is a good starting point, what sort of sensory data an arbitrary animal (or robot) might use. Knowing the stimulus leads to analysis of the sort of noise to which it is subject and what sort of limitations would arise in its use. This constitutes the *physics* of the sensory data. Usually the physics is well understood.

Given the physics, the challenge is to understand the *transduction and computational* processes, how the incoming signal, such as light or sound, becomes an electrical signal travelling along nerves to the brain. Even this first stage is still far from completely understood. Recent progress has been extremely rapid through the advent of genetic techniques now contributing across many areas of sensory research.

Frequently, detailed understanding is greatest at the periphery, where the data is collected, the eye or the ear. Often our knowledge of the pathways becomes increasingly vague as we progress higher up the information processing, but, again, the new brain imaging techniques have enabled rapid recent developments. Unfortunately, some knowledge of higher level processing is important, because there is extensive feedback and top-down control affecting sensitivity and discrimination. Several examples of recent research in subsequent chapters will show just how important these are, more Popper's searchlight than bucket approach to knowledge (Popper, 1972).

At this point we could then go on to look at the details of how the sensory information is captured and processed, but instead we switch to the viewpoint. In some cases the psychophysics matches the physical limits very closely. In others there is a huge gap.

The *psychophysical and behavioural* viewpoint is quite different. Sometimes the perceptual limits match precisely what the physical limits would predict. In other cases they do not. In most of the senses animals have got close to the physical limits. Yet, the encoding in neural systems is sometimes highly compressed and effective and at others seemingly wasteful. One important characteristic of information extraction is normalisation, or removal of absolute values. This process is normally referred to as the extraction of constancies. So, for example, in vision contrast is preferred over absolute light level, and colour is transformed to be approximately independent of spectral content of the illuminant.

To understand the information processing and streaming of sensory information requires both viewpoints, the physics/transduction/encoding view and the perceptual/ overall signal detection view.

Finally, we look at man's attempts to copy sensory processing systems in robotics and virtual reality. This is the fastest changing aspect of the whole book and web site references are provided where new developments are likely to appear.

The following subsections briefly describe the main content of each chapter.

1.1.1   General concepts

This book owes much to the work of Horace Barlow, direct descendent of Charles Darwin, the unquestioned pioneer of information and signal processing ideas in senses, and leans heavily on his book (Barlow & Mollon, 1982) with John Mollon another pioneer of visual science. Inspired by Barlow, Allan Snyder, Simon Laughlin, Srini Srinivasan and others at the Australian National University further developed the information theoretical framework by which animals capture and encode the environments in which they live in ways suited to their lifestyle. From these early pioneers, the field of sensory information processing continues to grow rapidly spurred by some of the new techniques referred to above.

1.1.2 Mathematical preliminaries

Before beginning our study of the senses themselves, some mathematical preliminaries are neceesary. Chapter 3 tackles the first of these, the idea of representation: how is sensory information transformed, split apart and sent to various parts of the brain. Linear systems are the starting point. It may come as a surprise that in such obviously non-linear systems as nerve cells, so many information processing ideas derive from linear models. Fourier analysis is thus a key technique, important to understanding both vision and hearing, and a useful baseline in the other senses. Section 3.9.3 tackles representation from a different perspective – that of information. How much information can be captured and transmitted and how does its representation affect transmission efficiency? Some of the qualitative discussion of neural coding can now assume more quantitative form.

1.1.3 Hearing

With the mathematical preliminaries out of the way, it is time to start on the senses themselves. Hearing comes first, because it is one-dimensional. The Fourier (frequency) representation is familiar to us in everyday life, particularly in the context of music, and is thus easy to understand.

1.1.4 Vision

A few years ago spatial frequency and resolution would have been obscure concepts to most people. But digital cameras have changed all that. Thus extending frequency analysis to two dimensions for vision sits on top of common sense knowledge of pixels in digital cameras. Likewise a few years ago, image processing operations were the province of a select few, requiring expensive hardware and specialised software. Now image processing software comes free with most digital cameras.

1.1.5 Spatial information and the correspondence principle

Both hearing and vision extract spatial information from having two eyes and two ears. This spatial information comes from establishing *correspondence* between signals from each eye or ear. Because of the different spatial location each eye does not quite see the same image and the same feature is imaged in slightly different places. Similarly a sound from not directly straight ahead or behind will arrive at one ear before the other. Thus the brain needs to establish which signals separated in time and space correspond to one another.

The same correspondence process occurs in tracking moving stimuli. Now the matching has to take place at successive time instants. Thus Chapter 7 treats stereopsis, binaural hearing and movement together.

1.1.6   Colour and texture

The visual system also has two specialised subsystems exclusively concerned with 'what', i.e. object properties. They are colour and texture, treated in Chapter 8. The streaming idea really took shape in the 1980s where it became apparent that colour contributed rather little to spatial processing, be it the delineation of object boundaries, movement or stereopsis. Parallel processing of information, in vision, and elsewhere in the cortex is discussed in detail by Zeki (1993).

1.1.7   Touch: the somatosensory system

Hearing and vision are definitely metric senses in that we have clear quantitative measures for, say, light intensity or frequency amplitude. Touch has strong metric components, but is so diverse and so anisotropic and heterogeneous throughout the body that the linear, Fourier techniques useful in vision and hearing are not quite so useful. Chapter 10 looks at the diversity of the somatosensory and vestibular system. Touch (or haptics) is one of the fastest growing areas of artificial intelligence and now an essential part of computer games.

1.1.8   Chemical senses

The last of the human senses to look at are the chemical senses. Now the metric properties are much weaker. Taste and olfaction are much more like associative memories. It is not even certain that there are any computational building blocks, although it would be very nice to have ways of representing any given scent – wine tasting online! Artifical noses and tongues and encoding of olfactory and tastant elements are open research areas at present.

1.1.9   Non-human senses

But humans do not actually possess all the known animal senses – or certainly not at any conscious or significant level. So this last sense-specific chapter takes a look at some of these oddball adaptations found around the animal world.

1.1.10   Integration

Our conscious experience is an integrated one. The last chapter considers how the senses integrate, sometimes with synergy, sometimes with conflict, sometimes with one cross calibrating the other (as happens with spatial and auditory circuits in the owl), in which part of the mystery of consciousness lies. At the time of writing, there is increasing belief that brain imaging and other techniques will unlock some of the secrets of consciousness. To stray into this dangerous philosophical territory would be foolhardy, but there are concrete, intriguing experimental results, from phantom limbs to out-of-body experiences.

The book covers a wide range of topics, from some elementary anatomy through to the latest results in sensory information processing. The overarching ideas, in addition to specific sensory details, include:

- Awareness of the way senses vary across the animal kingdom, how they are used in different ways and how humans are distinctive.
- Knowledge of the ideas of modularity and streaming in sensory information processing.
- The idea of sensory inhibition and what it is for.
- The major parts of the brain associated with sensory information.
- The stages in processing sensory data.
- The relevance of understanding animal senses to multi-media and computer games.

It will subserve a range of interests: the biology of the senses; the design of human–computer interfaces; the building of artificial sensors, from robotics to animats. All of these disparate elements will combine together in the virtual worlds and computer games of the not too distant future.

2 Understanding sensory systems

2.1 Introduction

This chapter takes an overview of sensory systems and some of the general principles needed to understand them. It also takes a very brief look at the brain itself. Although the focus of the book is really at the input and encoding end of the brain, a knowledge of its basic structure is helpful in understanding how the various information processing pathways fit together.

Before looking at each of the principle sensory modalities possesed by humans it is intriguing to ask what senses *might* exist. Have animals learned to exploit every possible known physical force or interaction? On the other hand are there senses yet to be discovered, where there are no established physical mechanisms? The first of these questions is tractable, at least in principle, and §2.2 offers a framework. The second lies outside the scope of the book.

By starting with the physics the limits to information processing (§2.3) become apparent (and in fact animals are pretty good). Two overarching methodologies arise out of the physics viewpoint: the representation of signals, the subject of Chapter 3, a key idea in Horace Barlow's early work, and information theory, the subject of Chapter 4. They form the theoretical core of the book. But perception is not a simple one way transmission of information from eye or hand to brain. It is highly conditioned by what we know and what we expect. Thus what happens at the periphery depends to some extent on what happens deep inside the brain.

The raw sensory data undergoes very sophisticated neural processing and pattern recognition, often still way beyond what artificial systems can do. Neural systems and biological brains are not the only computational means for processing sensory data, but they stamp their own character on what animals extract from their senses[1]. Section 2.5 provides a minimal overview of the nervous system and sensory parts of the brain.

Fundamental sensory research pursues knowledge for its own sake. But it is also useful, both for creating better ways of interacting with computers and machines (human–computer interaction, or, most recently, perceptual user interfaces) and for building artificial systems, robots or animats in virtual worlds. Although the primary focus here is not an engineering one, §2.7 introduces this perspective which reoccurs throughout the book.

---

[1] Noisy, limited dynamic range, massivel parallel.

Science often progresses by the development of new measuring techniques, and this is especially true for the senses. This book does not have a historical structure, thus the pioneering measurements occur only in context but §2.6 describes the techniques which crop up later.

2.2 Sensory diversity

The senses most familiar to us are naturally the human ones of vision, hearing, touch, taste, smell and vestibular, from which we build up a coherent model of the three-dimensional world around us and the position of our body and its parts within it. If one or more senses become damaged or do not develop properly, this model of the world can still develop.

But some animals lack some senses completely, or have others which we do not have or senses which we had but which have atrophied over time. Humans have close to the best overall vision of the animal kingdom. Raptors, such as eagles, have higher resolution. Nocturnal birds such as the barn owl are much better in low light conditions, while leopard can hunt in starlight. Insects frequently exploit ultra-violet light. Perhaps the most extreme visual adaption is the giant squid which lives in the very dim light of the bottom of the ocean. With a *very large eye* the size of a football, it aims to catch every photon which goes past, rather like an astronomical telescope.

We use sniffer dogs because they have have more sensitive smell than ourselves. In fact, most mammals have an additional olfactory system, the vomeronasal organ, which has disappeared in humans (§9.4.1.1). Just as in vision, there is more than one way in which dog might outperform human. They might be simply more sensitive – a bloodhound's *very large nose* certainly helps with sensitivity – analogous to leopard hunting at night. They might also be sensitive to a greater range of chemicals analogous to the way some animals have trichromatic olour vision but others do not. They might also have a better memory. Could most of us given a sniff of something (like somebody's clothing) then go tracking this scent? Tasters of fine food and wine take note too: humans have around 6000 tastebuds, the humble rabbit double this number and the catfish 100 000.

Some bat species process ultrasonic echo-location signals. In fact many mammals, cats, guinea pigs and even dogs, can detect higher frequencies than humans. Are our pet cats and dogs driven neurotic by the ultrasonic trash from CD players? Animals vary not only in the frequency bands to which they are sensitive but also in their overall sensitivity, sometimes helped by having very large ears (§7.3.2).

Some animals have taken touch to greater extremes. In the serious bushfires in the San Francisco and Berkeley hills in the 1980s a lot of pet cats survived, albeit with serious navigational difficulties. Cats were very effective at surviving the heat of the fires by hiding below ground level. But many couldn't avoid getting their whiskers burnt. The sense of touch in cat has developed into a finely tuned navigational system, very useful in the poorly lit nooks and crannies cat loves to explore.

Animals experience a variety of forces, notably electromagnetic radiation, magnetic, electrical, optical, sonic and physical impact, and have developed specialised sensors to make use of some or all of these. But different animals may use different parts of each of these stimuli, the wavelengths of light, the frequency of sound and so on. Some animals have additional senses which which we humans do not have, have never had, or have atrophied over time and sometimes they extract different or more precise information such as cat from its whiskers. Evolution has proved continually effective in finding solutions to environmental problems. Part of this variety probably emerges from neural information processing constraints: simple low-cost solutions prevail. The human brain consumes a huge amount (20%) of the body's energy relative to its size – neural computation is expensive, but even in *Drosophila* fruit flies neural memory appears costly (Mery & Kawecki, 2005). *This is a key idea for robotics, virtual reality and computer games.* We should not be preoccupied with building human-like senses but should look for minimally effective solutions.

Insects, as demonstrated by the world-leading insect vision group formed by Adrian Horridge at the Australian National University in Canberra, are an effective model for many robotic systems. One of the most recent is Srinivasan's development of tiny surveillance helicopters, one of many ideas spawned by insect vision (Srinivasan *et al.*, 1999).

Snakes and other reptiles have heat-sensing detectors, although no animals have achieved infrared vision, similar to military night glasses. Pit vipers have exceptional temperature sensitivity (to around 10^{-3} °C). Rattlesnakes have pushed electromagnetic sensitivity into the red with an infrared eye (§11.2.1). Such animals are cold blooded. Rodieck speculates (1973, p. 292) that in warm-blooded animals the thermal radiation within the eye would swamp any putative infrared detector. They also do not have ears, but have a primitive vibrational sensitivity, which presumably could become an ear with a few more eons of evolution.

Sometimes the sensory capabilities of animals have remained unknown long after they were first described. Birds, for example, were thought not to have a sense of smell until the last decade (§9.4.1).

We human beings have a range of senses well suited to the environment in which we live. But other animals have found ecological niches, where some of the physical stimuli are weak or missing. At the bottom of the ocean, in caves or underground, there is very little or no light at all. Bats cope using sonar. Deep ocean dwellers have responded by creating their own light through phosphorescence.

Elsewhere evolution has refined other senses: the barn owl (§7.3.4) has exquisite directional hearing; bats and dolphins create their own sonic probes for echo location (§11.4.1). Human hearing cuts off at around 20 kHz, cats go higher at around 40 kH, but dolphins trump terrestial mammals with a whopping 100 kHz cut-off. At the other extreme tigers and elephants use infrasound, sounds so deep we would not hear them (§5.4.2).

Similarly the touch sensors throughout the body are refined in whiskers and are further refined even further as the hair cells of the ear. Cats and rodents use their whiskers not just as mere collision detectors, but to build a coherent model of an environment when they cannot see very well if at all (§10.6.1). But an underground resident living in a world where vision and hearing would be pretty useless, the star-faced mole, has the strangest touch sensor on the planet (§10.6.2) with six times the number of mechanoreceptors of the human hand. For bats, cats, rats and star-faced moles these non-human sensors circumvent the lack of light in their hunting or habitation environments.

Other animals have gone even further and developed sensors for new stimuli. Platypus, *Ornithorhyncus anatinus*, a uniquely Australian monotreme[2] has an electric sensor, helpful in finding prey in murky river conditions. Electic sensors and electic weapons (stingray, eel) occur in other animals (§11.1), but they are almost all aquatic. Again, a simple information processing rationalisation springs to mind: the dielectric constant of water is much higher than air, so allowing better transmission of electric signals.

Magnetic sensitivity exists in birds for navigation, although birds get lazy – carrier pigeons follow motorways and even use the exits (Guilford *et al.*, 2004; Pilcher, 2004)! Recent work has shown magnetic navigation in turtles (Lohmann *et al.*, 2001) and mammals (Nemec *et al.*, 2001). When anything is discovered in a mammal, there is always the possibility that it might exist in man, and there are indeed conferences on magnetoreception in people.

Some human senses are vestigial or at the least non-conscious and magneto reception is one of the most contentious. Pheromones are another example. In animals they are well established and very important. They are the primary strategy ants use for cooperative foraging and navigational strategies. In moths and many other animals they are the means of attracting and locating members of the opposite sex ready to breed. What is less well known is that humans used to have a distinct sensory system for processing pheromones, the vomeronasal organ, still found in other mammals. Recent experiments have shown that humans are sensitive to differences in histocompatibility via odours (§9.4.1.1). They are processed by the vestiges of what is the vomeronasal organ in other mammals. However, this sense is *unconscious*, raising the issue of conscious awareness of sensory information, which will crop up again in §6.6.2. In this and in numerous other cases in this book, the question of human pheromones is not fully resolved at the time of writing.

Trade-offs of the different sensory options abound in the animal kingdom. For example, as colour vision evolved, it took over some of the roles performed by olfaction (§12.4.10), but whether this relates to the loss of vomeronasal sensitivity is not clear. It is interesting too that in studies of scaling of brain *cerebrotype*[3], Clark *et al.* (2001) found that the ratio of a range of areas including pyriform cortex and olfactory bulb (part of smell) have shrunk relative to the neocortex (where colour is processed).

Within a family of animal lifestyle, simple *scaling* effects are apparent. A bird which hunts from the air needs quite fine visual resolution. A bird which spends most of its time in vegetation has different needs. But within this subjective bandwidth, evolution has

---

[2] A mammal which lays eggs.

[3] The volume fraction for each component of the brain with total brain volume defined as 1 (Clark *et al.*, 2001).

tended to optimise information gathering. This makes sense. Once your environment, size and movement are set (and difficult to change), then you want maximum flexiblity within this setting, which will allow change of food, awareness of new predators and general adaptation to changing conditions. In an elegant study, Snyder and Miller showed that the focal length of the eyes of falconiformes (birds of prey) scales directly with the height at which they fly (and effectively their size: if you are as big as an eagle, you can dive faster, but you need to fly higher to avoid being seen) (Snyder & Miller, 1978).

But, although we might feel that we sweep up sensory data indiscriminately within this subjective bandwidth, human senses are still highly selective of the information they capture. Many existing standards in audio and vision already exploit such selectivity in the design of interfaces, storage and reproduction, and the technology is pretty mature. On the contrary haptic (touch) systems are still developing rapidly (§10.8.4) and maybe with a few years we will have online wine tasting and cooking courses replete with examples.

Artificial sensors are already commonplace and ubiquitous computing is a current trend. Miniaturisation and low power consumption necessitate capturing and processing the minimal information for the task at hand, from sentient clothing to internal body monitors, from surveillance monitors to smart dusts.

The designer of robots for the real world or *animats* for virtual worlds and computer games faces many sensory challenges. Compromises are essential to achieve real-time performance and thus there is a lot of interest in specific signalling as well as general purpose seeking out information from the environment. The richness of a real or virtual world comes from having very many different animats or plants all interacting and competing with one another. Not every animat needs to be a super animat. We need the insects and all sorts of primitive aliens, products of the mind of the designer. Thus the book underpins a library of possible senses which an animat might want.

2.2.1   Integration and lifestyle

It is important to see the different senses as both supporting and complementing each other, and how they suit the animal as a package – how senses fit into ecology (Dusenbury, 1992; Lythgoe, 1979). Thus a herbivore needs to be able to discriminate the food it eats (recall that humans have fewer taste buds than other animals such as rabbit and catfish) (§2.2). So vision is useful, colour probably particularly so. But smell and taste provide additional information. If it needs or chooses to eat at night (just simply to get enough food, to avoid predators or when it is cooler), then even moonlight may be absent and there is no light at all. So chemical sensors would be needed to distinguish food items, but exceptionally good, directional hearing, to detect predators would have definite survival value.

A predator needs long range vision or hearing and very good short range depth sensing (stereo or movement) for pouncing and fighting. But to hunt at night, vision may eventually not be enough. The barn owl can use hearing to locate prey very precisely in total darkness (§7.3.4).

2.3   Sensory information processing

A number of common patterns emerge from this diversity of sensory information processing:

- An animal has a number of senses which are partially *complementary* to one another: it can hear, but not see in the dark; it can detect odour traces from sources which are no longer present; although it may be possible to live missing one or more senses, survival is imperilled.
- The incoming signal first receives some preliminary, *non-neural filtering*, such as the optics of the eye (§6.4) or the ear canals (§5.3.2), which serve to restrict the incoming signal and better match it to the next stages.
- Each sense has a variety of receptors which detect different parts of the stimulus: in vision we have receptors sensitive to different wavelengths (which gives us colour) and sensitive to different intensity ranges (scotopic) versus daytime (photopic) vision; human touch has at least six different types of receptor (depending on how you count them).
- These detectors *transduce* external input signals, light, sounds, vapours, into an electrical signal.
- The different receptor signals may be multiplexed together in some way or they may travel in parallel pathways right into one or more parts of the brain; the M-cell visual pathway is largely colour free, but the P-cell pathway carries both colour and spatial information (§6.6).
- There is often a narrow channel through which the information is squeezed, a sensory bottleneck. Sometimes the physical limits create this, such as limits to vision imposed by the wavelength of light (§6.2). This may be more of an evolutionary accident than design or it may arise from some external constraint. In the case of vision, the bottleneck is effectively the optic nerve, where 10–100 million receptor signals are compressed into about a million ganglion cells. Too thick an optic nerve would impede eye movements. This bottleneck gives us a way into measuring the real information capture by a sensory modality.
- There is often substantial *convergence*, often into the information bottleneck. In the mammalian retina there are upwards of 100 million rods (§6.5.1.1) but two orders of magnitude fewer ganglion cells taking visual information to the brain. So very many rods map to each ganglion cell. On the other hand in the very centre of the eye, where vision is at its most acute (the fovea), there is around one P ganglion cell per cone photoreceptor. In olfaction some 50 million olfactory receptor neurons map to just 2000 glomeruli in the olfactory bulb (§9.3.4).

Colour vision illustrates the contrast between general and highly specialised sensory adaptation rather well. Humans have three spectral types, from which a vast range of hues may be discriminated, and surfaces and objects throughout our world are labelled by colour. At the other extreme, mantis shrimp has eight different spectral types, tuned, amongst other things, to pick up the fluorescence from other shrimps (§8.6.4). This distinction between a general processing strategy (abstracting (colour) reflectance properties of objects) from a more wavelength/phenomenon specific recognition has become an important psychological distinction of autism, something to which we allude very briefly in §6.7.1.2.

There is a frequent misunderstanding as to how the brain integrates sensory data. Subjectively we feel as if we have a unitary sense of the world across all domains. So indeed there is convergence at the highest world model levels. But most of the analysis and processing takes place in separate streams – so colour, for example, plays little role in edge detection, motion or binocular vision and the senses do not remain in exact temporal synchrony (§12.4.11). The last chapter looks at how sometimes this integration can go astray such as in syanaesthesia (§12.3.2) and lead to spooky events such as out-of-body experiences (§12.4.8).

Some of the senses have natural *metrics* associated with them, such as the amplitude of a sound wave in time or the magnitude of its frequency components. We can talk about the position of a point in an image and the relative magnitude of different components. All sensory data is metrical in this sense. But in some senses, notably the chemical senses, taste and olfaction, the metrics are far less clear: no theory as yet predicts how to represent any given odour in terms of a mixture of other odours, something we can do in precise ways for, say, colour, based on established physics. Ripe strawberries smell like ripe strawberries. Fruits contain esters which give them their distinctive fruitiness.

But a fine wine has a bouquet made up of many components in a delicate balance, and its decomposition is somewhat subjective. When a connoisseur detects a trace of oak for a wine matured in oak barrels, it is reasonable to assume that the chemicals which generate the taste or smell of oak are present. But when a wine is described as having lemon or floral characters, this could be the same chemicals, chemicals from the grapes which stimulate the same olfactory receptors or taste buds or just an analogy. The answer does not seem to be known as yet.

As discussed in Chapter 9 there are around 1000 receptor proteins, yet we can discriminate around 10 000 odours. So receptor signals *are* combined in some way to produce composite sensations, but it is not clear how.

Thus we refer to senses as metrical, where we have measurable ways of breaking up and constructing signals, and non-metrical, where we have more of a holistic pattern recognition. Chapter 3 goes into detail in how we represent metrical signals.

2.3.1   The information theory paradigm

A pivotal development in sensory research was the idea of *redundancy reduction* introduced by Barlow (1959) and subsequently shown to be applicable across many domains. Sensory information is compressed by removal of redundancy. To measure compression we need the tools of information theory discussed in §3.9.3. One of the key topics of this book is the information carrying capacity of sensory systems, the information carried and processed by nerve cells, or neurons. Snyder *et al.* (1977a) extended these ideas to vision in both invertebrates and vertebrates. The data

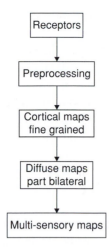

Figure 2.1 Flow of sensory information. The maps may have a real-world intepretation, such as the human body in touch, or have a more abstract nature, such as audio frequency.

which hits our senses contains much redundancy, simultaneous repeated encoding of data. Laughlin (1981) took the important step of measuring the characteristics of this incoming data, and its impact on information collection. A whole subdiscipline of *scene statistics* grew out of this early work.

2.3.2 Streaming

The broad structure of sensory information pathways is one of moving from precise topographic maps in the spatial senses (e.g. map of the world in vision, map of the body in touch) through to more diffuse maps and multi-sensory integration as shown in Figure 2.1.

It is no surprise to find the information from the different sense organs, the eyes and ears, travelling to the brain along different neural pathways. A little less obvious is the way information within a sensory domain is split up into separate streams on the way to the brain, raising a number of interesting issues:

- Why stream the information at all[4]?
- How is information from streams integrated into perception?
- Are there deviations from this strategy of necessity?

Streaming economises on processing – keep only task or function related information. Thus Barlow (1986) suggested that different visual areas would have cells clustered together of similar properties. Going deeper into the brain from initial sensory processing, these areas would become non-topographic, such as in specialisation for colour (§8.4.2).

---

[4] But see the discussion on stereopsis, where stereo information is extracted before any pattern analysis occurs (§7.2.3).

So, for example, the stereopsis and movement streams make little use of colour, so it makes sense to get rid of it from these streams. Each stream may need different temporal dynamics or different spatial resolution, allowing further optimisation. In the metric senses, the nature, structure and, to some extent, functionality of streams is apparent. But in the chemical senses, it is less clear. There are several taste categories (bitter receptors are more prevalent in some type of taste bud than in others (§9.5.1)), but in olfaction, apart from the split into pheromones and other odours, the streams, if any, are not clear-cut.

Barlow argued that limited connectivity would also drive local areas and almost three decades later the overarching network structure of the brain has not only become an area of considerable activity but the results seem to endorse his viewpoint (Laughlin *et al.*, 1998; Sporns *et al.*, 2004).

Although there are a variety of hierarchical streaming strategies, a dominant one is the separation into **'what'** versus **'where'**, the separation of identity of things in the world from where they are. The idea of just two pathways, *what* and *where* (Mishkin *et al.* in Zeki, 1993, p. 184). This corresponds, broadly speaking, to a broad subdivision of the brain into *dorsal* (top) and *ventral* (underneath)[5]. But some generalisations, such as this, are overly simplistic (Zeki, 1993, Chapter 20).

The problem of integration is a complex one, still only partially solved at the present time. But it's important to realise that our experience of an integrated coherent world does not necessarily imply a single area of activity in the brain. The different streams do not necessarily operate at the same speed. Three streams discussed in Chapters 6, 7 and 8 are colour, form and movement, processed in that order. There is as much as 80 ms difference between colour and motion (Zeki & Moutoussis, 1997), raising subtle philosophical issues.

Modular organisation is widespread through biological and many other systems. A relatively ordinary desktop computer has, at a simpler level, the same division of forces: file systems, web servers, FTP daemons all running pseudo-simultaneously, each performing their own tasks, each with their own communication protocols. Fundamental components such as the file system are themselves built of numerous subprocesses for reading and writing, organising, cataloguing and so on. In fact arguably the big breakthrough in software in the 1980s was object-orientation, followed by software agents in the 1990s where the modularisation of software approaches its zenith.

But there is still strong controversy about just how modular the brain actually is. Chapter 6 describes some very recent work on the recognition of faces and other objects and how these representations are *distributed* rather than modular. Even the streaming mechanisms are to some extent abstractions with an absolute demarcation sometimes being difficult to demonstrate experimentally. The anatomical realisation of modularity is not straightforward. Complex subsystems within subsystems abound, and as our understanding increases, so more and more levels of structuring emerge. Take the grandmother cell. Decades ago the speculation that there might be a cell somewhere

---

[5] It is easy to get confused here. Dorsal and ventral strictly refer to back and stomach, but in this context take their meaning from a quadraped. Back is confusing in the human head as it suggests a horizontal rather than vertical positioning.

in the brain responsive to grandmother went out of fashion. But it's back! Associative memory models of neural systems (§2.4.5) and simulations of neural pattern recognition and recall made such an idea distinctly unattractive. But now at the time of writing very specialised cells are now turning up (§4.3.1).

2.3.3 Theory of computation

Another well-known figure in vision was David Marr (Marr, 1982), who proposed a new way of thinking about how our senses work. He distinguished three levels: the theory of computation; the algorithms to do the computation; and the implementation of the algorithms within the brain. Since then many arguments and debates have raged about the distinction between these levels, especially between algorithms and implementation. Not Marshal McLuhan's *the medium is the message*, but the *neurons are the computation*. This book cuts across all three levels, but with more focus on what information is available and how it is used rather than the precise algorithms of sensory pattern recognition.

2.3.4 Mindsets and top-down processing

Over the last decade or so our thinking about the general strategy of sensory processing has been shifting in a profound way. Across many thousands of studies lies the assumption that information is processed sequentially, from sensors through to high level perception deep within the brain. Yet sensory systems abound in recurrent connections, but whether the initial phases of sensory pattern recognition involve feedback is controversial (Fabre-Thorpe *et al.*, 2001). In the first indications that the full story might be more complicated, Snyder and Barlow (1988) argued that sensory detail is suppressed by higher level concepts.

One of the subsequent manifestations of this has been the discovery in more than one sensory modality of detailed neurological mechanisms by which this occurs. Inhibition is one driving mechanism but there is now evidence that central *mindsets* can actually modify the firing of nerve cells in the cortical brain areas right at the beginning of the chain of sensory information processing in touch (§10.4.9) and vision (§6.6.2.1).

At the time of writing new results reflecting this different perspective are coming in, suggesting a possible paradigm shift is on the horizon.

2.4 Pattern recognition and computation by neurons

A brain is a giant network, much larger and more complex than even the internet[6] or the US telephone network. Its computation thus depends on the individual elements themselves (the neurons (§2.4.1)) and the properties of the network itself (§2.4.2).

---

[6] But the internet is catching up. At the time of writing there are about 200 million web sites, not far off the number of neurons in a human brain. The connectivity between web sites has a fractal nature rather similar overall to brain structure.

2.4.1   Neurons

The diversity of neurons is huge: they vary in what they look like, how big they are, how connected they are and many other things. They also vary in the way they communicate. The primary communication is chemical, via *neurotransmitters*, of which there are upwards of 20, from small molecules like nitric oxide to large organic molecules, such as dopamine, glutamate, serotonin and others[7]. It might seem strange that since the signal processing and transmission within a neuron is electrical that the communication between neurons should be largely chemical. But computer networks have a similar structure if we go to a high enough level of abstraction: internet packets have header information as to where they are they are going and the type of information they carry. Packets may be destined for specific ports on a destination machine and so on. Multiple neurotransmitters support intricate mixing of messages of different types.

The fine details of neural communication *may* matter in terms of what computations are possible, but they can be largely ignored in looking at the general principles of sensory information processing. But there are some interesting philosophical issues. The senses are driven by the world, by the impact of light, sound and so on. But to what extent has the neural machinery adapted to the senses, or ways of solving sensory problems been found for existing neural systems? Neurons are slow and noisy but operate in very large parallel arrays. They have limited dynamic range so cells have to adapt to the averaging input level and signal differences, which occasionally limits what they can do.

For present purposes it is useful to think of three broad neuronal categories:

1. Neurons which *transduce* a signal from some external form into an electrical signal; e.g. a photoreceptor transduces light (photons) into an electrical voltage.
2. Neurons which simply compute; they take inputs from other neurons and output to yet more neurons.
3. Neurons which control other body components such as muscles.

Sensory processing primarily involves only the first two of these. Digging deeper into the information processing behaviour requires knowing various characteristics – the threshold sensitivity, the dynamic range and properties such as degree of linearity and intrinsic noise. There are two general types of neuronal signal:

1. Cells which produce a continuously varying voltage output; photoreceptors do this, rather like the output from a photocell.
2. Cells which fire voltage pulses; by this we mean that the cell output is a series of pulses, or *spikes*, of fixed magnitude at varying rates; in some ways you might think of this as like a digital signal but its coding properties are quite different to the digital signals transmitted over computer networks, again a matter for §4.6.

---

[7] Little mention will occur of electrical transmission, but this is not an indicator of its importance. It may play a role in synchronisation of neurons and is abundant in the thalamo-cortical fibres. Electical synapses may also undergo long-term modulation (implying neural learning (Landisman & Connors, 2005)).

The abstract structure of a spiking neuron is quite simple. At the input end it connects to a number of other neurons, perhaps as many as 100 000 via connection points known as *synapses*. This input net is the *dendritic tree*. To first order, the inputs from all the other neurons are summed linearly, independent of where on the dendritic tree the input neuron makes contact. Most work in artificial neural networks assumes linear summation and such networks are capable of universal computation and function representation. For this book, this assumption will suffice, but there are complex computational processes which go on in the dendritic tree.

The neuron sums the incoming signals until *its activation* reaches a threshold (Katz, 1966). At this point it fires a voltage spike and the activation returns to zero. After a spike the neuron exhibits a refractory period during which another spike cannot occur. Thus a cell has a maximum firing rate and thus very strong input signals will saturate the output.

The neuron transmits its output along the axon, which may be a really long cable, as, say, in the case of a neuron in the spinal cord. At the other end of the axon is another network of connections to the tendrils of the input dendritic trees of other neurons.

Spiking neurons come in two types: myelinated and unmyelinated, comparable to optical fibres versus wires. A myelin sheath around the axon of a neuron enables it to conduct impulses much faster. The sheath appears white, thus myelinated neurons constitute the white matter of the brain, the unmyelinated neurons, the grey matter.

Obviously, the signal coming from a train of spikes is effectively positive, with a minimum of zero spikes. Thus to code positive and negative signals would require a *maintained firing rate*, where the neuron fires continually at the level corresponding to zero, with decreases in this firing rate corresponding to negative values. Although there are a few cells in sensory systems with such maintained firing rates, this wastes energy, a crucial resource in neural systems (§2.4.3). Thus cells frequently exist as on–off pairs, one responding to a signal turning on, the other to the signal turning off.

This is a *grossly simplified* version of neural behaviour! To begin with it completely ignores computation on the dendritic tree itself. Although this was first observed a while ago, it has recently become a hot topic of investigation (Branco & Häusser, 2010). Then there are numerous non-linearities, different mechanisms of neuron-to-neuron signal transmission and various sorts of global modulators. Excellent sources of more detailed neural information are books by Nobel Laureate Bernard Katz (1966).

2.4.2 Receptive fields and inhibition

A very useful concept in the metric senses, especially vision, is the *receptive field*. The first stage (§6.5) describes how receptors of some sort transduce an external input into an electrical signal. Many receptors together cover the input space – the retinal image in the eye, the frequencies we can hear, the molecules we can detect and the various impacts on the skin and body. Each receptor will sample only a part, usually a small part of the input space.

Neurons now collect this information from receptors, often pooling them, but each samples only a fraction of the input space – in vision a direction in real space, in hearing a band of frequencies and they may attach different weights to the different receptors from which they receive input. The receptive field determines this region from which the neuron receives its input and the weight attached to different points within it. It is what engineers would essentially think of as a filter.

Receptive fields, once thought to be fixed early in life, are now known to be much more dynamic; they change with experience, age and even over short time periods under certain stimuli (Fu *et al.*, 2002). As discussed in §2.4.3 neural computation requires significant energy. If a neuron were to signal both positive and negative values it would need to maintain an activity level of around half its maximum. Although some cells do have a *maintained discharge* many others operate differently.

The key factor here is *inhibition.* One of the attempts to provide an overarching framework for sensory information processing came from Nobel Laureate Georg von Békésy. His belief, eloquently explored in a book entitled *Sensory Inhibition* (Békésy, 1967), was that inhibition is a universal computational strategy of the nervous system:

> More and more it became clear that lateral inhibition is a common feature of all the sense organs . . . It was this generality that seemed to be important because it interconnected, at least in one respect, the fields of vision, hearing, skin sensations, taste and smell.

It now appears that inhibitory circuits decline with age, and thus receptive fields, themselves, shift. One profound recent experiment showed that loss of visual performance in old monkeys could be restored by the injection of inhibitory neurotransmitters, leading to hope for drugs to enhance cognitive performance in old age (Leventhal *et al.*, 2003; Miller, 2003).

2.4.3   Network properties of the brain

We can ask some of the same questions about the brain as we can ask of a telephone system:

- What is the global structure or hierarchy?
- What are the tranmission times?
- What is the connectivity like (is it dense, sparse, do some neurons have more connections than others)?

The modular and hierarchical characteristics of the brain are very evident in sensory pathways. There are strong analogies to an old-fashioned telephone system in which local exchanges connect to people's homes with low bandwidth cables; big cables connect exchanges; and ultimately huge bandwidth optical fibre on the ocean floor connects continents. Dense cortical connections are locally clustered into columns and hypercolumns, but larger connecting fibres run across the whole brain.

The last decade has seen a massive growth of interest in the structure of networks, particularly in networks which require only a small number of hops to get from one node to another but yet are far from fully connected. This small world structure, introduced by Watts and Strogatz (Watts, 1999), is common to many environmental, biological and social systems and basically features local connections with a small number of shortcuts. Scale-free networks, whose importance and prevalence was recognised by Barabási's group (Barabási, 2002), also have the short distance property (known as diameter in network parlance) but they get it in a different way. In such a network the number of connections, k, a node has follows a power law distribution:

$$P(k) \propto k^{-\gamma} \tag{2.1}$$

where $P(k)$ is the probability a node has k connections and γ is an exponent, with values around 2–3 for many systems. Scale-free networks have large hubs which creates short connection paths, an idea exploited by airlines. So to fly from Bathurst in NSW Australia (tiny regional airport) to Santa Fe in New Mexico you could go via Sydney (big hub) to Los Angeles (very big hub) to Denver (big hub) to Albuquerque (small hub) to Santa Fe (small regional airport).

The big interconnecting neurons are usually myelinated and under a microscope they appear white. Without myelin, neurons appear grey, as in the grey matter of the brain. Impulses are conducted more slowly, but neurons take up less space and more can be crammed into a limited volume such as the skull. The stripes of V1 (hence its name striate cortex (§2.5.2)) arise from the myelination of inputs to Layer 4. There is an exciting new area of research opening up in understanding the connectivity.

The space taken up by the neuronal axons and dendrites is considerable. In the grey matter, they represent 60% (Laughlin & Sejnowski, 2003) of the volume. The white matter itself occcupies 44% of the cortical volume. It scales as the 4/3 power with the grey matter over a large number of mammals from pygmy shrews to elephant, *five orders of magnitude in brain size*. This corresponds to a constant bandwidth of long-distance communication per unit area of cortex (Zhang & Sejnowski, 2000).

Studies of network structure and the volume occupied by white and grey matter collectively point to a cortical need to minimise energy use by cortical neurons (Laughlin & Sejnowski, 2003). Streaming of different types of information is consistent with this viewpoint.

These long range connections (which may be myelinated for speed) play a major computational role. In fact a new model of autism, *under connectivity* theory, suggests the condition may arise from a lack of this long range connectivity (§6.7.1.2). Not only the flow and integration of information depends upon this connectivity but so does the synchronisation of the brain as a whole, an issue considered in §12.4.11.

Not only do the different senses project to different brain areas, but individual senses, such as vision, themselves have numerous pathways. Since there is much variation among animals, most of the discussion will concern mammals. Although a lot of human specific data is now coming in, thanks to the advanced imaging methods (§2.6.2), much of our knowledge derives from studies of other mammals, notably cat and monkey.

But from the perspective of the animat designer or roboticist, a mammal centric view is limiting. Birds, insects, even moles, have their ecological niches wherein they compete successfully and may prove optimal low-cost models for some applications. The bulk of the discussion here revolves around mammalian architectures, allowing in other animals where distinctive strategies have evolved.

2.4.4   Self-masking

In some senses, such as hearing and touch, the animal itself can activate its own sensory systems – we feel our own touch and hear our own voice. But it would obviously save some computational effort if the sensory system were warned in advance of what is going to happen. Indeed this does seem to occur through *corollary discharge.* This is a general strategy used within the brain for supplying information about planned movements, such as saccadic movements of the eyes (§12.1.6), vocalisations (§5.4.4.1) and perhaps underlies the curious phenomenon whereby it is impossible to tickle oneself (§10.4.9)!

2.4.5   Associative memory

Our conscious minds often tend to think of concepts in a very analytical way. A chair is a thing to sit on; it typically has legs or some sort of support, a seat, which may be padded, and a back. Visually and conceptually it is a hierarchy, from the grain of the wood to the overall shape, building the concept of a chair. But a lot of sensory pattern recognition does not seem to be like this at all, particularly for the older senses such as olfaction.

Instead, it comprises associative memories. In the most abstract sense an associative memory is the pairing of a set of outputs with given inputs. Neural networks are particularly good at forming associative memories and one of the most important foundational papers in artificial neural networks (ANNs) was an associative memory algorithm by neural network pioneer, John Hopfield at Caltech (Hopfield, 1982).

In the Hopfield model, the ANN forms an associative, content addressable, memory for a set of input patterns. It does not have a very high capacity. The number of attractors (memories) is of a similar order to the number of neurons. It is a powerful abstraction, far removed from the complex dynamics we see in real neural systems. Whether the results hold for these more complex biological systems is not fully understood at the time of writing.

2.5   A tour of the brain

Understanding animal senses necessitates following the pathways from the sense organ, the ear or eye, right through to the brain. Evolution has generated some very complex and subtle ways of breaking information up into streams, Many man-made devices already use this very effective strategy.

In the brain, though, some of the organisational factors may be more contingent or dependent on factors such as energy supplies (§2.4.3) rather than driven by computational efficiency. Trichromatic colour vision, for example, appears relatively infrequently in mammals, basically in just primates such as monkeys and humans. So, the colour pathways and centres are in a sense grafted on. Evolution does not offer the options of a complete rebuild – every intermediate stage has to be able to survive in its own right.

Nobel Laureate John Eccles described the human brain as the most complex structure in the universe (Popper & Eccles, 1977). The parts of the brain associated with vision have at least a hundred anatomically distinct areas. Brain anatomy and function fill huge books, but a basic knowledge of organisation will suffice to gain deeper insights into sensory pathways such as:

- streaming from an information processing point of view;
- streaming and sensory capacities in an evolutionary context;
- and how top-down central control influences sensitivity and discrimination.

2.5.1 A thumbnail sketch

The brain sits on top of the spinal cord and in a very rough sense the oldest parts of the brain are here. The outer areas, particularly the cerebral cortex are younger in evolutionary terms. The spinal cord becomes as it enters the skull the *brainstem*, the oldest part of the brain. The brainstem feeds into the *diencephalon*, which itself feeds into the cerebral hemispheres comprising the *basal ganglia*, the *amygdala* and the *cerebral cortex*. The cortex will feature prominently throughout the book, the basal ganglia, which have important roles in learning, less so. The amygdala play a strong role in the processing of emotion, and various special pathways to them complement the main pathways for sensory pattern recognition[8]. The latter comprises four lobes, *occipital*, *temporal*, *parietal* and *frontal*.

The brainstem has many sensory relay areas such as pons and *reticular formation*. Some sensory nerves project in through these areas. The 13 cranial nerves come in to the reticular formation, some of which are sensory, some motor and some combined as shown in Figure 2.2 (Fields, 2007a; Saper, 2000)[9]. The pons and medulla transmit much of the somatosensory information through a subunit called the *medial lemniscus*.

The diencephalon has two areas, important in sensory information processing: the *hippocampus*, heavily involved in memory and the *thalamus*. The latter is one of the most important areas for sensory systems with at least 50 subdivisions: almost all senses pass through it as if it were the relay station for sorting and distributing sensory information. It is highly modular with units which are sense specific. There are two distinct pathways, with differing behaviour at a neurotransmitter level, the *lemniscal* or primary and the non-lemniscal or associative pathway (Mooney *et al.*, 2004).

From the thalamus sensory pathways project into the the *cortex*, the main information processing area of an animal brain. Each sense typcially has its own cortical area, but pathways may project to several different areas. Pathways from these sensory specific

---

[8] There is a host of different terms reflecting evolutionary and developmental perspectives, such as the term telencephalon for the cerebral cortex.

[9] The numbering starting at zero might be a bit confusing (although not to computer programmers), but this arose because the thinner cranial nerve zero (*nervus terminalis* was discovered in 1878 in elastobranch (shark) and then in 1913 in humans, long after the numberings of the other nerves had become entrenched (Fields, 2007a; Whitlock, 2004).

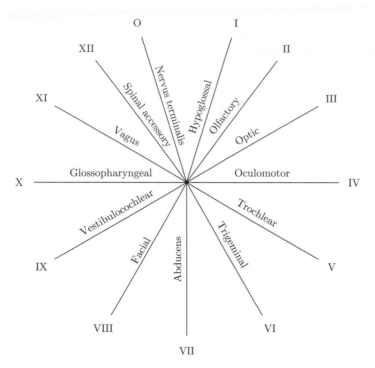

Figure 2.2 The cranial nerves. (Simplified illustration from Table 44.1 of Saper, 2000.)

areas subsequently feed to areas fusing information from multiple senses, discussed in the final chapter.

The cortex is itself like a generalised neural processing system, consisting of six layers, as shown in Figure 2.3. Pathways in and out and sideways tend to be associated with different layers across most of the sensory cortex areas. The cortex has many folds, referred to as *sulci* with pronounced ridges on the edges, referred to as *gyri*[10]. The thickness of the cortex is around 2–4 mm and does not vary very much among primates. These folded structures increase the surface area. Many cortical cells are unmyelinated (grey matter), but some cortical areas show distinct striations due to myelinated inputs as in the V1, the primary visual cortex (§2.4.3; §2.5.2).

The cortical areas are not particularly specified as vision or hearing. The kind of cortical area which develops depends on the input. Experiments, in which the inputs are misdirected, show that the visual cortex may, for example, develop where the auditory cortex would normally be, or where whisker barrels might appear (§10.6.1). It is as if the cortex receives inputs in the early development stages and develops an organisational, computational structure around them. The science fiction notion, akin to William Gibson's cyberdeck, of being able to add new sensors, computer databases, satellite position

---

[10] Pl. of sulcus and gyrus respectively.

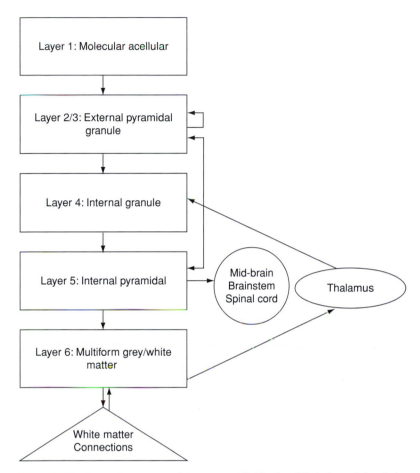

Figure 2.3 The layered structure of the cortex. Highly simplified view of the six layers of the cortex.

systems, whatever, may not be impossible. The very first examples of this appear in the well-established cochlear implants (§5.5.5) for hearing and the newer retinal implants (§6.10.2), methods of interfacing synthetic devices to hearing and vision to ameliorate eye or ear damage.

Section 10.6.1 shows that whiskers in animals such as mouse or rat have associated with them in the cortex distinct patterns of cells referred to as barrel fields. The basic organisation of this part of the cortex is what Rakic (2001) calls a *protomap* where a cortical organisation is generated without input from outside. Fukuchi-Shimogori and Grove (2001) show that for these barrel fields just one *growth factor*, FGF8, controls the development of these fields. If extra FGF8 is inserted in a different part of the brain a *second set* of barrel fields appears (Rakic, 2001)!

This is not to say that the wiring in the adult is not highly detailed and highly specific. There are more and more indications of how precisely evolutionary pressure has shaped the wiring. In late 2000 how specific wiring takes place in olfaction was established, and this is discussed further in §9.3.9. There are indications emerging of competition for brain space. At an individual level, brain areas can get reused. Visual and auditory cortex get reassigned in blind and deaf people respectively. At an evolutionary level, it is beginning to look as if the growth of one brain area (say for language) has to be at the expense of other areas (Gazzaniga, 1992). It has also been suggested that the reason the evolution of colour vision pushed out olfaction was the need for brain space (§12.4.10 (Holden, 2004b)). Finding room for colour means something else has to shrink. But colour and smell may often provide redundant information. Bright colours signal ripe fruit as well as the odorants wafting away. Thus some genes may become redundant as information becomes adequately covered by something else, in this case colour (§12.4.10). In other words it is hard to untangle the pressure for space from the atrophy from lack of use as more powerful systems subsume other roles.

2.5.2   Nomenclature

Labels for brain areas can be quite confusing! Homologous areas in different animals, even within a quite specific group such as primates, may have different designations.

Korbinian Brodmann in the late nineteenth/early twentieth century labelled most of the main brain areas in numerical sequence (Kandel *et al.*, 2000). Thus Brodmann area numbers reflect contiguity and can seem quite arbitrary, e.g. areas 17 and 18 are associated with vision but area 44 with language. Of course at that time he had very little to go on as to what each area did, so there is not that much in the way of logical ordering of different senses, let alone much else. As our understanding has deepened, so the need to split areas, such as 3a and 3b (which process different types of somatosensory information) has become apparent.

Some of the basic Brodmann areas now have additional designations. Thus the visual cortex of primates is usually described as V1 (the first cortical area to get information) followed by V2, V3, V4 etc. But sometimes these areas have other common designations. The primary visual cortex, V1, has distinct stripe patterns (§2.4.3) and is thus referred to as *striate* cortex. Other visual areas thus become *extrastriate*. Thus we shall try to keep the labels as simple and as explicit as possible. But readers following up the original references need to remember this diversity of nomenclature. For example, the labels V1, area 17, calcarine and striate cortex all refer to the early stages of visual processing.

Finally, it is easy to confuse nerves and neurons. The body has a number of high bandwidth communication channels, the nerves – the cranial nerves such as the optic nerve (Figure 2.2), and the 31 dorsal root ganglia (§10.4.3), each containing many nerve cells. These individual cells are also referred to as fibres and neurons. This book mostly adopts a single nomenclature, in which *nerve* refers to a big trunk cable containing many nerve cells. The nerve cells themselves we refer to as *neurons*. An assembly of neurons in the cortex, say, we refer to as a *neural network*. The use of neural networks in artificial intelligence is now widespread. We refer to them as artificial neural networks, abbreviated ANNs.

2.6 Investigative methods of brain science

Many techniques have contributed to our understanding of how the brain works, but as the twentieth century drew to a close, the range, power and resolution of these techniques was growing rapidly. Some of them have made a big impression on sensory processing research of one form or another.

One of the most exciting challenges of brain science is the way the brain works at many levels. Understanding how individual neurons work does not tell us everything about the large-scale dynamics. Thus a range of investigative techniques are necessary, from understanding how transmitters work between nerve cells, to the dynamics of the cell itself, to observation of whole brain dynamics. It is in this latter domain of system behaviour that gigantic progress has occurred in the last decade.

The oldest techniques are anatomical – slicing the brain up and elucidating its structure. Some nomenclature still dates from the early anatomical work (§2.5.2). The finest level of detail comes from looking at the behaviour of individual neurons, either through selective staining techniques or through *electrophysiology*, the recording of the electrical signals from the cell itself. Finally there are techniques which measure or modify the electrical and magnetic properties of the brain, giving us unprecedented access to brain dynamics.

Despite the long interest in sensory information processing, going back as far as the ancient Greeks, there have been big discoveries and radical changes in our thinking. They stem from new powerful measurement techniques:

- *Brain imaging and stimuli* (§2.6.2) keeps getting better and better and the latest technique, functional magnetic resonance imaging (fMRI), offers increased spatial resolution, but is slow; new results from this technique occur often in the book; magneto encephalography (MEG) improves on the resolution of fMRI but is still extremely expensive (§2.6.2). A new and completely different technique is *transcranial magnetic stimulation*, which allows part of the brain to be temporarily switched off (§2.6.2.1). Thus human subjects find their skills and performance vary dramatically as part of the brain is shut down (and not always for the worse! (§12.4.12)).
- *Genetic sequencing* (§2.6.3) is clarifying many issues and has had a huge impact in the chemical senses.

2.6.1 Anatomical methods

Neural tissue at first glance looks a tangled mess. Many different cell types may be intertwined together and sorting out their pathways is even worse than untangling ropes on a boat. Thus techniques for making one cell class or pathway stand out against another are very valuable.

To determine structures of pathways ideally requires structural features, their *morphology*, to be directly visible, such as different shapes and sizes. One of the basic methods for identifying anatomical structure is by staining with Nissel agents which show up the cell bodies and their different morphologies are thus distinguishable after staining. These staining patterns give rise to what is referred to as the *cytoarchitecture*. But where this is not possible, the contrast between features may be enhanced by selective staining techniques of which there are many.

Cytochrome oxidase is a key metabolic enzyme, active when the cell is busy. Staining cells according to the activity of this enzyme has revealed fine details of cortical architecture. We shall see, in studying visual area V1 (§6.7.1), how useful such anatomical techniques are in revealing diversity of function. In fact Zeki (1993) argues that:

If any of the anatomical methods reveals a specific architecture, in which adjacent cortical subregions within an area differ obviously and repetitively from one another, it is certain that sooner or later a functional difference corresponding to the architectural heterogeneity will be found.

Even more is possible by adding tracers, which are taken up selectively by some cells and may sometimes show entire patterns of connectivity.

The most detailed analysis of brain function is to go right down to the level of an individual cell and see how it behaves. This is the domain of electrophysiology. Tiny electrodes inserted into the cell record neural activity. Now, clusters of electrodes can record simultaneously the activity of sets of cells and examine collective behaviour.

2.6.2 Dynamic imaging, stimulation and blocking

There is nothing, though, quite like seeing the living brain in action. Over the years a variety of methods have been tried and throughout there is a trade-off between speed and spatial resolution. Techniques looking at individual cells may have high temporal resolution, but can only look simultaneously at a small number of cells.

Some, for example, involve injecting voltage sensitivity fluorescent dyes into neural pathways. Thus any electrical activity in the nerve cell and concomitant voltage change generates light. The response is fast.

The electrical fields on the skull yield coarse-grained but useful indicators of brain activity (§2.6.2.2), but the real breakthrough (in fact a recent Nobel Prize) has been magnetic brain imaging. Results are coming in so thick and fast that many of our models and preconceptions are under threat.

The first technique revolutionising our understanding is *magnetic resonance imaging* (fMRI), which allows us to see activity in the brain at high spatial and reasonable temporal resolution. It is a somewhat indirect technique, measuring the blood oxygen level dependence (BOLD) arising from neural activity. When cells are active they need energy, their metabolic activity increases and this shows up in the oxygen they take from the blood. Oxygen has two unpaired spins in its outer shell of electrons which enables it to be detected by magnetic resonance. Because the enhanced metabolic activity does not coincide exactly with the neural activity, the temporal resolution of fMRI is not so good as some other techniques, such at positron emission tonography (PET) or electroencephalography (EEG) (§2.6.2.2).

PET is an expensive, somewhat invasive technique, with good spatial and temporal resolution. The subject breathes in a radioactive isotope of oxygen O^{14}, which appears in the brain within less than a minute (and has a half-life of around two minutes). Each positron emitted by the isotope eventually decays into two gamma ray photons travelling straight out in opposite directions. By catching these photons it is possible to work back and discover where the disintegration occurred, and thus develop a map of the most metabolically active regions of the brain.

2.6.2.1   Trans-cranial magnetic stimulation: tweaking the brain

Another technique which has great promise is *trans-cranial magnetic stimulation* (TMS) which enables localised areas of the brain to be transiently enhanced or inhibited in activity. Thus we can now observe how the brain behaves if we (reversibly) disable or stimulate small parts of it. An even newer technique is *trans-cranial direct current stimulation* (DCS) which uses electrical current rather than magnetic fields and is less invasive and requires less medical supervision for its use. The potential here is just enormous and far from fully realised at the time of writing (Snyder *et al.*, 2004).

2.6.2.2   Electroencephalography

Neural activity in the brain generates electrical activity on the scalp. Sensors attached to the skull record this EEG activity. Needless to say, not all of the activity of 10^{10} neurons is accessible this way. But EEG does have quite characteristic patterns for specific behaviours such as attention and sleep. The advances in signal processing driven by increasing computing power have enabled more and more to be extracted from a number of sensors, now up to several hundred. EEG does have advantages over other more spatially precise methods – it's fast, non-invasive and now will operate wirelessly.

2.6.2.3   Magneto encephalography

The latest advance is magneto encephalography (MEG), the most expensive, with the resolution of fMRI, but a 1 ms temporal resolution, three orders of magnitude better. Whereas fMRI uses an indirect measure of neural activity, based on blood flow, MEG measures the magnetic fields generated by the electrical activity of the nerve cells themselves.

A subject in an MEG machine looks just like somebody in a dryer in a hair dressing salon. But there the resemblance ends. The multi-million dollar costs of setting up MEG involve building special rooms with very high insulation from ambient magnetic fields.

2.6.3   The impact of genetics

The human genome programme and the surge of interest and results in genetics have had a considerable impact on sensory information processing. In vision we now know the precise location of the genes which code for colour vision (§8.1) and have some idea of how they vary within populations and across species. In hearing, very recently the genetics of hair cells have thrown light on just how the enormous sensitivity of mammalian hearing is achieved.

But it is in the chemical senses that genetics have dramatically changed the way we think. In discovering the genes which code for the receptor proteins, we now know the dimensionality of these senses in a way we did not before. Particularly interesting is taste, the poor relation of olfaction. But what was once just the generic taste bitter, turns out to have as many as 80 genes coding for bitter receptors. So our discrimination of bitter is really quite powerful. Numerous other examples occur throughout the book.

2.7 Senses in computer games and virtual worlds

The 1990s saw an explosive growth of the graphics and sound used in computer games. From iconic two-dimensional (2D) maps, three-dimensional (3D) became a requirement for almost any successful game. Surround sound, better resolution, faster updates make the visual experience an ever more exciting one. However, there is a downside as noted by James Dunnigan, a designer of military strategy games (Dunnigan, 2001). The demands of the eye-candy are now so great that much of the game budget goes on the visual, auditory and tactile elements, and only a tiny fraction gets spent on the strategy designs! Games now cost millions to develop, mainly because of the high interface quality demands.

But over the last decade games have been entering a new phase. A game such as chess is a totally deterministic game: what is possible at any time, even the eventual winner assuming best play is given (although the players do not usually know this). So despite the visual and tactile aesthetics of some hand-crafted chess pieces, the abstract structure of the game is the key thing. Game designers have heretofore sought total control of the game at any point. Evolution or learning within the game has the disadvantage that it may go into unknown or undesirable regions.

One promising future direction which may redress the current imbalance of sensory impact over strategy is the move towards virtual worlds. There are now games, such as Second Life, where people play in large numbers within a virtual universe, a separate existence occupying large parts of their time. The business model of these games as one of regular subscription and maintenance of the customer base (against other competing immersive worlds) is crucial and the game interface must be impeccable.

A prime requirement of a really successful game interface is *consistency*. Bugs creep into games where characters fall through walls. Objects in a game appear, but none are interactive. So, it will not be able to pick up an object on a table and throw it through a window, if this particular action has not been programmed into the game. Thus the player's choices are limited and the environment is unnatural. However, if we move towards building realistic environments, in which all the objects have physical/biological characteristics, then we eliminate all these problems at once. Any entity in the world then has to sense the virtual world around it and move accordingly. We are some way off achieving this on a game scale, but how fast we are advancing and how far is evident from this perspective of the early 1990s from eminent philosopher, Daniel Dennett (1991):

It is a familiar wall these scientists have hit; we see its shadow in the boring stereotypes in every video game. The alternatives open for action have to be strictly – and unrealistically – limited to keep the task of the world-representers within feasible bounds. If the scientists can do no better than convince you that you are doomed to a lifetime of playing Donkey Kong, they are evil scientists indeed.

But it might be held up as the holy grail of virtual world and game design. Of course any rich simulated environment needs many insects and other simple animals as well as the high level non-player characters.

3 Introduction to Fourier theory

Time present and time past
Are both perhaps present in time future
And time future contained in time past.
                    Burnt Norton (from *Four*
                    Quartets) by T.S. Eliot

3.1 Overview

This theoretical chapter examines core ideas for the whole book, the ideas of linear systems, vector spaces and functional representation. All the sensory modalities sample the incoming signal in some way, producing a discrete representation. They then transform and split this signal up into streams. Along the way the data is usually compressed. This chapter deals with how to sample a signal and represent it in different ways. The section on information theory (§3.9.3) then takes on the topic of compression.

One fundamental representational example is Fourier decomposition, crucial to vision and hearing. It is one of the most important and powerful ideas in the whole of engineering and communications and it is essential to understanding sensory processing. The mathematics is rather complicated, but the emphasis herein is on the ideas rather than the detailed formalism. For the interested reader, excellent books describing the mathematical content in much more detail are *Linear Systems, Fourier Transforms and Optics* by Gaskill (1978) and *The Fourier Transform and its Applications* by Bracewell (1999).

We already have an intuitive feel for these ideas. We know how somebody's voice changes if they have a cold, when they talk over the telephone; we know how pictures can be blurred by a dirty lens, or can vary in contrast from soft portraits to sharp lithographs. All these common phenonema are examples of *filtering*, of changing the *frequency balance* in a signal. We already have an everyday knowledge of what frequency means in sound, but we shall see that frequency is useful in vision and to some extent in the somatosensory system. The analysis of signals in terms of their frequency composition was the work of eighteenth-century mathematician, Jean Baptiste Fourier.

As a precursor to Fourier theory two general concepts are necessary: linear systems theory and the concept of superposition (§3.2). They are helpful in understanding the behaviour of filters and provide the tools to look, say, at the effect of the lens of the eye on the visual image or the way the ear encodes sound.

3.1.1   Superposition, frequency and filters

Even somebody who has not studied any mathematics, computing or engineering probably already has a good intuition for frequency. Many people know the notes of the musical scale. Each key of a piano corresponds to a particular frequency. Middle C is 262 Hz, the A to which orchestras tune is 440 Hz. In days gone by, before the advent of electronic tuners, the device for tuning a musical instrument was the tuning fork. You can almost see the tuning fork vibrating as the forks go backwards and forwards at a *specific frequency*[1]. But then, if middle C on a piano has an equivalent pitch on the violin, why do they not sound the same?

In fact the notes of musical instruments are made up of many frequencies all mixed together, or *superimposed*. It is this range of *harmonics* (simple multiples) of the *fundamental frequency (pitch)* that gives each musical instrument its distinctive sound or timbre. The sensation of a distinct pitch occurs because all the harmonics are simple multiples of the fundamental, but the exact algorithms which map harmonic structure to pitch are complex (§5.3.5).

Now, the really interesting thing is that we can take signals of all kinds, not just musical notes and decompose them into a mixture of frequencies – except that now the frequencies are not simply related to one another. Even signals with sharp edges, signals made up of pulses or signals with a very irregular structure can be decomposed. Chapter 5 explains that this is exactly what the ear does.

But apart from knowing that we can decompose signals in this way, what use is it? Firstly, it's extremely useful, because it simplifies the mathematics of many complicated processes. Three examples will be the idea of *filters*, the sampling theorem and the analysis of information rates in §3.9.3.

Again, filters are an everyday concept, but whereas we think readily of mechanically filtering out contaminants from solutions, or extracting caffeine and flavours from coffee beans, what we do in sensory systems is to filter out or modify parts of signals, a little more abstract maybe. When a trumpeter puts a mute on the front of her instrument, the sound loses some high frequencies and becomes more mellow. Now imagine when you've got a bad cold, or you've just landed in an aeroplane, and your ears are partially blocked. Everything sounds different, fuzzy – filters again.

What happens to the sound is easy to explain in the *frequency domain* – a lot of the high frequencies have been filtered out. In the same way, if you look through a dirty window, what you see is blurred, again high *spatial* frequencies have been lost.

When we listen to music on a CD, what we hear is transformed and degraded from the original signal, but in fact the CD is actually not bad as a source and Chapter 5

---

[1] A question you might want to think about, and will be able to solve after reading about movement detection in Chapter 7, is why we cannot normally see the vibration forks of a tuning fork. To what note would the fork have to be set for us to be able to track the arms of the fork?

will show later how a CD is matched to the performance of human hearing. More of the degradation comes in the later stages of reproduction, particularly the loudspeakers, but there are even bigger errors in the spatial representation of the sound, an issue we take up in Chapter 7. A lot of the theoretical underpinnings of sensory information processing can be accomplished with a subclass of filters, notably *linear filters* or filters in linear systems. Linearity is a prerequisite of simple Fourier domain filtering.

The second, extremely important use is in the Nyquist sampling theorem (Nyquist, 1928), the cornerstone of digital signal processing. Take a continuous signal, say the voltage output from a microphone. One might think that if you approximate it by a series of samples taken at regular time intervals that some information would be lost. Not so. In a profound and far-reaching result Nyquist showed that if the signal samples are sufficiently close together, it is possible to reconstruct the original signal *exactly* if there are no errors in the sample values. Listening to music from a CD has involved digitising the orginial musical signal, storing the digital data on the disc and then converting it back to a continuous signal to drive the loudspeakers. How often is it necessary to take samples to ensure no errors? Section 3.6 gives the details.

In vision, the decomposition of images into *spatial* frequencies is less intuitive to most of us. But we are already familiar with two-dimensional wave patterns, such as ripples on water, caused by the wind or some other disturbance.

Frequency analysis is also useful in thinking about touch. We shall see in Chapter 10 that there are populations of different touch receptors, some sensitive to very fast vibrations, others to much slower displacements of the skin.

In the non-metric chemical senses, taste and smell, many theoretical issues are less clear. But one idea, which we shall come across in this chapter, the idea of *degrees of freedom*, is useful. We can now estimate the degrees of freedom for sense and taste, but this work is really new, driven by information from the human genome project.

So the ideas are very simple: we want to understand how we analyse signals into different frequencies all added together; we want to see how adding, removing, weakening or strengthening some frequencies relative to others changes the signal; and we want to see how to digitise a signal without losing information. Linear systems and Fourier Analysis could easily occupy entire books and they do (Blahut, 1985; Gaskill, 1978), but our needs are a basic understanding of a few key points:

- Linear systems (§3.2) and superposition (§3.1.1);
- Sinusoids, or sine and cosine waves in other words (§3.1.1, §3.2.4);
- Convolution (§3.4), a fundamental linear operator, the Fourier Transform (§3.5);
- The way filters work in *reciprocal domains*, such as time and frequency (§3.5);
- The Nyquist sampling theorem for digitising signals (§3.6);
- Windowed frequency representations, where sinusoids are damped out by a window which goes to zero at plus and minus infinity, creating a localised *wave packet* (§3.8);
- And finally a topic which leads into the information theory of the next chapter, representations which are based on statistical properties of signals and images (§3.9).

3.2   Linear systems and operators

Linear systems are the basis of all we are going to discuss in this chapter. They are pervasive, although Richard Feynman, Nobel Laureate, and one of the great expositors of scientific ideas, had a rather pragmatic view (Feynman *et al.*, 1963, pp. 24–25)[2]:

> ...why *linear* systems are so important. The answer is simple: because we can solve them! So most of the time we solve linear problems. Second (and most important) it turns out that the *fundamental laws of physics are often linear*... *That* is why we spend so much time on linear equations: because if we understand linear equations, we are ready in principle, to understand a lot of things.

To make these ideas more precise, we need a clear definition of the somewhat vague all encompassing term of a system. For our purposes a system can be almost anything, a machine, a computer, an animal population, an environment and so on. Because of its extremely general nature we often refer to it as a black box – we don't worry about what's inside but only about how it behaves. We only know that if we put a particular thing in, we get something out – we know the *input/output relationship*.

Thus a system may be defined as the result of an *operator*. In mathematical terms operators are quite diverse. They may be simple multiplicative factors, but may also include derivatives (differential operators) or integration. We represent operators by a special calligraphic notation, enclosing their operand in curly brackets {}. The mathematical formalism we do not need to worry about and we discuss one useful operator in detail in §3.4.

An operator often represents a *functional*, which is a little different to what you would normally think of as a function[3]. A function takes a point in space or part of a space (called the *domain*) and delivers a value within the *range* at that point. It is a *mapping* from the domain to the range.

A functional is a little more complicated than a function. Functions are easy to understand. The volume, V, coming from the stereo is a function of the position of the volume control, p_{vc}, which we could write as $V(p_{vc})$. The amount of light coming into a camera is a function of the shutter speed or the aperture of the lens. A functional, however, takes not a value or set of values as input, but an entire function, and spits out another function! So, a stereo amplifier takes an input signal, $f(t)$, a voltage as a function of time, and delivers a new, bigger signal, $F(t)$. If it's a good amplifier, the signal coming out will just be a replica of the original, at a higher level, but if it's a bad amplifier, considerable distortion of the signal coming out may occur, making it non-linear. Thus an operator (functional), S takes as input a function from some set,

---

[2] Around the time Feynman wrote this, Lorenz was publishing his work on climate modelling, which is now regarded as the first real study of chaos. The advent of powerful computers has made it possible to solve many non-linear problems, and in particular to study chaos. But the thrust of Feynman's argument remains strong. We shall really need non-linear ideas only occasionally, such as when we discuss associative memory in olfaction.

[3] Gaskill (1978) notes that we can think of a functional as a system with memory. Because it takes a function as an input it, we can think, in the case of a system defined in time, as it requiring knowledge of the function through past and future to compute the output at any given time.

$f_1, f_2 \ldots$ and generates for each an output $g_1, g_2 \ldots$

$$g_i(x) = S\{f_i(x)\} \tag{3.1}$$

An optical example will make this more concrete. The camera used to look just like a black box, although one of the famous early cameras was called the Box Brownie! It takes a pattern of light intensities and forms an image on the sensor. So, if you imagine taking a photograph of a painting, each point on the painting becomes a point on the sensor. Ideally there would be an exact match, but there will be some loss of colour accuracy, some loss of sharpness and some distortions due to the lens, all of which make the output a degraded version of the input. So we could represent each point on the painting by a flux of light coming from it, $P(x, y, \lambda)$, where x, y are the coordinates on the painting and λ is the wavelength of light[4]. The operation of the camera converts this P to an image in the camera back, $I(u, v, \lambda)$ where u, v are the coordinates on the camera film. So we can write the operation of the camera in terms of an *operator*, C, like this:

$$I_1 = C\{P_1\} \tag{3.2}$$

Suppose now that, using a mirror we superimpose the light flux from two paintings, P_1 and P_2 the resulting image is now given, approximately by

$$I_{double} = C\{P_1\} + C\{P_2\} = I_1 + I_2 \tag{3.3}$$

In other words we have calculated the camera image separately for each input and added them together. For imaging of light through a perfect diffraction (limited (§6.2.3)) this holds, but would not hold exactly for practical lenses.

The general definition of a linear system then is given arbitrary inputs $f_1(x)$ and $f_2(x)$, which under the action of operator, S, give $g_1(x)$ and $g_2(x)$, then for any constants, a_1 and a_2 the operation of S on af_1 is ag_1 and on the sum is:

$$S\{a_1 f_1(x) + a_2 f_2(x)\} = a_1 g_1(x) + a_2 g_2(x) \tag{3.4}$$

This equation encapsulates the *principle of superposition*.

Many systems are linear only over a limited range: they become insensitive at the lower end and saturate at the upper end of this range, i.e. eqn 3.4 is only valid for some restricted range of values. In the camera example above if we take the image on the sensor in a digital camera, the signal recorded by each pixel will be approximately linear in the light incident on the pixel, but at some point the pixel will *saturate* and its output will no longer increase as the light level increases. A microphone, which may have quite a linear response, has some threshold sensitivity, some minimum sound level before it will produce any output at all, and some maximum level at which it blows apart!

---

[4] We will have a lot to say about wavelength and colour later; they are *not* the same thing, but the distinction is not important for the present discussion.

3.2.1 Discrete and Continuous Systems

A central theme throughout the book is that of sampling (§3.6) where a signal comprises a set of samples of a continuous signal. Eqn 3.4 seems like a discrete summation, but it may equally well be continuous. The *integral operator*, $\mathcal{L}$, is a linear operator. It transforms one function f into another g via a *kernel* $b(v; x)$ according to

$$\mathcal{L}\{f(x)\} = g(v) = \int f(x)b(v; x)dx \qquad (3.5)$$

using the variable v to indicate that the domain of g could be different to that of f. The kernel, $b(v; x)$ is also referred to as the Green's function. The Green's function results from application of the system operator to the Dirac delta function (§3.3, §3.3.3).

3.2.2 Illustrative examples of operators

A very simple linear operator, $\mathcal{C}\{f(x)\}$, would be multiplication by a constant, say ξ. So $\mathcal{C}\{f(x)\} = \xi f(x)$ but squaring a function, SQ, giving $SQ\{f(x)\} = f^2(x)$ is obviously not linear. But in each of these cases we would calculate the function $f(x)$ first and then carry out the operation. But other operators have a much more radical effect and the original function is left nowhere to be seen. Two cases useful to us in the book are differential and integral operators. The differential operator $\mathcal{D}\{f(x)\} = \frac{df}{dx}$ is linear since:

$$\mathcal{D}\{(af_1(x) + bf_2(x)\} = a\frac{df_1}{dx} + b\frac{df_2}{dx} \qquad (3.6)$$

Integral operators might be a little less familiar and require a few mathematical caveats on the functions being suffficiently 'well behaved' which are beyond the scope of this dicussion. Instead of thinking about a set of functions f_i (as in eqn 3.1), basically a linear operator (which we are going to look at in more detail in §3.4) such as $\mathcal{Z}\{f(x)\}$ has a *kernel* $z(x)$ so that:

$$\mathcal{Z}\{f(x)\} = \int f(x')z(x - x')dx' = g(x) \qquad (3.7)$$

Such an operator is also linear since:

$$\mathcal{Z}\{af_1(x) + bf_2(x)\} = \int (af_1(x') + bf_2(x')z(x - x')dx'$$
$$= a \int f_1(x')z(x - x')dx + b \int f_2(x')zx(x - x')dx'$$
$$= ag_1(x) + bg_2(x) \qquad (3.8)$$

3.2.3   Vector spaces and eigenfunctions

The principles of linear systems and superposition provide a simple way of applying an operator to a signal. If the signal is made up of the sum of several components and we apply some scaling operation to each of the components without mixing them, then the new signal is just the weighted sum of the results on the components added together. The components essentially remain independent. In Fourier theory these components are waves, sound waves, oscillating patterns of light intensity or anything you like.

The space in which these components are defined may be one-dimensional, as for sound, two-dimensional as for vision or even higher if we combine space and time. However, this representation idea can go further, where the points in space are actual functions, which has a similar feel to it to the idea of functionals. The commonplace idea of a vector is a set of numbers giving the coordinates in space. However, we can consider a space to be made up not of coordinates but of *functions*, such that any point in the space is a function, a *vector space*.

Associated with any linear operator there are special functions we call *eigenfunctions*, which *span* a vector space. In other words any function within the space can be represented as a linear sum of the set of eigenfunctions, or in fact, any set which span the space.

Conceptually the idea is a simple one, though. When the system operator gets an eigenfunction as input, the output is the same eigenfunction, perhaps with some scaling. *Eigen* is the German word for 'own' and it simply means that these functions are the operator's 'own functions'. Thus for an operator $S\{f(x)\}$ its eigenfunctions, $\phi_k(x)$ satisfy the equation:

$$S\{\phi_k(x)\} = \lambda_k \phi_k(x) \qquad (3.9)$$

where λ_k is a real or complex number, the *eigenvalue*.

Consider all the functions we can define as a function of two real variables. Mathematicians call the *space* in which these functions live $\mathcal{R}^2$, or, in general, for n variables, $\mathcal{R}^n$. We can now define *orthonormal basis sets* to represent any function which lives in such a space. Thus we can represent $\psi(x)$ in terms of the basis set ϕ_i as:

$$\psi(x) = \sum_{i=1}^{\infty} a_i \phi_i(x) \qquad (3.10)$$

where a_i are numerical coefficients. The orthonormal constraint is the following:

$$\int_{-\infty}^{\infty} \phi_i \phi_j \, dx = \delta_{ij} \qquad (3.11)$$

where the Kronecker delta, δ_{ij} is zero unless $i = j$ in which case it has the value one.

It's now fairly easy to see how we can calculate the coefficients in the expansion:

$$a_i = \int_{-\infty}^{\infty} \psi \phi_i \, dx \qquad (3.12)$$

There are many possible basis sets. But for any given operator, $\mathcal{S}$, which maps any function in the space into another function, the eigenfunctions, ξ_i are unchanged by the operator as in eqn 3.9.

3.2.4 Linear shift-invariant systems

There is one special class of linear system which occurs frequently in signal and image processing, the linear *shift-invariant system* (LSI). This is just a system in which as an input function shifts, the output shifts too, but doesn't change shape, i.e.

$$\mathcal{S}\{f(x - x_0)\} = g(x - x_0) \tag{3.13}$$

for any x_0. The kernel $z(x; x')$ becomes simply $z(x - x')$. It is quite easy to see what the eigenfunctions of this equation are. Try $f_k(x) = e^{j2\pi kx}$ so that:

$$\mathcal{S}\{e^{j2\pi kx}\} = g_k(x)$$

Then:

$$\mathcal{S}\{f_k(x - x_0)\} = g_k(x - x_0) = e^{-j2\pi kx_0}\mathcal{S}\{e^{j2\pi kx}\} = e^{-j2\pi kx_0}g_k(x)$$

i.e.

$$g_k(x - x_0) = e^{-j2\pi kx_0}g_k(x)$$

This equation is satisfied if:

$$g_k(x) = \lambda_k e^{j2\pi kx} \tag{3.14}$$

i.e. the eigenfunctions are complex exponentials, i.e. complex mixtures of sine and cosine waves. The eigenvalues, λ_k depend on $\mathcal{S}\{\}$.

The usefulness of eigenfunction analysis now becomes clear. The complex exponentials, $e^{j2\pi kx}$ comprise sine and cosine functions, elementary wave-like functions, and are the eigenfunctions of LSIs, a point to which we return in §3.9.1. This is not surprising. A cosine function looks exactly the same if you shift it by one full period; shift it by a quarter of a period and it becomes a sine. Analysis of signals into frequency components began with Fourier in the nineteenth century and the methods of representing signals by sine and cosine waves bear his name, the Fourier Transform and Fourier Analysis.

The summation in eqn 3.10 is over a discrete set of functions. We also have integral transforms and operators, where the sum becomes an integral. In eqn 3.12 the basis functions become one continuous *kernel*, defined everywhere, and the coefficients, become the *transform*, again, defined, everywhere. For the sensory analysis of this book, we can slide from one to the other without serious risk of error or ambiguity, albeit sliding over some of the more formal mathematics.

3.3   Important functions

There a few special functions which are very useful in working with linear systems and each will crop up later on.

3.3.1   The Kronecker delta function

The **Kronecker delta** function, δ_{ij}, has cropped up already. It is zero unless $i = j$ in which case it has the value one. It crops up frequently in operations on matrices or summation over indices. It is different to the Dirac delta function (§3.3.2) and is distinguishable by its subscripts and lack of an argument.

3.3.2   The Dirac delta function

The **Dirac**[5] **delta** function (not the same as the Kronecker delta (§3.3.1)) is easy to understand conceptually and exceptionally powerful. But it is rather nasty to make mathematically rigorous, because it is essentially infinite at one point and zero everywhere else! Thus rather than with a 'normal' function, where one defines it at every point in space, the Dirac delta function is defined in terms of its properties under integration:

$$\int f(x)\delta(x - x_0) = f(x_0) \tag{3.15}$$

The delta function is symmetric and has unit area, thus it scales according to:

$$\delta\left(\frac{(x - x_0)}{b}\right) = |b|\delta(x - x_0) \tag{3.16}$$

One way to think of the Dirac delta function is to imagine it as the limit of a symmetric pulse, say a Gaussian. As the pulse gets narrower and narrower, keeping the area constant, it gets taller and taller, eventually becoming infinite.

3.3.3   The impulse function

If we replace the filter convolution in the convolution formula (§3.5) with a delta function, then the function I is unchanged. Such a filter does not degrade the input signal at all. If we replace I with a delta function, the result is now the filter itself. In other words, stimulating a linear system with the delta function enables us to determine how it will behave for any input. In this viewpoint, we call the delta function, the **impulse** function. The system response to the function gives the Green's function (§3.2.1). We can now find the response to any input function, by imagining the input to be made of a weighted series of delta functions. Because the system is linear, we can now add together the series of Green's functions to get the system output. Later in the book, we shall use the

---

[5] Dirac was one of the most important theorists of quantum mechanics. Amongst other things he formulated the relativistic Shrödinger wave equation, thereby discovering the positron, antimatter and netting a Nobel Prize.

two-dimensional impulse function to an optical system, which is a point source of light, $h(x, y; \mu, v)$, to get the image intensity distribution, $\psi(x, y)$, from an input intensity distribution, $I(\mu, v)$, given in eqn 3.17:

$$\psi(x, y) = \int \int I(p, q), h(x, y; \mu, v) d\mu dv \qquad (3.17)$$

3.3.4 The comb function

Section 3.6 discusses *sampling*, the representation of a signal by its value at a discrete set of points. To get the sample values we need a relative of the Dirac function called the **comb** in Gaskill's notation (Gaskill, 1978). It is really just an infinite string of regularly spaced delta functions at some sample interval, x_s (Gaskill, 1978) (eqn 3.18). When we multiply a signal by the comb function, the result is just the sequence of values at the sample points multiplied by a delta function at each point (eqn 3.19):

$$comb\left(\frac{x}{x_s}\right) = \sum_{n=-\infty}^{\infty} \delta(x - nx_s) \qquad (3.18)$$

$$f(x)comb\left(\frac{x}{x_s}\right) = \sum_{n=-\infty}^{\infty} f(nx_s)\delta(x - nx_s) \qquad (3.19)$$

The comb function can also be two-dimensional, in which case it becomes a grid of delta functions and it can be shifted using the same shift as for the delta function itself.

3.3.5 Rectangular, cylinder and sinc functions

Two fairly ordinary functions, also important to sampling (§3.6), are the **rect** (rectangular) and **sinc** functions, both of which could be considered to go to delta functions as they become very thin (Figure 3.1).

rect is just what its name implies:

$$\text{rect}\left(\frac{x}{b}\right) = 1 \quad \frac{-b}{2} \geq x \leq \frac{-b}{2}; \quad 0 \; otherwise \qquad (3.20)$$

whereas **sinc** is:

$$\text{sinc}(x) = \frac{\sin(\pi x)}{\pi x} \qquad (3.21)$$

rect serves to chop out a piece of a domain (in the sampling case a piece of the frequency domain), while **sinc** is used for interpolation between sample points as discussed in §3.6.

Closely related to **rect** is its two-dimensional analogue, **cyl(r)** the cylinder function, $\text{cyl}(\frac{r}{d})$, obtained by rotating the rectangle function – exactly what it sounds like, a cylinder in two dimensions of radius d and unit height.

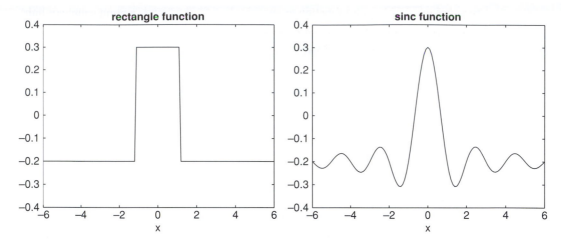

Figure 3.1 Rectangle (left) and sinc (right) functions.

3.4 Convolution

In the painting example (§3.2) one of the ways in which the image might be a degraded version of the original is that it might be slightly blurred. Each pixel in the image has got bits of its neighbours in the painting mixed in with it. So, sharp edges have got smoothed a little and spread out into neighbouring pixels. This spreading operation is approximately independent of position in the image[6]. So this makes it a shift-invariant operation, as we discussed in §3.2.4. We can express this as an integral as in eqn 3.17, for a blurring operation, $\mathcal{B}$:

$$P(x, y) = \int_{-\infty}^{\infty} \int_{-\infty}^{\infty} B(x', y')I(x - x', y - y')dx' dy' \qquad (3.22)$$

just as in eqn 3.17. We shall use a special symbol, $\otimes$ to denote convolution, viz.:

$$P(x, y) = B(x, y) \otimes I(x, y) \qquad (3.23)$$

Convolution is essentially a *filtering* operation and we could consider the blurring operation B to be an example of a filter. At each point x, y we look at all the neighbouring points of $I(x, y)$ where the filter B is not zero. Then we multiply the value of I at each neighbourhood point by the *reflected* value of B at that point and add all the results together to give the new value at x, y. In effect we are mixing in I with all its neighbours, weighted according to B.

The convolution is shift-invariant because the blurring function, B, is the same everywhere, it does not change with position. In many physical systems this is an idealisation. We shall see later that the image quality of the human eye falls dramatically away from

---

[6] This of course is not completely true for real lenses, which suffer from various kinds of *aberrations*, some of which create distortion at the edges of an image. The human eye suffers badly from aberrations at large pupil sizes, rendering the image quality at 6 mm pupil worse than at 2 mm, where it could in principle be three times better.

the straight-ahead position. In such cases it is often convenient to work in patches, where the function, B, is approximately constant.

It is fairly easy to see that convolution is both associative (eqn 3.25) and commutative (eqn 3.24):

$$g(x) = q(x) \otimes p(x) = p(x) \otimes q(x) \tag{3.24}$$

and:

$$g(x) = (p(x) \otimes q(x)) \otimes r(x) = p(x) \otimes (q(x) \otimes r(x)) = p(x) \otimes q(x) \otimes r(x) \tag{3.25}$$

The computational complexity of convolving two images of size $N \times N$ increases with N^4, making convolution a computationally extravagant operation. As we shall see shortly, there is a faster way of doing it, resulting from a discovery by Cooley and Tukey (1965). The *Fast Fourier Transform* scales as $N^2 \log N$, a big improvement. Since then other improvements have been made, discussed in detail in the book by Blahut (1985). These days, it is possible to patent algorithms and receive royalties from them. In their day it was not. Had they been able to patent their algorithm it would have the been enormously profitable, and perhaps made them the first computer millionaires.

3.5 Fourier Analysis and the Fourier Transform

Finally we get to the details of how to analyse a signal into its frequency components, or more generally from one domain to another, *reciprocal domain*, usually frequency. Fourier Analysis and the Fourier Transform do essentially the same thing. In the first case the domain is finite, akin to the strings on a violin and the frequencies occur in discrete, finite steps. In the latter, the domain is infinite and frequency is a continuous variable. Closely related to the continuous transform is the *Discrete Fourier Transform*. Most computational work in audio or image processing uses this discrete version. However, the ideas which are important for understanding senses come from the continuous transform and the sampling of continuous signals, so these are the foci of the present chapter.

The Fourier Transform (FT) then, is the method of finding the frequency components of a signal. There is a variety of different ways to define it, and we're going to have to leave the mathematical niceties to the many excellent books specialising in the area, such as Gaskill (1978) and Bracewell (1999). We will use what is sometimes called the real form. It's actually not the mathematically most elegant, but it has the advantage of relating directly to the equations for a real wave form, for a real function, $f(x)$[7].

---

[7] x will often be either time for an auditory signal, or space for an image; we refer to the domain of $f(x)$ as the time or space domains and the transform as the frequency domain.

The simplest form uses complex exponentials as the frequency components, where $F(v)$ is the Fourier Transform at frequency v in eqn 3.26:

$$\mathcal{F}\{f(x)\} = F(v) = \int_{-\infty}^{\infty} f(x)e^{-j2\pi vx}dx \tag{3.26}$$

which can be written as a sum or real, R, and imaginary, J, parts as in eqn 3.27, since $e^{i\theta} = \cos\theta + j\sin\theta$:

$$F(v) = R(v) - jJ(v) \tag{3.27}$$

where j is $\sqrt{-1}$ and R and J are:

$$R(v) = \int_{-\infty}^{\infty} f(x)cos(2\pi vx)dx \tag{3.28}$$

$$J(v) = \int_{-\infty}^{\infty} f(x)sin(2\pi vx)dx \tag{3.29}$$

where v is the *frequency*. We shall use a special symbol to indicate Fourier Transform pairs, viz. $f(x) \Leftrightarrow F(v)$ and we call the function $F(v)$ the *spectrum* of $f(x)$. So, this is the key definition, a measure of the strength of each frequency component, F. Note that although f is real, F is *complex*[8]. But the transform can't generate new information in any sense. So since $F(v)$ has two components for the one $f(x)$ there must be some symmetry or redundancy. It is easily shown that if $f(x)$ is real, then $F(v) = F^*(v)$.

There are a couple of simplifications we can look at. Firstly, if $f(x)$ is an even function, i.e. $f(x) = f(-x)$ then, because sin is odd, only R is non-zero and the FT is real. Similarly if $f(x)$ is odd, $f(x) = -f(-x)$ then only J is non-zero and the FT is imaginary.

The *power* at each frequency, Z, is obtained as the square of the amplitude (a basic result from wave theory):

$$Z = R^2 + J^2 \tag{3.30}$$

and is a real measurable quantity. The absolute value of $F(v)$ is the *magnitude* and is the square root of the power. The distinction between the real and imaginary parts will rarely be important in the book, but the concept of frequency will come up time and time again. Going back to our original musical instrument example, we can now talk about the power or magnitude of each harmonic making up the notes on a violin or piano.

---

[8] Some readers may not have come across complex numbers. They form a very big part of mathematics and it is far too big a field for us to even touch on here, hence we have adopted the real form. The trouble starts with this quantity j – we cannot take the square root of a negative number! Two negative numbers multiplied together give us a positive result. Complex numbers are created by adding a second, so-called imaginary part to an ordinary (or real) number. This imaginary part is a real number multiplied by j. Packages such as Matlab will keep track of the real and imaginary parts for us automatically and we will often perforce skip over the details.

Finally the *Discrete Fourier Transforms*, used in computer signal and image processing is:

$$F_i = \sum_{k=0}^{N-1} e^{\frac{j2\pi kl}{N}} f_k \tag{3.31}$$

Comparing eqn 3.31 to eqn 3.26 the integral has been replaced by a summmation over N (equally spaced) components f_k. Further discussion of the DFT occurs in §3.7.

3.5.1 Useful properties

The Fourier Transform has a number of useful properties, which we will describe here without proof:

Scaling: if we scale one domain, the other scales inversely, i.e.

$$f(x/b) \Leftrightarrow |b| F(bv) \tag{3.32}$$

It is easy to see what happens here. Imagine slowing down the tape in a tape recorder, the pitch of the recording goes down. This relationship says that the frequency falls in a linear relationship as time is stretched. In the neurons which go from the eye to the brain, the area of the world captured by the cell varies inversely with its spatial frequency bandwidth (Chapter 6).

Shifting: if we shift $f(x)$ to $f(x - x_0)$ then the FT is multiplied by $e^{j2\pi vx_0}$ referred to as a *complex phase factor*:

$$\mathcal{F}\{f(x - x_0)\} = e^{-j2\pi vx_0} F(v) \tag{3.33}$$

Linearity: this should come as no surprise:

$$a_1 f_1 + a_2 f_2 \Leftrightarrow a_1 F_1 + a_2 F_2 \tag{3.34}$$

Central ordinate: the zero of the FT is equal to the area under the original function. We often refer to this as the DC term by analogy with electrical circuits:

$$F(0) = \int_{-\infty}^{\infty} f(x)dx \tag{3.35}$$

A strategy used by sensory systems, especially vision, is to get rid of this DC term, which represents the total signal energy coming in. The *relative* values of signals are usually more valuable for pattern recognition and analysis.

Convolution: last but not least comes the most useful property of all, a powerful motivation for studying the FT in its own right. The FT of a convolution of two functions, f_1, f_2, becomes the product of the transformation, F_1, F_2:

$$F_1 F_2 \Leftrightarrow f_1 \otimes f_2 \tag{3.36}$$

This relationship enables us to do filtering operations in the Fourier domain by multiplying spectra (which is easy) rather than carrying out integrals (which is usually hard). Neurons carry out Fourier-like operations in hearing, vision and to some extent in touch.

Table 3.1 Some simple Fourier Transform Pairs. All may be scaled and shifted according to eqns 3.32 and 3.33 respectively.

Function	Transform
$\text{rect}(x)$	$\text{sinc}(\nu)$
$\text{sinc}(x)$	$\text{rect}(\nu)$
$e^{-|x|)}$	$\frac{2}{1+(2\pi\nu)^2}$
$e^{-\pi a x^2}$	$e^{-\frac{\pi\nu^2}{a}}$ (Gaussian function)
$delta(x - x_0)$	$e^{-j2\pi x_0 \nu}$
$\text{comb}(x)$	$\text{comb}(\nu)$
$cyl(r)$	$\frac{J_1(\pi r)}{2r}$

3.5.2 Fourier Transform of common functions

Table 3.1 lists some functions which have an analytical form for their transform. It is far from complete but includes those which will be useful later.

The **comb** function transforms to just another **comb** function. This is a little tricky to see because of the unusual behaviour of the delta function itself. The transform seems to be an infinite sum of harmonics of the reciprocal of the comb spacing, δx, at frequencies of $\frac{1}{\delta x}$. These reinforce at integer values and become infinite, but cancel everywhere else. and another series of Dirac delta functions results. This transform is useful in understanding sampling and interpolation in §3.6. Thus the comb function, like the Gaussian, is its own Fourier Transform.

3.5.3 The optical and modulation transfer functions

Diffraction, as we discuss further in §6.2.3, imposes a cut-off in the maximum spatial frequency, ν, in the image. This cut-off is different for coherent light, such as that emitted by lasers and used in holograms, and the incoherent light we get from the sun and other common light sources. It is incoherent light which is relevant to us here.

For incoherent light we get a gradual attentuation of contrast in each spatial frequency until we reach the cut-off. This attentuation function we refer to as the *Optical Transfer Function*, or OTF, $p(\nu)$. For a circular pupil of diameter D (Gaskill, 1978):

$$p(\nu) = \gamma_{cyl}\left(\frac{\lambda\nu}{D}\right) \tag{3.37}$$

where:

$$\gamma_{cyl}(r) = \frac{2}{\pi}[\cos^{-1}(r) - r(r - r^2)^{\frac{1}{2}}]\text{cyl}\left(\frac{r}{2}\right) \tag{3.38}$$

where the *cyl* is the cylinder function (§3.3.5).

The OTF operates on the Fourier Transform, which is a complex quantity in general. Hence a useful practical alternative is a simplification, the *Modulation Transfer Function*, or MTF, which is simply the modulus of the OTF.

The value of the MTF at any frequency is the change (usually the loss) of amplitude of a sinusoidal wave of the frequency passing through the system. The wave loses modulation or, in other words, contrast. Thus we can measure the MTF in a simple way. We put a sinusoid into a system and measure for every frequency the ratio of how much the amplitude is attentuated. Experiments of exactly this kind have been carried out for human vision, resulting in MTFs for the human eye.

A lens, for example, may pass low spatial frequencies with almost 100% contrast, yet may show a progressive loss of contrast as the frequency increases. It gets softer in photographic parlance. Similarly, the human hearing system passes some frequencies without significant loss, but loses everything below about 20 Hz and above about 20 kHz.

If the system is shift-invariant (§3.2.4), then a sine or cosine function passes through without change in or mixing of frequencies. Thus to measure the MTF we simply pass sinusoids through our system under investigation and measure the amplitude on the way in and on the way out. The lens test chart for a camera does this, amongst other things, with patterns of bars of different spacing and orientation. Similarly we measure the visual performance of the eye by measuring the loss of contrast at different spatial frequencies.

If the system is not shift-invariant, then this procedure does not work. So, if the spread function varies with position (as it does in the human eye), then a single sine or cosine will be turned into a mixture. But quite frequently useful information can be obtained in local regions and for photographic lenses, for example, it is common to measure the approximate MTF in different zones of the lens, such as the centre and periphery. Such mixing can also occur if the system is non-linear, as is often the case with amplifiers in the sound (music) modality.

The OTF has one further property. If we take its Fourier Transform it gives us the *point-spread function*, or PSF, $a(\theta)$ which describes the degree of spreading of a point of light as it passes through an optical system. For the circular pupil case of eqn 3.37, the PSF (measured in radians) is:

$$a(\theta) = \frac{J_1^2(\frac{D\theta}{\lambda})}{\pi \theta^2} \tag{3.39}$$

where J_1 is the Bessel function of first order.

3.5.4   Autocorrelation and power spectra

In assessing system performance and information capacity, random signals (noise) play an important role, discussed in detail in §3.9.3. For an ensemble of signals, $s_i(t)$, $s_j(t')$ the correlation between them, R_{ij} is:

$$R_{ij}(t, t') = \mathcal{E}\{s_i(t)s_j(t')\} \tag{3.40}$$

for signals s_i and s_j, with the operator $\mathcal{E}\{\}$ giving the expectation value or average over an *ensemble* (simply speaking the set of all signals).

An important simplification now yields an important result from the Fourier Transform. Physical systems are said to be *ergodic* if the time average is the same as the ensemble average, i.e.

$$R(t - t') = \mathcal{E}\{s_i(t)s_i(t - t')\} = \int_{-\infty}^{\infty} s_i(t)s_i(t - t')\, dt \qquad (3.41)$$

which is the convolution of s with a reflected version of itself.

The power spectrum, $P(v)$ is the Fourier Transform of the autocorrelation function. Since the effect on any signal of a linear system is to multiply the signal transform by the OTF, it is easy to show that the effect on the power spectrum is:

$$P' = M^2 P \qquad (3.42)$$

where M is the MTF.

3.6   The sampling theorem

We finally come to one of the great ideas in communications theory. It is sometimes referred to as the Shannon–Nyquist sampling theorem. In fact it probably dates back even before Nyquist (1928) to Hartley (1928), but in the pre-digital days, its uses were maybe limited. It was with Shannon's sampling communications theory that it became important, and people refer to the Shannon paper (1948) (where he does quote Nyquist), hence the duality. It often happens in science that the person who popularises an idea becomes associated with it most strongly. Like many great ideas, it's easy to understand after the fact which makes the discovery seem less interesting or less profound than you would expect.

Suppose we want to digitise a continuous image, such as a photograph print. You might think that we could never do this perfectly: no matter how close together we make the pixels, there might always be sudden changes in-between which might get missed. What Nyquist showed was that if the image was *band-limited*, then in fact *perfectly accurate* reconstruction between the sample points was possible for finite sample spacing[9].

The assertion that a signal is band-limited implies that the spectrum is finite only in a finite band and is zero everywhere else. In optical systems the diffraction of light creates a finite spatial frequency cut-off in images (Chapter 6). In sound, the cut-off is imposed by mechanical properties of the ear in biological systems and by direct filtering in digital sound reproduction.

To simplify the discussion, we will assume, as in many biological applications, that the finite band extends outwards from zero to a finite *cut-off frequency*, v_{co} as shown in Figure 3.2.

---

[9] However, this analysis ignores noise. Yet biological sensory sytems are plagued by noise and some compromises are necessary. We look at how noise enters into information content in §3.5.4.

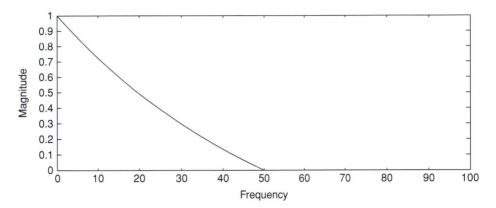

Figure 3.2 Example of a band-limited function. This is typical of the spatial frequency cut-off arising from diffraction through a finite aperture.

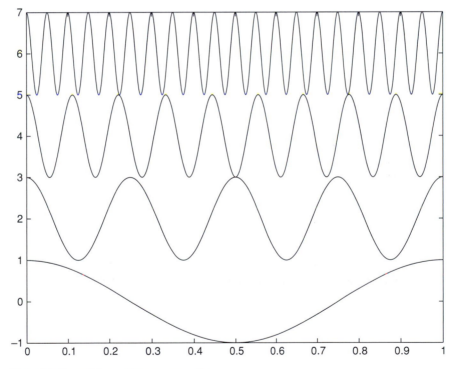

Figure 3.3 Several frequency components.

Figure 3.3 shows some of the frequencies which make up the spectrum of a square wave. As the frequency gets higher, the rate at which the function changes increases. But if the spectrum has a finite cut-off, then that limits the maximum rate of change. So we can see that we might not want to take samples too many times, because the possible change between sample points is limited.

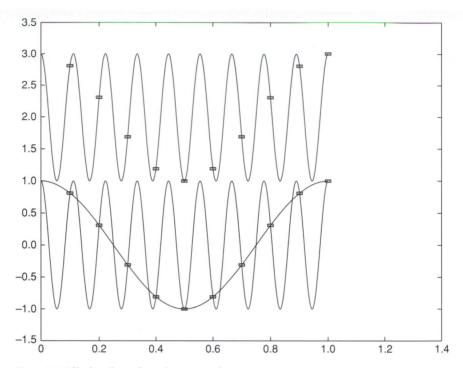

Figure 3.4 Aliasing through too low a sample rate.

It turns out that to reconstruct the signal perfectly we need to take sample points at just two per cycle of the maximum frequency in the signal. If we take samples further apart, *aliasing* occurs.

Computer graphics fans will already be familiar with aliasing artefacts – things such as jagged lines. But in fact the original meaning of alias was an assumed name, i.e. having more than one name for the same thing. In signal processing terms it meant one frequency masquerading as another. We show this in Figure 3.4. In the top half of the figure, we show the sample points as taken on a frequency, which is too high for the sampling rate. In the bottom figure, we show that there is another, lower frequency, which goes exactly through the same sample points. Thus the two frequencies are confused and errors result.

Now most images do not have simple periodic structures and contain mixtures of many frequencies. Aliases now appear at all these frequencies. The same errors occur as at the frequency level but they manifest themselves now in the jagged edges and pattern distortions with which we are familiar in the computer games world.

3.6.1   Reconstruction of the signal

Having sampled the signal, maybe stored it or sent it over the internet, we probably eventually want to reconstruct it as an analogue signal. This is the process of *digital to analogue conversion.* It is an essential step in the transformation of the digital signal on a CD to the sound waves which reach the listener's ears. Modern home theatre systems often have digital inputs for signals from DVD players for movie sound tracks as well as CD. But the signal which drives a loudspeaker is (mostly) a continuously varying voltage. So the series of pulses, the digital signal, has to be decoded into the signal amplitude at each sample time. Then these values have to be interpolated or smoothed, before they are sent to the loudspeaker[10].

The reconstruction process is simple. We multiply each sample value by an *interpolation function* and sum up the results, i.e. we convolve the series of sample values with the interpolation function. The exact form of this interpolation function comes from studying the Fourier Transform of the signal.

Sampling a continous signal creates a new sampled signal whose spectrum is the original signal replicated endlessly at intervals of twice the sampling frequency. Now, because the signal is band-limited and the sampling rate is high enough, these replicants do not overlap. Hence simply applying a brick wall filter restores the original spectrum. In the space or time domain this function is a *sinc* function, given in Table 3.1.

How does this transform replication come about? If we think of the signal in one dimension, stretching from $-\infty$ to ∞, sampling consists of extracting the value of this signal at regular intervals. We get this by multiplying by the comb function (§3.3.4).

Now, we have to think what this sampling process corresponds to in the Fourier domain. The multiplication becomes a convolution, which (§3.3.4) is just another comb function. But recall that the transform of the comb function is itself a comb function. The further spread out the teeth are in one domain, the more widely spaced they are in the other. It should now be possible to see that when we convolve a function (the Fourier Transform of our original function) with a comb function, we get the transform replicated forever centred around the teeth.

If we could now extract the original Fourier Transform from this endless set of images, we could recover the original function exactly. If we have sampled at the Nyquist frequency or above, these images do *not* overlap. Thus if we multiply by a **rect** function the size of one image, we can chop out all of the images except the original and get back our original Fourier Transform without error. Multiplying by a function in the Fourier domain corresponds to convolution in the time domain and the convolution is with the transform of a rect function. That transform is a **sinc** function. Thus we see how sinc interpolation really works.

But suppose that we have not sampled at the Nyquist frequency, but have *undersampled.* Now the images overlap. No matter what function we multiply by, we always get some horrible mixture. The mixing in of other images creates the errors which become the aliasing artefacts.

---

[10] At the time of writing a few manufacturers, such as Tact, are selling digital, or Class D, amplifiers which do not convert analogue to digital in the normal way at all, but merely send the loudspeaker a modified series of pulses. Such amplifiers are capable of very high accuracy as nothing is lost in the conversion process.

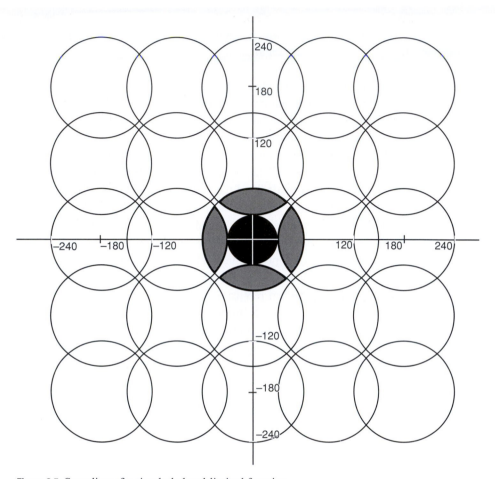

Figure 3.5 Sampling of a circularly band-limited function.

3.6.2   Two-dimensional sampling

The straightforward generation of sampling to two dimensions is to just turn our comb function into a brush. We have a grid of spikes extending in both x and y directions. Exactly the same arguments apply. The images in the sampling domain now correspond to points in all directions and the interpolation function now chops out a rectangle in the centre.

But for the pupil of the human eye and other optical systems with circular pupils, the cut-off is the same in all directions. Hence the Fourier Transform of the sampled signal appears as in Figure 3.5.

If the pupil is circular, there will always be some oversampling at the corners, where there is no signal. So a square sampling array is not perfectly efficient. If the circle has radius r, the ratio of image to sampled area is $\frac{\pi r^2}{4r^2} = \frac{\pi}{4} = 79\%$ efficient. Hexagonal sampling is an improvement. Simple geometry gives a sampled fraction of:

$$\frac{\pi r^2}{\frac{6r^2}{\tan 30}} = 0.91$$

Hexagonal sampling is common in animal visual systems. In the man-made domain, Fuji's SuperCCD has an octagonal sensor, the efficiency of which is left as entertainment for the reader. Looking at Figure 3.5, the empty space where the Fourier Transform is zero implies that the function which selects out just one spectrum could wander around in this region. Thus the interpolation function is not unique. It could be a Bessel function (the transform of a circle, Table 3.1), which is hard to calculate or it could simply be a two-dimensional sinc function corresponding to the tranform of a bounding square.

Another possible sampling strategy is to use irregular or pseudo-random arrays. These can be made isotropic, but they are more susceptible to noise.

3.7 The discrete Fourier Transform

The Fourier Transform is easier to understand in its continuous form (the CFT), but in practice in computing and engineering we usually work with its *discrete* form (the DFT). This is somewhat more messy algebraically, but a quick overview will help us to understand the idiosyncracies of FT software.

We take the Fourier Transform of a finite set of numbers. There are all sorts of details about periodic sequences buried here, and Nyquist sampling of a continuous signal always creates an *infinite* set. However, within this finite domain the properties of filtering and convolution still hold. We get for a function f sampled at N points, where $j = \sqrt{-1}$

$$F(v_k) = \sum_{\alpha=0}^{N-1} f(x_k)e^{\frac{-j2\pi v_k x_\alpha}{N}} \tag{3.43}$$

The input sequence of points is discrete and the output sequence is also discrete. We can make this easier to read and remember by just using the integer counters, k and α, i.e.

$$F_k = \sum_{\alpha=0}^{N-1} f_\alpha e^{\frac{j2\pi k\alpha}{N}} \tag{3.44}$$

3.7.1 The zero frequency term

If we set $k = 0$ then the exponential term has the value 1 at all times so we get

$$F_k = \sum_{\alpha=0}^{N-1} f_\alpha \tag{3.45}$$

which is often referred to as the DC term as for the CFT. In a time signal, this would be the sum of values for all times. For an image it would be the sum of all pixels. In general this is a large number which has to be handled with care when plotting the DFT.

3.7.2 Phase shifts

The DFT has similar phase behaviour to the CFT. If we shift all the f values by some constant, x_0, then we get:

$$F'_k = \sum_{\alpha=0}^{N-1} f_\alpha e^{\frac{-j2\pi v_k(x_\alpha - x_0)}{N}} \tag{3.46}$$

$$F'_k = e^{-j2\pi x_0} \sum_{\alpha=0}^{N-1} f(x_\alpha) e^{\frac{j2\pi v_k \alpha}{N}} = e^{-j2\pi x_0} F_k \tag{3.47}$$

showing that we have the original DFT multiplied by the *phase factor* $e^{-j2\pi\beta}$.

3.7.3 Negative frequencies

Negative frequencies are unfamiliar but are just simply an extension of the definition of frequency. Alternatively our common experience of frequency is just one aspect or domain of the Fourier Transform. For the discrete case we can rewrite eqn 3.44 to show the negative frequencies clearly:

$$F_k = \sum_{\alpha=0}^{\frac{N}{2}} (f_\alpha e^{\frac{j2\pi k\alpha}{N}} + f_{N-\alpha} e^{\frac{j2\pi k(N-\alpha)}{N}}) - 0.5 f_{\frac{N}{2}} e^{\pi jk} \tag{3.48}$$

But $e^{j2\pi\alpha}$ is just 1 throughout, so this becomes:

$$F_k = \sum_{\alpha=0}^{\frac{N}{2}} (f_\alpha e^{\frac{j2\pi k\alpha}{N}} + f_{N-\alpha} e^{\frac{-j2\pi k\alpha}{N}}) - 0.5 f_{\frac{N}{2}} e^{\pi jk} \tag{3.49}$$

showing that the negative frequencies start at the end and work backwards. Note that the highest frequency has value $k = \frac{\frac{N}{2}}{N} = 0.5$ which fits in nicely with the idea of the Nyquist theorem sampling the highest frequency at twice per cycle.

3.7.4 Real signals

The signals we deal with in sensory information, such as sound and light, are all real valued. Yet the Fourier Transform is complex. But if we start with N numbers, we can't just end up with $2N$ distinct values because we have moved to the complex plane. This is the same degrees of freedom argument as §3.5. There has to be a relationship between the Fourier values. If we take eqns 3.49 and 3.45 we see that the DC terms and the maximum frequency terms are both real. Thus we must have $N - 2$ remaining independent values. For these other values $F_k = F_{N-k*}$ where the asterisk denotes the complex conjugate.

3.7.5   Real Fourier Transform

We might wonder if there are any circumstances in which the DFT of a real signal is itself real. Again, using eqn 3.49, if we put $f_\alpha = f_{N-\alpha}$, i.e. alternatively make f symmetric about zero, then the two first terms of eqn 3.49 become complex conjugates of each other and the DFT values are all real.

3.8   Other frequency related representations

We have looked at the canonical sampling strategy dating back to the pioneers of communcations theory, Shannon and Nyquist at the beginning of the twentieth century. There are more complex strategies, which are more common in animal senses. Denis Gabor was a prolific theorist and inventor. Amongst his greatest practical achievements is the invention of the electron microscope. But he also introduced the eponymous function class (Gabor, 1946), which is an example of a whole family of sampling strategies. In fact, wavelets (§3.8.2), the latest incarnation, are now a major part of image processing and unerlie image compression standards such as JPEG 2000 (§7.6.2).

3.8.1   Gabor functions

The Fourier Transform of Dirac delta function, $\delta(x - x_0)$ becomes just a single complex frequency component $e^{j2\pi x_0 \nu}$ (§3.5.2). Following the scaling principle (§3.5.1), if we widen the delta function, then we convert the single frequency into a narrow band of frequencies. The first idea in the Gabor Transform is to introduce a spreading function, in fact a Gaussian.

The Gaussian, $g(x) = e^{-\pi x^2}$ has the easiest of all Fourier Transforms: it is the transform of itself, i.e.

$$g(t/a) \leftrightarrow g(a\nu) \tag{3.50}$$

So if we transform a narrow Gaussian function of time we get a broad, Gaussian spread of frequencies.

Gabor proposed that instead of sampling a signal as a series of point samples, where we take the value of the signal at specific, regular intervals, we sample with a Gaussian window. Gabor then showed that we can replace sampling at the Nyquist frequency by sampling at a lower frequency but taking several different widths of Gaussian window at each sample point. The Gabor functions are thus:

$$g_{mn}(t) = e^{im\Omega t} G(t - nT) \tag{3.51}$$

where $G(t)$ is the Gaussian window. Figure 3.6 shows this function.

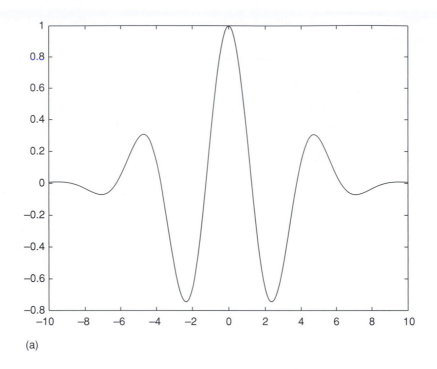

(a)

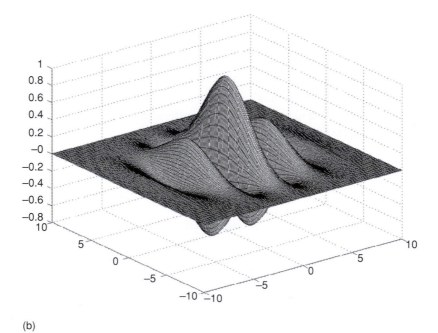

(b)

Figure 3.6 Gabor functions: 1D (a) and 2D (b).

In effect what we are doing is splitting the frequency spectrum with each Gaussian window and separately sampling each band. The bands overlap of course because the Gaussian is a smooth function without sharp edges. Gabor showed that such sampling strategies are exact providing we keep the number of sample points the same. So if we sample at a quarter of the Nyquist frequency, then we need four different Gaussian windows at each point. The *number of degrees of freedom* remains the same. Stepan Marcelja (1980) introduced Gabor functions as a model of the properties of cells in the visual cortex and numerous follow-up papers exploited their properties.

It turned out subsequently that there is nothing special about the Gaussian window. More or less any smooth function will do and there is an entire class of such transforms, known collectively as *Wigner transforms* (Wigner, 1932). Figure 3.6(b) shows a typical 2D Gabor function similar to the receptive fields of the primary visual cortex.

In creating the Gabor functions we start with a window function, in this case the Gaussian. We then shift it in space and modulate it with a sinusoid[11], which has the effect of shifting it in frequency. Recall that the Fourier Transform of the convolution of two functions is the product of the Fourier Transform of each function independently. A single frequency is like a point in the space or time domain, so the transform of the sinusoid is a spike, or Dirac delta function in the Fourier domain. The Gaussian is its own transform. A fat Gaussaian in the space domain transforms to a thin one in the Fourier domain and vice versa. So, the transform of the product of the convolution of the Gaussian and the delta function is just the Gaussian shifted out to the frequency of the sinusoid.

This process is quite general. We can pretty much use any reasonably behaved window function which we translate and modulate, for the Wigner Transform. In effect the window function is just like a spread out version of the delta function which we used to build the comb function for traditional sampling.

In the Fourier model the Nyquist theorem told us exactly how often it is necessary to sample to achieve exact reconstruction without noise. With these hybrid functions, localised wave packets, there are again criteria for how many functions we need to take. Gabor solved this problem with a visual analogy: for one-dimensional functions we plot frequency against time and consider the Gabor functions as tiles of dimensions Ω (frequency) and T (time). To get a *basis* the tiles have to be sufficiently small, which occurs when $\Omega T = 1$. In fact the basis is numerically unstable at this minimum number of functions and, with these wave packet function sets, *frames* are more common. Frames *oversample* and create a measure of *redundancy* in the function set.

3.8.2 Wavelets

The Gabor function is tricky to work with numerically (Bossomaier & Mackerras, 1992). Hence other window functions which were faster and easier to compute appeared. The

---

[11] We should really be talking about complex exponentials here, rather than sinusoids, since a sine or cosine transforms to two delta functions, one at positive frequency, the other negative. However the general argument is unchanged.

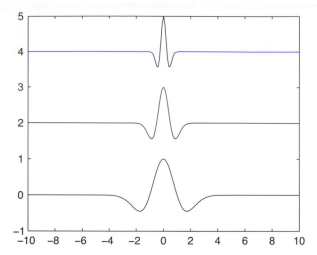

Figure 3.7 The first three dilations of the Mexican hat wavelet.

most important of these is the *wavelet* (Strang & Nguyen, 1996), which has become very important for practical applications. The wavelet transform runs faster than the Fourier Transform in many situations.

Whereas in the window transforms we modulate and translate, in the wavelet transforms we *dilate* the window function and translate it. In the wavelet decomposition we start from a single wave pulse, $w(t)$ which we then translate and *scale*. Translation produces the set of functions $w(t - k)$ and scaling changes the width of the function according to a scale factor of a^j. Thus our wavelet basis is:

$$w_{jk}(t) = a^{\frac{j}{2}} w(a^j t - kT) \tag{3.52}$$

The translations occur over a 'period', T with a dilation factor, a. The product $T \log a$ is important just like the product ΩT for the window functions. Again the basis-frame transition occurs at $T \log a = 1$. Common values are $T = 1$ and $a = 2$.

Again, the wavelet function can be pretty much any well-behaved function, although we shall remain a bit coy about what well behaved really means. One popular form of wavelet is the Mexican hat, used in computer vision. It is the second derivative of the Gaussian:

$$w(t) = (1 - t^2) e^{-\frac{t^2}{2}} \tag{3.53}$$

Figure 3.7 shows the first few dilations of this wavelet.

Wavelets really became important in the 1990s and a great many variations in window functions and scaling properties came under scrutiny. They now form the basis functions for the latest image compression standards such as JPEG.

Biological systems are in general fairly loose in their implementation of any such rigid formalisms. The ideas, though, of wavelets and scaling will be very useful in the book.

3.8.3   Finiteness

The Fourier Analysis and Fourier Transform operate assuming infinite time or space dimensions. Cosine or sine waves go on forever. Yet many practical systems are finite. A variety of mathematical tricks enable us to get around this problem.

Using wavelets, or some sort of local transform is a practical solution: we can just assume everything outside some region is zero and thus the contributions of wavepackets outside the region will be zero too.

A more powerful method is to look at eigenvectors corresponding to finite boundary conditions. This takes us somewhat beyond the depth of this book, but we do briefly analyse an example of this when we cover vision. For finite domains, instead of talking about a sampling *rate*, we move to talking about a total number of samples in the whole domain. We call this number the *degrees of freedom*. A different set of eigenfunctions (since these systems are no longer shift-invariant) takes over, *prolate spheroidal wavefunctions* (Slepian & Pollak, 1961). These have the property that they are band-limited in one domain and maximally concentrated in the other.

3.9   Statistical methods for finding representations

Fourier methods are fundamental to much of optical and general signal processing. But the representations used in animal neural systems are not perfect Fourier representations. Although Fourier Analysis is a good start in matching sensory detectors to the world, it is not perfect, and with advances in measuring techiques (such as digital cameras) and more powerful computers, other techniques have come to the fore. These often involve statistical or entropic concepts and thus overlap somewhat with §3.9.3 which discusses entropy. Two of them are *principal component analysis* and *independent component analysis*. The first centres around Gaussian approximation of signals. The second, the reverse, signals which are minimally Gaussian in some way, which we shall briefly discuss. The distinction is quite subtle. In the first case we find a set of functions, like the complex exponentials of Fourier Analysis, such that there is no correlation in the expansion coefficients, the a_i in eqn 3.10. The expansion functions are orthogonal. In the second the basis functions are *statistically independent*, but the coefficients are not necessarily so.

A third idea, linked to both to some extent, is the idea of sparse coding and minimum entropy codes, which we consider briefly in §9.3.8.

3.9.1   Principal component analysis

Principal component analysis is one way of describing a particular representation, but it goes under other names too. The *Karhuhen–Loéve Transform* is essentially the same thing, while *singular value decomposition* is closely related. The representation is formed from the eigenvectors of the correlation matrix.

Formally, for a set of signals, $x(t)$, the Karhuhen–Loéve is formed from the integral eqn 3.54:

$$\int R(t, t')\phi(t') \, dt' = \lambda\phi(t) \tag{3.54}$$

for an expansion such as eqn 3.10. Glossing over some of the mathematical details, such as the exchange of integration and averaging steps, we can see intuitively how to obtain this equation.

For independent coefficients:

$$\mathcal{E}\{a_i, a_j\} = \delta_{ij} = \mathcal{E}\left\{\int \phi_i x(t) \, dt \int \phi_j x(t') \, dt'\right\} \tag{3.55}$$

But $\mathcal{E}\{x(t)x(t')\} = R(t, t')$ giving:

$$\mathcal{E}[a_i, a_j] = \int\int R(t, t')\phi_i(t)\phi_j)\phi_j(t') \, dt \, dt'$$

But since the ϕ_i are orthogonal, then eqn 3.54 gives us:

$$\mathcal{E}[a_i, a_j] = \int \lambda\phi_i(t)\phi_j(t) \, dt = \lambda\delta_{ij}$$

as required.

Section 3.2.4 described how the complex exponentials are the eigenfunctions of shift-invariant systems. In such a system, $R(t, t') = R(t - t')$. From the shift theorem of Fourier Transforms:

$$\int R(t' - t)e^{-j2\pi vt'} \, dt' = e^{-j2\pi t} S(v)$$

where S is the spectral density, the Fourier Transform of the autocorrelation function. This equation matches eqn 3.54.

If instead of using a continuous signal $x(t)$ we use a set of samples, then the integral equation reduces to a standard matrix eigenvalue equation:

$$\mathbf{R}\,\vec{x} = \lambda\vec{x} \tag{3.56}$$

which is the form in which principal component analysis is usually expressed. This discrete form is particularly useful in much biological sensory analysis because of the discrete nature of receptors and nerve cells.

Note that we have implied nothing about the relationship between the basis functions, ϕ_i, except that they are orthogonal. Independent component analysis takes an alternative approach and makes the basis functions the focus of attention.

3.9.2 Independent component analysis

Independent component analysis (ICA) is a much more recent technique, dating from the early 1990s (Comon, 1994). The domain usually used to get across the idea of ICA is the cocktail party phenomenon (Hyvärinen, 1999). In this scenario we have a number of people all talking away independently and we want to separate one conversation from another, and maybe concentrate on the most salacious. So in trying to untangle these signals, a reasonable starting point is that they are statistically independent. Now the central limit theorem in statistics tells us that as we add independent random variables together the aggregate distribution gets more and more Gaussian. Thus finding the independent components boils down to finding the maximally non-Gaussian set of components.

Thus we assume that our fragments of conversation (or signals in general), which we denote by $x_i(t)$ are made up by the superposition weighted sum of a number of N independent conversations, $s_j(t)$, with coefficients a_{ij}:

$$x_i(t) = \sum_{i=1}^{N} a_{ij} s_j(t) \tag{3.57}$$

We know neither the matrix a_{ij} or the components themselves, $s_j(t)$. Thus a procedure will be to find both via numerical optimisation. The first requirement is a measure of non-Gaussian character. A very brief description follows in §3.9.3, but this is a statistically difficult area beyond the scope of this book. A fuller description is given by Hyvärinen (1999). The optimisation is usually numerical rather than finding some functional representation. A series of signals $x_i(t)$ are collected from the domain of interest and digitally processed to arrive at a set of signals expressed as a series of digital samples, much in the manner of many filters for digital signal processing.

3.9.3 Measuring deviation from Gaussian

It will come as no surprise that there is no unique measure of how far a distribution deviates from a Gaussian distribution. Some of the statistically best methods turn out to be very difficult to estimate in practice. So, ICA contains a considerable amount of judgement and variability in its application. It has, nevertheless, proved to give useful insights into sensory information processing.

Broadly there are two approaches to measuring the deviation from Gaussian. The first uses the shape of the probability function directly, usually using a property called *kurtosis*, $K(x)$, defined by eqn 3.58:

$$K(x) = \, < x^4 > -3 < x^2 >^2 \tag{3.58}$$

where the angle brackets denote expectation values.

The alternative is to use a property of the Gaussian distribution discussed in §4.2.8. Of all distributions, the Gaussian has the highest *differential entropy*, $H_D(x)$, for a given variance. Thus we can define a quantity *negentropy*, $J(x)$, which is how far short a distribution falls short of the maximum as eqn 3.59 shows:

$$J(x) = H_D(x_g) - H_D(x) \tag{3.59}$$

where x_g corresponds to the Gaussian distribution with the same variance as x.

There are many numerical twists and turns in using either of these approaches. Only those used in particularly sensory applications will be considered as we come across them.

4 Introduction to information theory

Yet herein will I imitate the sun
Who doth permit the base contagious clouds
To smother up his beauty from the world
Then when he comes
He wished for comes
And nothing pleaseth like rare accidents.
King Henry IV Pt. 1,
William Shakespeare

4.1 Overview

Information must be one of the most used words of today's world. Computers, media and the internet dominate many of our lives. The volume of information we collect and store increases annually, so much so that energy efficiency for data storage and manipulation is becoming a serious issue (Ranganathan, 2010). This omnipresence has given us everyday measures of information, bits, nips, nibbles, bytes, kilobytes, megabytes, gigabtyes terabytes, petabytes, exabytes... But what do these quantities really mean?

Entropy is effectively average information, and the quantity which defines the capacity of a transmission channel or storage medium. It is this second of the two fairly mathematical chapters addressing three interrelated concepts: information and entropy (§4.2, §4.2.1); noise (§4.4); and bandwidth, which links back to the previous chapter on Fourier Analysis. The emphasis is, again, on the concepts and how to visualise what happens rather than detailed physics and mathematics. Bialek (2002) gives a very detailed and thorough account of the mathematical framework.

Information has a companion concept, entropy (§4.2). Whereas anybody would offer you a definition of information, fewer would hazard a guess at entropy. Yet together they are amongst the most powerful of all concepts in science and engineering. In statistical mechanics, entropy is a measure of the disorder of a system and corresponds to the number of microstates a system can occupy. Without delving too deeply into the physics, which space does not permit, the relationship to computer systems can be explained fairly simply.

Information (§4.2.1) is a very widely used term, varying from the very precise (e.g. size of a computer hard disc) to the very fuzzy (information on a web site – amount of space on a disk, useful data relative to other sites, unique repository of facts or what?). We are going to look at the precise, quantitative side. In particular we want to know how to estimate the information content of a *sensory signal*. We can use this in data compression, human–computer interface design or building an animat.

Chapter 3 dealt at length with signals and their representation. Complementary to the signal is the idea of a *channel* to transmit signals in the most effective way. A channel can be an electrical wire or an optical fibre. It can also be a nerve cell or biological receptor and is a core theoretical concern of sensory systems.

It is quite useful to think of a memory as like a channel too, where we have the same concept of capacity and use the same unit, the bit. Once we know the information in the signal we know what sort of *channel capacity* we need to transmit this signal and we know what size storage medium to record it (§4.5). Again, channel capacity is an everyday concept. Most people who connect to the internet from home will know the speed of their modem in kilobits per second. Each animal sense transmits information from the sensor, the eye, ear, nose, to the brain and this book tries to achieve a measure of this data transmission similar to the way we rate a modem. This is immediately useful. We can say what modem (or internet speed) we need to completely satisfy the data capactiy of our senses, and thus define the requirements of artificial systems.

The Nyquist theorem states (§3.6) that in the absence of noise it is not necessary to take samples really close together to get perfect reconstruction of a signal – if the signal is band-limited in the right way. So it is also with channels. One of Shannon's surprising and exceptionally powerful results was that a channel does not need to be completely free of noise or interference to send a signal *with arbitrary accuracy* provided the signal is limited and encoded in the right way (§4.2.6). Imagine, for example, that we have bags which will hold up to 1 kg of rice, but the rice is contaminated with 5% of fine sand. We can still get 950 g of rice in the bag and can use a filter to separate the rice from the sand afterwards. It is possible to drill a 3 mm hole in a music CD and still reconstruct the information *exactly.*

Today, like the Nyquist theorem, this might seem obvious. At the time it was a profound insight. It is sad to note that during the writing of this book Claude Shannon died. Shannon's paper is still a benchmark in clarity of technical exposition and profound insight (Shannon, 1948).

Light, sound and electrical signals are all continous on a macroscopic scale. But for all practical purposes discrete systems are more useful in the study of the senses. The reason is noise, which effectively chops up continuous signals into discrete chunks. Making this idea precise is what is tricky, but it is easy to understand at an intuitive level. Suppose, that the petrol with which we fill up our car has been mixed with 10% of ethanol. We could be pretty sure that we have around 90 litres of petrol if our tank holds 100. But it would be pointless to say we had 90.0 because this level of accuracy is greater than the likely variability in the ethanol contamination. Section 4.4 takes up the central issue of noise in more detail.

Noise is always with us. It determines how accurately we can measure things. In fact once we have determined the capacity of a channel in the presence of noise, we can then, in effect, ignore the noise, using the same limits on the rate at which we can feed information down it with arbitrarily small errors. At the simplest level this is obvious. If we allow enough room for the signal and the noise, then we can transmit signal and noise – providing that they do not get inextricably tangled.

In the computer age these ideas are commonplace: storage devices include redundancy to allow them to recover from read errors or damage. But animal sensors, eyes, ears etc., are analogue systems creating continuous (§4.2.9) rather than discrete signals[1]. Digital systems are discrete. We talk about so many distinct bits on a hard disc, so many transmitted along a wire, or in many areas of signal processing or computing. Continuous systems are more like the printed image[2].

In some images you can actually see the noise. An X-ray in a hospital exposes the patient to dangerous radiation, which can itself cause cancer. Doctors want to minimise the radiation dosage to reduce the risk to sickness from the diagnosis itself and thus make the flux of X-rays as weak as possible. But X-ray photons are highly energetic (which is largely why they are dangerous), so, compared to a light image, the X-ray image is made up of a much smaller number of much bigger photons than is an optical image. This causes a very coarse granularity, referred to as quantum mottle.

After having studied entropy, information and noise, it is possible to combine them into channel *bandwidth* and encoding. The transmission and analysis of sensory information takes place using neural networks. Biological neurons have their own limitations and speed and noise and these issues as they relate to senses are taken up briefly in §4.6. Again in our computer literate world, this is now an everyday concept, but the goal is to compute it in sensory systems where the information is not digital in the manner of computers and telecommunications.

4.2   Entropy and information

Entropy is one of the most abstract concepts with which to come to grips. It is often described as a measure of the disorder of a system, the lower the entropy, the higher the order. There is a common sense link to how easy it is to find things within a system. The author has a high entropy office where searching for things is a major undertaking.

Entropy is a powerful idea which has spread from thermodynamics where it originated, to many other fields, including chaos, and of course, computing. So, the second law of thermodynamics tells us that the entropy of a system *always* increases – you can never win out against disorder as no doubt many of us have experienced in family, home and office. We shall find that entropy serves as a sort of measure of capacity so it will be useful to keep this idea on hold.

---

[1] This is not strictly true at the smallest scales where signals transfer from one biological cell to another as ions or tranmitters, but this microscopic scale is outside the scope of the book.

[2] We should not be too pedantic here; in fact even a photograph is discrete, composed of silver crystals in the negative or dye clumps in a print.

4.2.1   Information

The storage capacity of a computer disc is a convenient place to start. How many different files could you store on a 1 G bit disc (available storage after formatting etc.), each of which completely fills it? There are 10^9 bits[3], each of which may be 0 or 1 making 2^{10^9} possible files. The information capacity, or entropy H in this case, in bits is simple the logarithm to base 2 of the number of possible files, N_f which could be stored[4]:

$$H = \lg N_f \qquad (4.1)$$

The choice of the logarithm as a measure of information was not in any sense a unique one. But Shannon (1948) argued that it has a number of convenient properties. For example, it makes information additive. So when we add an extra hard disc we would like the total information to be the sum of each individual disc. We get that with the logarithmic formulation.

This equal likelihood is tricky. On a computer disc there is no difference between one bit pattern and any other, but this is not true in general. One way we can think about it is in the way states clump together. Suppose we think about the mixtures we can make with six letters. We plot them on six axes, each point along the axis corresponding to a letter. If we take say, 1000 arbitrary mixtures of 3 letters, they will form a uniform cloud. But if we take English words, this cloud will be clumpy. Some combinations of letters are very common, others never occur. Qantas, the Australian airline, is one of the few well-known words in which 'q' is *not* followed by a 'u'. So we can see intuitively that the space of English words is not quite so random as the first case and it has a lower entropy.

We can rewrite eqn 4.5 in terms of the probability $\frac{1}{N}$ for each state as:

$$H = -\sum_{1}^{N} \frac{1}{N} \lg\left(\frac{1}{N}\right) \qquad (4.2)$$

which allows us to see how we deal with states of varying likelihood:

$$H = -\sum_{r=1}^{N} p_r \lg(p_r) \qquad (4.3)$$

The quantity $\lg(p_r)$ is the *information* and thus S is its average over the set of states.

Figure 4.1 shows the relationship between entropy and information. The relationship is now an exact one. A number of people, notably Richard Feynman (Feynman, 1984)

---

[3] The name *bit* comes from J.W. Tukey, who we came across in the previous chapter as one of the inventors of the Fast Fourier Transform.

[4] This is very similar to the basic definition of entropy. S, in statistical mechanics, where, for a system with N states, the entropy is given by the same equation but with a multiplier k, Boltzmann's constant, and the logs are taken to base e and where all the states are assumed to be equally likely. Without going into detail, essentially all this is saying is that there are numerous states all *equally likely* which are indistinguishable. So in a gas, very many molecules travel in all directions with wide variation in energy. A swap of the speeds and directions between any two molecules does not in anyway affect the temperature and pressure of the gas.

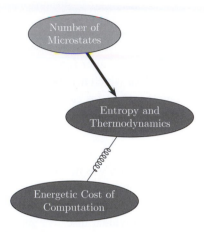

Figure 4.1 Entropy and information.

at Caltech, and Charles Bennett (Bennett, 1982) and Rolf Landauer (Landauer, 1961) at IBM, asked just how much energy computation requires. The eventual result turned out to be zero! Computation itself does not require energy. But the destruction of information does, and many algorithms destroy information along the way. The energy required depends on the temperature, and it turns out to be precisely **kT,** where **T** is absolute temperature and **k** is Boltzmann's constant.

An interesting consequence of such thinking is to ask how *thermodynamically efficient* biological neural computation is. Laughlin and others laid the foundations for doing this by studying the costs of neural spikes and synaptic transmission. It turned out that it requires 10^4 ATP molecules to transmit a bit at a chemical synapse and up to 10^7 for neural transmission by spikes with each ATP conversion requiring approximately 25 kT (Laughlin *et al.*, 1998). Compared to a man-made computer it is surprising just how efficient biological computation actually is.

Despite the relative thermodynamic efficiency of neural computation, the brain still uses a lot of energy, around 20% of the total at rest and relatively far more than any other part of the body. Laughlin and Sejnowski (2003) argue that these voracious demands for energy constrain brain structure and function to a significant extent. The importance of this for sensory information is in the idea of sparse codes, discussed in §4.3.1.

Let us now look at information from the probability of events. Information is a reduction of uncertainty. The less likely an event, the more information we can glean from it, as shown in Figure 4.2.

Sources of information are a dominant feature of our lives, from newspapers to television, from email to the world wide web. With the advent of so much information being delivered to us by computer, we have a general awareness of how information is measured. We know that downloading a 10 Mbyte image takes longer than a 10 Kbyte text file along the same connection. But to understand how these measurements come about, we need to go back to the statistics (and entropy) of the *information source*.

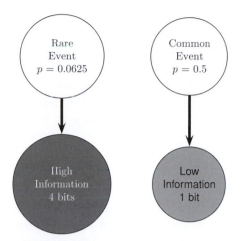

Figure 4.2 Uncertainty and information.

If we toss a coin and are told the answer is heads, we reduce the uncertainty of the outcome. Since there were two options, with equal probability one half, we say that the information gained is $H = -log_2 p$ (Hartley, 1928) which is just 1 bit as we would expect. As the probability of an event becomes more unlikely, the information goes up, again, just as you would expect.

So, if you know that your next door neighbours come home at 5 pm on a Wednesday, then go out to a movie at 6 pm, coming home around 9 pm, then we expect to see the car in the driveway between 5 and 6 pm and then not until 9 pm. So car not around at this time conveys little information. But if the car is there between 6 and 9 pm, we do learn something. If they quite often do something different, maybe watch a movie on TV instead, then it doesn't tell you very much. But if it's unusual for them to break this pattern, we get a stronger message. Perhaps one of them is sick, or the car is giving trouble.

This notion of rarity leads to the definition of information, i_r, from an event r implied above:

$$i_r = -\lg p_r \tag{4.4}$$

for an event, R of probability, p_r. This is a *statistical* concept, defining the information relative to some set of objects or events. For the study of the encoding and transmission of sensory information this is the concept we need. In some ways it would also be attractive to think about some sort of intrinsic information of an object. But this intuitive concept is mathematically somewhat intractable. But it has become mathematically precise quite recently with the invention of *algorithmic information theory* (§4.3). We get the entropy by averaging over all possible events according to their likelihood as in eqn 4.3. So the entropy is the average information of the source from eqn 4.3:

$$H = -\sum_{r=1}^{N} p_r i_r \tag{4.5}$$

Once we have the entropy of a signal we are in a position to determine the *channel capacity* to transmit it without error.

A useful analogy in thinking about channel capacity is the number of seats on a bus. The bus can carry a maximum number of people, but seats may be occupied by luggage, and the number of people already on the bus is an uncertain quantity.

The extension of entropy to two or more variables is easy. The joint entropy, for two random variables, X and Y, consisting of sets of events x_i and y_i is:

$$H(X, Y) = -\sum_{i.j} p(x_i, y_j) \lg p(x_i, y_j) \tag{4.6}$$

If events x_i and y_j are independent, then $p(x_i, y_j) = p(x_i)p(y_j)$ from which it follows that the joint entropy is just the sum of the entropies of the components. They are independent and do not interact, so the entropy is additive; the information from two sources is additive if they are independent of each other.

If one set or variable is broken up into subquantities, the entropy is simply obtained by adding up the subcomponents one by one. Consider a simple shopping list for a fresh produce market, consisting of carrots, broccoli and fruit. The number of possible shopping lists is eight and if each item occurred equally likely on the list 3 bits would be needed for the list. But suppose that fruit, f, gets bought more often, say, $\frac{5}{12}$ of the time, with vegetables, v, $\frac{7}{12}$, carrots, c, $\frac{3}{7}$ and broccoli, b, $\frac{4}{7}$. Now, because fruit is there nearly half the time, there is less information to encode. Section 4.2.6 describes how to do this. The total entropy, $H_{shoppingList}$ 1.555 bits, which is also equal to:

$$H_{shoppingList} = -p_{fruit} \lg p_{fruit} - p_{veg} \lg p_{veg} + p_{veg} H_{veg}$$
$$= 0.526 + 0.454 + 0.575 = 1.55$$

In the case where they do interact, the combined entropy must be lower, because knowing something about one tells us something about the other. For this we need the *conditional entropy*. If we are given the value of one random variable, X, we can ask about the entropy of some other variable, Y, given X. This is the *conditional entropy* denoted by $H(Y|X)$ which requires the conditional probabilities, $p(y|x)$. For example, a perfect typist intending to type an a will type a, and nothing but a, but most of us make mistakes occasionally so we might end up with an s or some other letter. Thus we can define the probability $p(s|a)$ as the probability of getting an s given an attempt to type a.

To express the conditional entropy in terms of event probabilities, first consider just one event, x_i, and ask what is the entropy of Y given this value. This might be the range of possible outputs given a single input value. Thus the conditional entropy of Y given x_i is:

$$H(Y|x_i) = -\sum_j p(y_j|x_i) \log_2 p(y_j|x_i) \tag{4.7}$$

To get the conditional entropy of one distribution against the other, we just have to average over all the events x_i, i.e.

$$H(Y|X) = -\sum_i p(x_i) \sum_j p(y_j|x_i) \log_2 p(y_i|x_j) \tag{4.8}$$

The joint entropy may now be expressed in terms of the conditional entropies:

$$H(X, Y) = H(X) + H(Y|X) = H(Y) + H(X|Y) \tag{4.9}$$

So start with either the entropy of X (or Y) and add the entropy in the other after taking out the variation due to the first. If X and Y are independent, then $p(y|x) = p(y)$, eqn 4.7 gives $H(Y|X) = H(Y)$ and we get back to the sum of the indendent entropies as required.

These ideas of entropy and information are really confusing when people first come across them. But, despite the mathematics of this chapter, the ideas can be understood in a really simple way.

Imagine that we have some 'scraethies' made up of pictures of Hollywood stars of the golden age of cinema. We have 256 pictures and we want to know the star hidden underneath the coating. This is a completely arbitrary collection, not an A-list or a list of Oscar winners or anything else systematic. We might find Kathrine Hepburn but not Cary Grant. So when we are told that a picture is, say, James Stewart, we have gained precisely 8 bits of information. To put it another way, we need a byte of information to specify all the film stars in our collection.

First we scratch away the top and find that our star has black hair. Gentleman of course prefer blondes, so really black hair is not so common in our collection, and in fact there are just 16 stars with black hair. So, the probability of black hair is $\frac{1}{16}$ and we have gained $log_2 16 = 4$ bits of information. We keep scratching and now find something really interesting: our star has black skin. There are just two black stars, Sidney Poitier and Louis Armstrong. Our probability of drawing a black star was thus 2 in 16 and worth 3 bits. The 1 bit left is the choice between the two remaining stars.

4.2.2 Mutual information

One of the most widely used statistical and information theoretic techniques in use at present was almost lost to humanity! The Reverend Bayes, an eighteenth-century vicar, published a small number of theologically oriented works during his lifetime. However, on clearing up miscellaneous papers after his death, an unpublished manuscript was discovered. It contained a theorem, now known as Bayes theorem, now of enormous importance, having virtually spawned an entire field of statistics now spread into network theory.

Bayes theorem states (Papoulis, 1984):

$$p(x_i|y) = \frac{p(y|x_i)p(x_i)}{\sum_k p(y|x_k)} \tag{4.10}$$

$p(x_i)$ here is the *a priori* probability of of x_i, before we know anything. $p(x_i|y)$ is the *a posteriori* probability of x_i the value after we know y. The idea is that probabilities are conditioned on our prior knowledge. A really interesting demonstration of how the brain effectively uses Bayes theorem in practice came from Summerfield *et al.* (2006). It illustrates the complex dynamics of sensory perception, an issue to which we return briefly in the last chapter. Essentially they set up three categories of images, faces, houses and cars, contaminated by noise making them very hard to identify. Subjects were then given a 'mindset' as to whether there would be faces. Much other work elsewhere had demonstrated the sorts of areas in the prefrontal cortex which are active as it processes faces. These areas lit up during the face mindset regardless of whether a face image was present or not!

An extremely important concept in measuring the communication of information is a symmetric combined entropy, the *mutual information*. Given two variables linked in some way it measures how much the probability distribution contains about the other. The following discussion uses entropy, but alternative approaches define it in terms of the Kublick–Liebler divergence, referred to briefly in §4.2.2.1, between the (conditional) distributions.

When a communication channel is used to transmit a set of messages, X, which are received as some set Y, the entropies of these messages can be used to calculate the channel capacity. Intuitively, a good channel will have a very close or perfect matching between input and output, and the the mutual information of the two will be very high, or, alternatively the channel will not generate any additional entropy.

From the point of view of the receiver, the input is uncertain, given the output. This is $H(X|Y)$, sometimes also referred to as the *equivocation*. From the sender's point of view, the input X is known, but the output may not be, and we call the quantity $H(Y|X)$ the *prevarication*. In general the two are not equal.

Imagine Karen in one department sends a set of coloured forms to Keith in another. Each form may be red, green or blue. Jill knows the colour of the form she sends (women are not often colour blind, whereas men quite often are (§8.3.1.4)). If the probability of red, green and blue are, respectively, $\frac{1}{4}, \frac{1}{3}, \frac{5}{12}$, then the source entropy is 1.55 bits. But Keith *is* red–green colour blind. So he sees blue forms and some intermediate colour, let's say, brown, with probabilities $\frac{7}{12}, \frac{5}{12}$. The mutual information is simply whether the form is blue or brown (red/green). The prevarication is zero, so the mutual information is just 0.98 bits – the sender knows what she sent and that her colleague is colour blind. The equivocation is non-zero, because the receiver cannot distinguish red from green[5].

In most situations we are not interested in some absolute information, but the information one thing tells us about another. We call this the *mutual information*, which we usually denote by $I(X; Y)$ for the mutual information about X given Y[6]. It is a symmetric

---

[5] This example is a little provocative. Section 8.3.1.4 argues that such mistakes would be rare in practice for colour blind individuals.

[6] Strictly speaking this common nomenclature is inaccurate; I is a function of the probability distributions of the random variables X and Y, not the variables themselves.

Table 4.1 Vehicle probabilities. (4WD Toyota = 4WDT; Other 4WD = 4WDO; Other Toyota = T; Any other = O.)

Vehicle	4WDT	4WDO	T	O
Probability	0.2	0.1	0.3	0.4

Table 4.2 Vehicle entropies.

Entropy	$H(X)$	$H(Y)$	$H(X, Y)$	$H(X\|Y)$	$H(Y\|X)$	$I(X, Y)$
Value	1	0.88	1.85	0.85	0.97	0.03

quantity. Its value is:

$$I(X; Y) = H(X) - H(X|Y) = H(Y) - H(Y|X) = H(X) + H(Y) - H(X, Y)$$

(4.11)

This looks at first, somewhat confusingly, like the joint entropy, but if we substitute for $H(X|Y)$ in the case when the variables are independent, we find that the mutual information is zero.

Let's consider a simple example. Suppose Xak and Yentl both keep logs of traffic along their road, each for their own reasons. Xak works for the local Toyota dealer and is worried about the sales revenue in the area. Yentl is worried about the recent reports which say that most children injured on the road or in driveways are victims of four wheel drives (4WDs) and is getting together a petition to the council on urban speed limits. Thus Xak watches Toyotas while Yentl watches 4WDs. Since Toyota makes Landcruisers and other 4WDs, clearly, Xak and Yentl have some information in common. $H(X)$ is the total information Xak collects. $H(X|Y)$ is the information left after we've taken out the 4WDs. So the mutual information is the 4WD Toyotas. On the other hand, $H(Y)$ is the total information Yentl collects about 4WDs while $H(Y|X)$ is the information left after we've taken out the Toyotas. So again, the mutual information is the 4WD Toyotas.

Table 4.1 shows some illustrative probabilities.

From the above equations we then find the entropy values in Table 4.2.

4.2.2.1 Comparing probability distributions

The Kullback–Liebler divergence, D_{KL}, is defined by:

$$D_{KL}[p||q] = \sum_x p(\xi) \lg \frac{p(\xi)}{q(\xi)}$$

(4.12)

for two probability distributions p, q. It is called a *divergence* because it is not symmetric and therefore not a metric. If we now let p be the joint distribution and q the marginal distributions, we get:

$$D_{KL} = \sum_x p(x, y) \lg \frac{p(x, y)}{p(x)p(y)} \tag{4.13}$$

4.2.3   Alphabets and codes

Discrete channels (digital telecommunications, computer buses, discs . . .) operate with *alphabets* of *symbols*. In the case of the vehicles we had just four different vehicles, in other words an alphabet of four symbols. Channels or storage systems may get these symbols mixed up, creating errors or noise. There are very many different codes with different design considerations. They are not going to be particularly important in this book.

Codes and alphabets can be anything: the theoretical framework only needs the probabilities of the events which can occur. For example, an ancient custom might use small bouquets of just three flowers exchanged between dating couples to indicate the strength of their interest. Three carnations indicates friendship and interest, three roses extreme passion. By convention flowers are sent by a go-between. There are four messages, corresponding to zero to three of either flower.

Jack and Jill, however, are not having a smooth ride. Their go-between, Jeremiah, has his own affairs to run and is always short of roses! He feels he can swap a carnation for a rose without too much risk of discovery.

Jack is a fickle character and the bouquet he sends weekly to Jill varies randomly across all five options, each having an equal likelihood. Half the time Jeremiah does a swap, of which Jack and Jill are aware, but they put up with it. Good go-betweens are hard to find.

We can see straight away that there is asymmetry in their viewpoints. When Jack sends three roses he knows that Jill might receive only two, but when he sends three carnations, that is definitely what Jill will get. But from Jill's point of view, she knows that if she received three roses, Jack is head-over-heels in love. But three carnations could still harbour some signs of passion.

In this scenario, the symbols are not the individual flowers, but the bouquets. The flowers are an *encoding scheme,* just like the representation of letters by the 7 or 8 bit binary codes defining ASCII text.

Neural codes have now become of great interest as neural measurement techniques get better and the sensory systems of insects have been excellent test cases for elucidating encoding strategies (Rieke *et al.*, 1996). Some are discussed in §4.3.1. Bandwidths of invertebrate cells are comparable with the values we get for mammals. The complexity and power of mammalian brains comes from the number and connectivity of the neural networks rather than in increased performance of the individual neuron. Section 5.4.3.1 gives an example of information extraction in crickets close to the theoretical maximum.

Table 4.3 Numbers of students in team sports.

Sport	Female	Male	Mixed
Tennis	6	12	9
Football	15	18	0

Table 4.4 State transition matrix for gender mix in sports.

0.2857	0.4000	1.0000
0.7143	0.6000	0.0

4.2.4 The state transition matrix

The mutual information is most often used in the context of communication channels. Data at the input from some alphabet, X, gets mapped at the output into some alphabet Y, which may in fact be the same. If the communication is perfect, then the entropy of the input and output will be identical. But channels are frequently corrupted in one way or another as we discuss in §4.4.

Every symbol in the input, x_i, in a perfect world would correspond to one and only one symbol, y_i, at the output. In practice, other symbols may occur instead. So, we can define what we call the *state transition matrix, $Q_{j|i}$* which is the probability that symbol y_j was received if the input was x_i. If there are no errors then $Q_{j|i} = \delta_{ij}$[7].

We can now expand the entropy in terms of probabilities:

$$I(X;Y) = \sum_j p_j \sum_k Q_{k|j} \log_2 \frac{Q_{k|j}}{\sum_i Q_{k|i} p_i} \tag{4.14}$$

It is fairly easy to show that this is equivalent to eqn 4.11.

Example

A survey of a group of 60 athletic students reveals that they play football and tennis, sometimes in mixed and sometimes in same-sex teams as shown in Table 4.3.

We want to find the mutual information in gender and sport. The entropy in sport, $H(S)$, comes out to be 0.993. Calculating the conditonal entropy for sport, given male, female and mixed come out to be $H(S|female) = 0.863$, $H(S|male) = 0.971$ and $H(S|mixed) = 0$. Mutliplying by the probabilities for male, female and mixed, we arrive at a mututal information of 0.205.

Alternatively we can determine the state transition matrix from gender mix to sport, which turns out as in Table 4.4.

We now insert this into eqn 4.14 to again arrive at a value of 0.205.

---

[7] δ_{ij} is the Kronecker delta which takes the value one if i and j are the same, and the value zero otherwise.

4.2.5   Predictive coding

Bayesian methods provide a general framework for signal detection, encoding or decoding against prior knowledge and we saw an example above (§4.2.2) of how the brain seems to use such methods. We make assumptions about what we are going to perceive and filter the incoming data appropriately. The precise details of how this works still remain unresolved. But at a more local level, the use of *predictive coding* has been demonstrated in the early stages of vision (§6.6.1.1).

In many real-world signals, each signal value exhibits correlations with its neighbours, recent intensities in sound, or adjacent pixels in images. One very effective compression method is to exploit this correlation on a pixel by pixel basis in predictive coding. The concept is simple. Because of the correlation between pixels, it should be possible to guess a pixel value based on its neighbours – in a blue sky, most pixels will be bluish. Thus measuring and encoding the prediction error will in general be more economical.

In classical methods, the coding is done on the fly. For a one-dimensional signal, the previous pixels are used. For an image, scanning from left to right, the pixels to the left and in the row above can be used. Obviously, if the whole image were available in memory (now much easier to achieve than when these algorithms were designed), then one can get a better estimate by using the neighbours in all directions.

Given that we have decided upon the neighbourhood, R, of pixels, p_i, to be used for the prediction, p_{est}, we now have to choose an estimator. A linear approximation is common (Gonzalez & Woods, 1992, p. 358):

$$p_{est} = \sum_i \in R w_i p_i \qquad (4.15)$$

where w_i are weighting coefficients.

Predictive coding not only forms the basis for image compression but also in a strategy used by human vision (§6.6.1.1 (Srinivasan *et al.*, 1982)) gives an example of such encoding in vision.

4.2.6   Huffman codes

The transmission of signals down noisy channels is probabalistic. On average we can pack a given amount of information down the channel. But optimal packing requires some flexibility in coding. Using a bus analogy, to keep a fleet of buses full all the time, without leaving people behind is not easy with buses of a fixed size if we want them to depart at fixed times. But if we have a range of bus sizes, from minibus to doubledecker, then we can get closer to keeping the buses full all the time.

This is how one of the fundamental coding techniques, which chunk symbols together, works. But if chunks are of varying size we need either to send start/stop signifiers (wasteful) or to employ some clever coding tricks. *Prefix codes* are such a trick. The Huffman code is an example and is optimal within the set of prefix codes, i.e. for a long message the length in bits is equal to the entropy of the signal.

Table 4.5 A Simple Huffman code.

Symbol	Probability	Code word	Information	Entropy (bits)
x_1	0.20	10	2.32	0.46
x_2	0.32	00	1.64	0.53
x_3	0.33	01	1.60	0.53
x_4	0.01	1100	6.64	0.07
x_5	0.04	1101	4.64	0.19
x_6	0.10	111	3.32	0.33
Total		**Entropy**		**2.10**

Given a set of symbols, $\{x_i\}$ with probabilities p_I we want to assign codes to the symbols such that for rare symbols we have a long code word and for common symbols we have a short code. This will give us the short sequence for a random selection of symbols. In the early days of radio transmission, Morse code used sequences of dots (short) and dashes (long) to encode the letters of the English alphabet. Thus a single dot represented e (most common letter) and three dashes z.

A prefix code is set up in such a way that no element of the code is the start of another element. Thus the code words do not need start and stop codes. Starting at the beginning, just keep on reading bits until the sequence matches a code word. Once a match occurs, there can no longer be a code word which could also match. Thus we decode this sequence and continue reading.

To construct a suitable code, we start with our set of symbols and look for the two lowest probability symbols, x_a, x_b. The higher the probability of the symbol, the later it will appear in the procedure and thus the shorter will be its code word. We combine these two symbols together, assigning them values of 0 and 1, to produce a new set $\{t_0, x_i \setminus \{x_a, x_b\}\}$ where the tree t_0 has replaced the two symbols with probability equal to their sum. We now recursively apply this procedure until we just have one large tree.

Table 4.5 shows a simple code for six symbols. The average code word length is 2.2 which is just a little larger than the entropy. Figure 4.3 shows how the code is constructed. Note that this code is not perfect, but with a small number of symbols this is inevitable. Improvements lie in taking the symbols in blocks and coding block by block. Such a strategy is also important if there is any correlation between symbols. For example, in the English language, with a few exceptions such as the Australian airline Qantas, the letter 'u' always follows 'q'. Hence we do much better by coding the probabilities in pairs of letters or even larger groups.

4.2.7 Dictionary methods

Another very efficient way of chunking symbols together is a so-called dictionary code. Here commonly occurring fragments are collected and encoded as units. Thus in English letter combinations such as *ton* are common, whereas others such as *gzx* are not. The best of these methods are typified by the so-called LZ codes devised by Ziv and Lempel, which underlie compression utilities such as *gzip* (Witten *et al.*, 1994), in which the dictionary

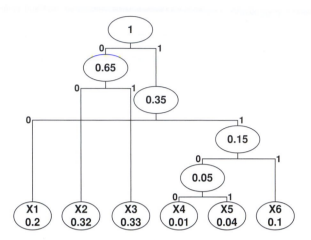

Figure 4.3 Construction of a Huffman code.

is built dynamically as coding progresses. LZ codes underlie GIF image compression. At a conceptual level, the nervous system uses dictionaries extensively through receptive fields and sparse codes. But the implementation details are quite different.

4.2.8 Entropy in continuous systems

So far we have considered discrete sytsems with finite alphabets of symbols. Yet most natural signals are effectively continuous. The outputs of many animal sensors (e.g. photoreceptors) are also continuous. Thus we also need to consider probability distributions rather than discrete event probabilities. The sum of eqn 4.4 becomes infinite if the probability distribution becomes continuous, which is not too helpful. But there is an alternative concept, which turns out to be very useful, the *differential entropy*. Note that although it *looks* like a simple generalisation of the discrete case it hides some nasty mathematical details:

$$H(X) = -\int_{-\infty}^{\infty} p(x)ln(p(x))\,dx \tag{4.16}$$

for the differential entropy, H of random variable, X. One mathematical issue is that this particular form, eqn 4.16, is not independent of the coordinate system. So, if $y = g(x)$ has a unique inverse, then:

$$H(Y) = H(X) + E\left\{ln\left|\frac{dg(x)}{dx}\right|\right\} \tag{4.17}$$

(whereas for discrete random variables $H(Y) = H(X)$). In practice this does not matter, because *differences* in entropy are (typically we want difference between signals and noise). Thus, because we are usually working with such differences, we can leave this issue moot.

The distinction between discrete and continuous systems can get a little philosophical. At the lowest, quantum mechanical level, everything is pretty much discrete. Light, vibrations, everything comes in discrete quanta, albeit with wave-like properties. But at the macroscopic level we rarely see this quantisation. We see it in X-ray systems (§4.1), and in very low light levels at night. It is pretty easy to calculate the differential entropy for the Gaussian distribution, or variance, σ^2, i.e.

$$G(x) = \frac{1}{\sqrt{(2\pi)}\sigma} exp \left(-\frac{(x - x_0)^2}{2\sigma^2} \right) \tag{4.18}$$

resulting in:

$$H(X) = \frac{1}{2} \log_e(2\pi e \sigma^2) \tag{4.19}$$

The Gaussian distribution has an important property. Its entropy, as in eqn 4.19, is the highest of any distribution for a given variance. Thus if we represent noise as Gaussian, we know that the loss of information is the highest possible for a given noise variance.

This leads to another useful quantity, the *negentropy* discussed in §4.2.10, where the differential entropy is subtracted from the Gaussian distribution of the same variance, creating a quantity which has to be positive (or zero if the distribution is in fact Gaussian).

4.2.9   Entropy of multivariant Gaussian

It is straightforward to generalise the one-dimensional case to multiple dimensions and this will be very useful later on.

In the Fourier domain each frequency is considered to have a Gaussian distribution in Gaussian noise. Thus we end up with an information in bits defined by:

$$H \approx 0.72 \sum_i \lg \left(1 + \left(\frac{S_i}{N_i} \right)^2 \right) \tag{4.20}$$

where S_i and N_i denote the *power* of the signal and noise at frequency i.

If we have N independent sensors each with a signal to noise ratio of ζ, then information capacity is given, following eqns 4.19 and 4.28, by:

$$H_N = 0.72N \lg(1 + \zeta^2) \tag{4.21}$$

But if the signals are correlated, then the capacity is reduced, since we are using some of the available range of the sensor in partially duplicating its neighbours. This is analogous to the difference between two large numbers – we rapidly lose precision as the numbers get closer together.

An N-dimensional multi-variate Gaussian distribution with a correlation matrix, R, has an entropy given by:

$$H_R = \ln(\sqrt{(2\pi e)^N |R|}) \tag{4.22}$$

where $|R|$ is the determinant of R.

Information theory has been quite a successful tool in optical image analysis, from photography to vision and this relationship becomes important when we come to discuss sampling arrays, in vision and in digital cameras.

In a seminal paper describing the application of information theory to optical signals, Fellgett and Linfoot (1955) used a Gaussian model. In the frequency domain, each frequency is considered as statistically independent of every other one. In any set of images there will be local correlations, because objects are a finite size. But each frequency component has contributions from all the objects in a scene. So if we imagine collections of objects of varying shapes and sizes distributed randomly across many images, then it is reasonable that there will be little correlation between frequency components. Each frequency component is now considered to be an independent channel with Gaussian distributed signal and noise as in eqn 4.21.

4.2.10   Negentropy

The differential entropy, eqn 4.16, is not invariant to a transformation, eqn 4.17. But a useful quantity is the negentropy, $J(Y)$, which as a difference of two entropies is invariant. Since the largest differential entropy of any random variable of a given variance is Gaussian, then the negentropy is defined by:

$$J(Y) = H(Y_g) - H(Y) \tag{4.23}$$

where Y_g, the Gaussian random variable of the same variance as Y, is always positive.

4.3   Algorithmic information

There is another form of information theory, much later than that of Shannon, invented more or less concurrently by Chaitin, Kolmogorov and Solomonoff. Whereas Shannon information is concerned with probabilities across sets of symbols, the idea behind the newer form is to look at the information content of a single event or entity, rather like the information content of a book or a map.

More formally the algorithmic complexity is the length of the shortest program which will output something or some description of it. This might seem to depend upon the language chosen for the computer program, but in the limit of large programs this is not the case. The computer program can be preceded by a translator (of fixed size) which translates the program into any language (of equivalent or higher power of course) we choose.

Figure 4.4 shows the relationship between the concepts.

There is a story of the great British philosopher Bertrand Russell taking somebody to task for remarking that some number plate, say ABC123, was very unusual. All number plates were equally likely – although perhaps this was before personal number plates were all the rage. But our intuition is still that this number plate is unusual, because it has a simple pattern. It is this notion of pattern and the complexity of the algorithm we need to describe something which is the essence of algorithmic complexity.

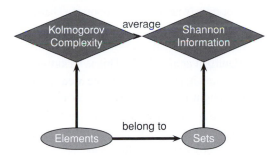

Figure 4.4 Relationship between Shannon and Kolmogorov information.

This ought to be a very useful concept in sensory information processing because what we are ultimately looking for is ways to describe things. In vision and hearing the pattern analysis is extremely sophisticated. In the chemical senses, the analysis is much more primitive. A great deal of our memory is constructive. We fit our experiences into patterns, we use expectations based on history and frequently our memory is unreliable as a result.

Although there might be some analogies, algorithmic information theory has had rather little impact in sensory analysis. It does give rise to principles such as *minimum description length* which are used in artificial intelligence, data mining etc. (Witten & Frank, 2000), but it has a bit of a problem. It is not Turing computable: to be able to compute the algorithmic information of a string turns out to be an equivalent problem to whether or not a Turing machine will stop. If we can't compute it, its attractiveness as a fundamental measure dwindles.

Nevertheless, we can impose bounds and it is still a very useful qualitative idea to work with. There are a couple of useful properties which help us here:

1. The algorithmic information is asymptotically independent of the language used; essentially we can prefix any program with a fixed length translator, but this means that we may not see the invariance until we get to very large programs.
2. The algorithmic information is related to the Shannon information. Li and Vitányi describe the relationship thus. If p is a Turing computable probability distribution, then the entropy $H = -\sum_x p(x)log P(x)$ is asympotically equal to the expected complexity $C = \sum_x P(x)C(x)$ (Li & Vitányi, 1993, p. 142). This is intuitively very satisfying. We add up the algorithmic information of every object and we get the Shannon information for the set.

4.3.1 Features and sparse coding

In a way neural systems tend to maximise algorithmic information, but in a somewhat indirect way. One of the ancient and, until recently discreditied, ideas, is that of the grandmother cell, a cell somewhere in the brain dedicated to grandmother, a term coined by Lettwin following a proposal for *gnosis* neurons by Konorski (Gaschler, 2006). But it now seems that neurons in various parts of the temporal lobe are highly specific for individuals, responding say to various diverse images of an actor such as Halle Berry and also to the text of her name (Quiroga *et al.*, 2005). This is an extreme form of what is generally known as *sparse coding,* and itself is similar to the idea of minimum entropy codes. Minimising energy use is thought to be an advantage or possibly the driver of sparse coding (§4.2.1).

Instead of having a representation, like the Fourier Transform, where a small or minimal number of functions which combined together give us a wide range of pictures or other stimuli, sparse coding goes to the other extreme, with many features all represented individually. The chemical senses, notably olfaction, seem to use sparse coding (§9.3.8) (Friedrich & Laurent, 2001) and other senses such as vision may do so too (§6.7.4). Songbirds, such as zebra finch (*Taeniopygia guttala*) also use ultra-sparse codes in the generation of their songs (Hahnloser *et al.*, 2002), while sparse codes have been demonstrated in the auditory cortex of rat (Hiromádka *et al.*, 2008).

Take face recogntion, for example. If we compress face images using any of the standard techniques, such as JPEG, we can get rid of a lot of statistical redundancy in the images and achieve tenfold compression with little loss. But that doesn't seem to be how we process faces ourselves. We think of them as made up of eyes, mouths, noses, with variations in complexion and colour. A police identikit builds up faces out of such components.

It would be possible to have many neurons devoted to all sizes, shapes and colours of eyes. Where brain development has gone off course, something like this seems to happen – the prodigious memories for details of autistic savants, for example (Snyder *et al.*, 2004). But even simple artificial neural networks can make do with far fewer cells. Training a feed-forward network to recognise eyes independent of size, shape and colour is quite easy (although computationally demanding) to do.

So, the code which would come closest to the code of lowest algorithmic complexity would combine feature detectors for eyes, but with a range of transformation operators, to change size, colour and so on, to generate the whole range of possible eye images. We do not yet know the full details. As we shall discuss in Chapter 6, there is still debate over holistic versus synthetic representations.

4.3.1.1   The cost of action potentials

Spikes are used through much of the brain and for the trunk communication lines, taking signals from the sense organs to the brain or control signals back to the muscles and organs of the body. But the recoding of the signal from a continuous receptor potential to a spike train comes at a cost.

A wide range of experiments give varying figures for the information loss. For example, Juusola and French (1997) measure a loss of almost an order of magnitude from around 1400–2000 bps for the graded potential to only 200 bps for the spike train.

4.4 Noise

With more and more of us living in cities, noise from traffic, sirens, building work, arguments next door, is all around us. But in information terms, noise is not only sound but unwanted signal fluctuations of any kind. In real physical systems we can never escape noise. Even ignoring quantum mechanics[8], at any temperature above absolute zero there is thermal noise.

But noise may take many different forms, each characterised by some statistical distribution. Most of communications theory is built around Gaussian noise, i.e. the variation in amplitude of the noise follows a Gaussian (or normal) probability distribution, for reasons we discuss below.

Thus voltages from, say, photocells are essentially continuous at normal light levels. But they are noisy: partly the noise is intrinsic device noise, part of it is actually quantum noise from the light signal itself. Now because of the noise, we can effectively think of the signal as discrete! Suppose we try to measure the voltage coming out of a sensor. We might get a single consistent reading, say of 2.4 volts. But the last decimal place might vary, say between 2.35 and 2.47 volts. Thus there wouldn't be much point in measuring the voltage any more accurately, because the noise has effectively limited the accuracy to the second decimal place – it has effectively imposed a quantisation on the signal.

But it's not a clean quantisation, like the digitisation of a signal of engineering solutions: the noise voltage fluctuates according to some continuous statistical distribution. It turns out that by far the most important distribution, and thus the only one with which we shall be concerned here, is the Gaussian distribution. It's important for several reasons:

- Common noise sources, such as photon noise, are approximately Gaussian if the captured flux exceeds around 50 photons;
- According to the Central Limit Theorem[9], if we have several independent sources of noise in a system, the sum will be approximately Gaussian;
- There are some interesting mathematical properties, which we will not delve into, one of which is that of all possible distributions of a given power, Gaussian has the largest *differential entropy* (§4.2.8); stated another way, Gaussian noise is the nastiest.

To put these abstract ideas into more concrete form, let us look at the types of noise which appear in each of the sensory modalities. In each case, the primary transduction stage sets the limits for everything which follows.

4.4.1 Photon noise

Quantum mechanics allows us to treat light as a wave or particle flux. The particle, or photon, viewpoint is useful when we want to look at the inherent noise in optical signals.

---

[8] In fact even at absolute zero there is still some residual noise referred to as zero-point vibration, which does not appear in classical/Newtonian mechanics.

[9] The Central Limit Theorem is one of the basic statistical theorems, which states, very broadly, that if we sum together a number of independent distributions, possibly all different, the result is Gaussian. As few as six will often produce a quite Gaussian result.

If we have an average flux of, say, N photons per second, and we capture a second's worth, then we do not get the same number of photons on every measurement. The distribution of the number we capture follows a Poisson distribution, with variance N:

$$p(k) = \frac{e^{-N} N^k}{k!} \tag{4.24}$$

For significant numbers of photons, this distribution rapidly becomes very close to Gaussian.

This is intensity noise. We shall see in Chapter 8 that wavelengths, red versus green, may also get confused and effectively contribute noise.

In the limits of low light levels biological photoreceptors are pretty good, in some cases able to detect single photons. But at higher light levels, other sources of noise within the (cone) photoreceptors may come to dominate – random flips in the pigment itself or thermal fluctuations in the cell membranes.

Another impediment to sensing the electromagnetic spectrum is background noise. We shall see in §11.2.1 that some animals have developed quite effective infrared vision. But they are cold blooded. Man would have great difficulties with thermal radiation from the body itself.

4.4.2   Hearing and touch

Hearing and touch have similar mechanics, although the transduction of hearing is a lot faster and more accurate. The noise here is vibrations induced by random thermal fluctuations in the sensors or membranes. At thermal equilibrium, each degree of freedom of a dynamical system will contain energy of kT, setting a useful fundamental lower limit. There will always be noise from the random bombardment of air molecules, thus there is no need to have sensitivity much below this level.

4.4.3   Chemical senses

There are at least two different kinds of noise. The delivery of molecules to receptors is stochastic: turbulence and concentration variations are significant along with the transport mechanisms for taking molecules to the receptor sites. There is also a form of *aliasing,* whereby many molecules may activate the same receptor to a greater or lesser extent. We wouldn't want to mix up benzaldehyde (not very toxic) and cyanide (§9.3.3).

4.4.4   Signal conversion

Even when the active agent, a photon or a chemical molecule, has reached its target all is not yet over. The process of conversion of the signal to an electrical impulse, *transduction*, is itself a discrete process, creating chemical pulses of varying kinds. The sizes of these pulses is also variable and thus a noise source. Further on, neurons are themselves noisy, and since the principles of information transmission along neurons are common to all the senses, we shall consider them in this chapter (§4.6).

4.5   Transmission along channels

Engineering and telecommunications have many different sorts of channel, discrete or continuous, with or without noise, all with different properties. So, how do we get from the properties of a channel and the noise to which it is subject, to the commonplace metrics of hard disc space and modem speed?

Consider our city bus as a communication channel. It can hold a given number of people sitting, some more standing and has space to put luggage. If we try to fit more than the maximum number of people in, some will get squashed and flustered, and may not get out in the same state in which they got in. If a group of, say 20, people wants to go into town to play bingo, then they just need to know that there will be 20 seats free. If the bus holds 25 people, but 5 usually get on first at the railway station, then on average they will all fit. But there will be some days when the bus is overcrowded and some days when it has empty seats. The capacity as seen by the bingo players is 20.l.

The key idea is thus *channel capacity*. If we try to transmit more data than the channel capacity we get errors. Seems simple enough, but Shannon's[10] expression of this in quantitative terms was a major breakthrough. Remember that entropy is effectively average information. So intuitively we would expect that the entropy of the source should not exceed the channel capacity and this is indeed the principle Shannon formulated. Since the mutual information between two distributions X and Y is the entropy in one given the other, then it seems as if channel capacity should be the mutual information between input and output.

There is just one consideration left to get the channel capacity – the nature of the source itself. It would make sense to choose the maximum capacity, i.e. the source which maximises the mutual information. The challenge for use of the channel is encoding the input in such a way that it has the same statistics as the optimal source. Hence channel capacity is:

$$C = max\, I(X; Y) \tag{4.25}$$

4.5.1   Capacity of a noisy Gaussian channel

When the channel is noisy, there is no certainty about the data transmitted. But we can get arbitrarily close. With a modified channel capacity, we essentially have the same capacity theorem as for the noiseless case. Shannon now defines the capacity as:

$$C_N = Max\{H(X) - H(X|Y)\} \tag{4.26}$$

which we know from eqn 4.11 is just the mutual information.

---

[10] Claude Shannon is the father of information theory, a theory which underpins almost everything in telecommunications, and is of widespread importance to computer science and artificial intelligence. His 1948 book, *A Mathematical Theory of Communication* was enormously influential and is still a highly readable classic today. Shannon died in February 2001.

We take the maximum over all possible information sources to the channel, i.e. to make best use of it we encode the data in the most appropriate way. So, in our bus we do not want luggage on the seats, as this will reduce the maximum number of passengers the bus can hold.

$H(X)$ is just the capacity of the channel without noise. But the capacity is then reduced by the equivocation, the uncertainty in the input given the output resulting from noise. If the channel is noiseless, the equivocation, $H(Y|X)$ is zero.

4.5.2   Capacity of a Gaussian channel

To use eqn 4.26 we need expressions for the entropy, but for Gaussian signals and noise we can get these from the variances as in §4.2.8, the noise variance, N^2, and the signal variance, S^2. The mutual information in the output of the channel, $S + N$ (signal with added noise) and the signal is $I(S + N; N)$, i.e.

$$I(S + N; S) = H(S + N) - H(S + N|S) \tag{4.27}$$

but since the signal and noise are independent, this reduces to, using eqn 4.19,

$$I(S + N; S) = \frac{1}{2} \ln \left(1 + \frac{S^2}{N^2} \right) \tag{4.28}$$

where we call the quantity $(\frac{S}{N})^2$ the *signal to noise ratio*, often written as **SNR**. Signal and noise tend to be used rather loosely as either the signal fluctuations themselves or the signal power. To relate eqn 4.28 to the idea of noise acting as the discretisation of the signal is easy. If the signal is much bigger than the noise, then we get simply $\ln(\frac{S}{N})$ which is just the log of the number of states.

This concept we shall use again and again in our thinking about sensory processes.

4.5.3   Waveform channel with noise

In the final example, we consider a channel in which the input is a continuous waveform, to which Gaussian noise is added by the channel. The capacity now depends on the bandwidth of the channel and is given in bits per second by:

$$C = W \ln \left(1 + \frac{S}{N} \right) \tag{4.29}$$

In Chapter 5 we shall see how to estimate this for a compact disc. But elementary computing arguments tell us the information rate of a compact disc (one channel) is the frequency 44.1 kHZ times the number of bits per sample, 16, i.e. 705 kbps. To avoid errors, we must band-limit the signal before it goes in. We shall see later that biological systems do not rigidly adhere to such a strategy.

4.6   Neural coding

As we saw above there are two broad categories of neural output, continuous and spiking. The framework for treating neural information, particularly for continuous outputs, was laid down in seminal papers by Barlow (1964, 1982), Laughlin (1981), Snyder (Snyder & Srinivasan, 1979; Snyder *et al.*, 1977b) and others and much of this work still holds good. The story for spiking neurons is still incomplete. Although the last decade has seen a lot of progress in measuring and modelling the information transmission of single spiking neurons, there is still a lot to be learned about the way spikes synchronise with one another across neural populations (Maas & Bishop, 1998).

In vision and perhaps hearing, we have very good answers to many coding efficiency questions. In the other senses a lot less work has been done and answers are less precise or non-existent. What we need are some rules of thumb which can help us decide what sensory systems to use and what demands they will make in different situations.

Within the trillion or so cells which make up a human brain, the complexity and role of interconnections is extremely diverse. But in the more limited task of understanding sensory information processing, coding is just a little bit simpler. One of the key ideas here is that of the *receptive field*, which is particularly useful at the interface between transducers and the neurons which take information away from the sensory organ and off to the brain.

As subsequent chapters will delineate, neurons apply some sort of filter to the input space. In vision they may look out in a particular direction, be sensitive to a particular range of wavelengths or have particular temporal properties. In hearing they may be sensitive to particular frequency bands. This variation across the sensory domain, along dimensions such as position and wavelength in vision, is known as the receptive field.

One very common characteristic of neural coding, which is particularly important in understanding receptive fields, is its adaptivity and gain control. Firstly, to cope with the huge range of possible signals which may come in, from the almost minimal sound of a cat stalking its prey, to the massive transient of a thunderclap, a general property is for cells to adapt to an average level and signal variations from that level. Predictive coding, discussed in §4.2.5, describes the computational efficiency which arises from such a strategy. Another feature is for neurons to signal only positive quantities. For a cell to swing both positive and negative would require continuous activity at the average level. But neural activity is expensive. The human brain uses around 20% of the body's energy at rest. So cells appear in on/off families.

To implement such a strategy requires *inhibition*, where one cell blocks the activity of another. Some examples of this will occur later on, but many details of the precise mechanisms are still unclear. However, one key component is that the the inhibitory cells are usually different using different transmitters. A single cell does not both excite and inhibit. A very exciting recent result, discussed in §6.7.2.2, shows that there may be differential ageing amongst cell populations with very interesting implications for senescence.

4.6.1   Continuous output

When a cell produces a continuous voltage output, the calculation of its information rate is similar to many engineering systems. These ideas were successfully applied by Snyder and others to determine the information capacity of animal photoreceptors.

A simple way to look at the information rate, which we can apply when the output is continuous, is the digital argument of §4.5.2. We take the maximum possible output (voltage), say, V_{max} and ask what the just detectable difference in voltage is, say, ΔV. Thus we have a number of states equal to $\frac{V_{max}}{\Delta V}$. In a sense this is how the number of bits was determined for a compact disc as we shall see in Chapter 5.

In fact the just detectable difference is an average quality, arising from noise fluctuations. But as we saw above (eqn 4.28), this relationship becomes exact for a Gaussian signal and noise (where both voltages are now standard deviations).

4.6.2   Spike output

When the output of the neuron is a series of spikes, the theory gets a whole lot more interesting. Many of the diverse range of neurons in the body communicate by firing voltage impulses or spikes. A pulse code has considerable advantages. Detection of a pulse is all that matters, its size is irrelevant. Hence transmission over long distances can be much more accurate. The issue becomes then not of whether the pulse gets attenuated during transmission, but more of whether its arrival time is affected. In the early days of CDs, one of the deficiencies in the sound quality arose not from the information content of the digital stream, but temporal jitter at the decoding stage.

For almost as long the average firing rate was thought to be the method of coding. A high firing rate represented a strong signal in a more or less linear way. So a rapid flurry of spikes indicated a strong signal not firing anything. So the noise would be calculated as the variance in firing rates. But neuron firing is very irregular, and it would seem that the information transport is rather low, or so it seemed for several decades.

The picture is slightly more complicated in that cells are often configured to fire when something changes, turns on or off, rather than to fire continually. Part of the reason for this is metabolic – it costs energy for the cell to keep firing even when nothing is going on. The brain in fact consumes about 20% of the body's energy when at rest, so neural costs are not insignificant. But also important is that it is often more useful to be monitoring changes in the environment, waiting for interesting things to happen.

But firing rate is not the whole story. Electrophysiologists are able to record the signals from individual neurons. The signals sound random, an irregular series of clicks or buzzes, each corresponding to a spike, the *spike train*. There is a limit to the maximum firing rate of a cell. It requires energy to build up a voltage spike and a recovery time afterwards is required. This so-called refractory period is rarely less than a few milliseconds. So no cells exceed a thousand spikes per second, and most operate at a hundred or below.

Now, if the spikes are random, what mathematicians refer to as Poisson point processes, then to get the firing rate we have got to average over quite a few pulses. Suppose we average over 100. That would give us an error of about 3%. But this number of spikes is going to take around a second at maximum rate. So one cell would not be able to provide us with the rapid processing times we need. Reaction times and motor coordination are down around a tenth of a second or below.

We could get this accuracy by averaging over lots of cells, and, for a long time, everybody assumed that is exactly what happened. After all we do have a lot of cells. Georgopoulos *et al.* (1986), for example, demonstrated that cells determining the movement of a limb in the monkey cortex operated as a population: the average movement direction across all the cells was used. On the contrary, other work, again on monkeys, suggested that a number of behavioural tasks could be pinned down to the activity of just a single or very small number of neurons.

But could we do better? Can we get more from the spike train than just the average firing rate? Opticon and Richmond (Rieke *et al.*, 1996, pp. 151–152) in a significant but somewhat underexploited series of papers demonstrated that one *could* do much better than the firing rate. But there were still conflicting viewpoints.

It was some years later that Rieke *et al.* (1996) and others studied the information transfer properties of a well-known neuron, H1, in the insect visual system. They came to the powerful and remarkable conclusion that every spike counts. So, the pseudo-random firing of a cell carries information right down to the variations in timing between the spikes. In that sense it is hardly random at all. Moreover, the H1 neuron performs better on natural than constant stimuli, and its efficient information transmission is consistent with the fly's visual system operating close to the limits set by photoreceptor noise for motion detection (Steveninck *et al.*, 1997). Similar results are found in the retinas of mammals and reptiles (Berry *et al.*, 1997).

To do this required a shift in the way of thinking. Instead of thinking about coding the value of something out there, such as a a sound or light intensity, by firing some number of spikes per second corresponding to the intensity, we think directly about the time variation of the signal itself. If we have a signal which varies in time, this will produce a varying series of spikes. Now instead of trying to estimate the average firing rate at any given time, Bialek *et al.* (2001) went even further. They deduced a filtering scheme to convert the series of spikes back to the original continuous input signal.

Perhaps somewhat surprisingly, this filter turned out to be linear. In other words, interpolating between the spikes with a linear convolution kernel returns the original stimulus to a high level of accuracy. In a further refinement Steveninck *et al.* (1997), improved the kernels for directionally sensitive neurons. The spike trains for stimuli in the preferred direction and time reversed stimuli (to simulate the opposite direction), were given values of plus and minus one respectively, are combined and used to calculate the kernel against the composite. A more accurate result applying this kernel to the original stimulus results.

Reich and colleagues (Reich *et al.*, 2001; Richmond, 2001) have recently made similar measurements in the visual cortex area V1 in macaque monkeys. They show that the information carried by neighbouring neurons is largely independent and that averaging over groups loses information. This holds at short time intervals (< 10 ms) even if the response properties of all the members of the group are very similar.

If we know how to estimate the original signal from the spike train, we can measure the error, hence the noise and hence the information transmitted by the spike trains. For the estimation procedure to work, the spikes have got to occur at the right time. The irregularity now becomes a property of the coding, rather than noise! So in turn, the errors in individual spike timing become the crucial factor in limiting information transfer. Victor (2005) gives a recent review of spike timing models, both theoretically and across most sensory domains. The temporal precision is usually down to 4 ms but with some possibly lower values.

Thus Rieke and his colleagues measured the spike timing and were able to deduce very precise estimates of information bandwidth of sensory neurons. The limit seems to be about 100–200 bits per second. It does of course vary according to the specialisation of each cell. An alternative way to think of this is in terms of bits per spike, which seems to be around 2–3 in most cases. In subsequent chapters we shall give some examples of bandwidths where they have been measured and calculated precisely.

Usually these measurements involve artificial random stimuli, since these are easy to standardise across animals and senses. But the senses have not evolved to be optimum for random signals. We shall see later in §5.4.5 that where natural signals have been used the coding efficiency goes up even further! Detailed discussion of stimuli and measurements will take us too far from our main theme. The interested reader can find a thorough discussion in the book by Rieke *et al.* (1996).

The figure of 1–2 bits per spike seems to hold across a very wide range of animals, invertebrate and vertebrate. Combining this with a firing rate of 100 spikes per second gives us a rule of thumb to be able to compare the senses. We will assume a 100 bits per second, identify the 'bottleneck', count the neurons and thus get our information capture rate. We can then relate this to what we would need in an artificial system, in virtual reality or in a computer game. There are examples where the firing rate goes higher than this.

4.6.3   Synchronisation

The analysis of information transfer of individual neurons begs consideration of what happens when neurons fire together as assemblies. The pioneer of *mass action* ideas was Walter Freeman (see, e.g., Freeman, 1999) and the last decade has seen many papers on the properties of synchronisation, where neurons fire together, or they show synchronisation of oscillations in their firing rate. Synchronisation has been invoked as a mechanism of attention, for binding together attributes belonging to individual objects, interaction between motor cortex and spinal cord (Schoffelen *et al.*, 2005) or even for visual awareness and consciousness (Stryker, 2001).

A persuasive current model for synchronisation is that it subserves attention. With many possible distracting stimuli, one component, one object, one sound source gets singled out. Synchronisation of the neuronal signals from this source strengthens transmission to higher levels, ultimately blocking out distractors (Fries *et al.*, 2001).

If many signals become temporally correlated, the dimensionality, or number of degrees of freedom, of the signal decreases significantly, so this has a significant impact on the information transmitted. But in sensory processing, there is a transition from transducing and transmitting to computation. The information theory framework really only takes us to the transition and is less useful beyond. This marks the place where we leave the analysis to other books.

4.6.4 Stochastic resonance

One respect in which noise is not entirely useless is a phenomenon known as *stochastic resonance*. Here adding noise can actually *increase* sensitivity. An easy way to see this is to think of a signal as having to get over some threshold. If there is no noise, below threshold is bad news. No signal is detectable. But now add noise. Sometimes the noise will add to the signal and push it over the threshold, referred to as the *iceberg* effect (Volgushev & Eysel, 2000), thought to operate in lowering the threshold of human hearing (Gebeshubera, 2000). We shall see a couple of examples later in the book, in enhancing touch in elderly people (§10.8.5) and in the orientation tuning of cells in the cat visual cortex (§6.7.2.1).

4.7 Envoi

Information theory has had a profound effect on the study of sensory systems. In many animals, from moths to man, from crabs to cats, we shall see examples of optimisation out to the physical limits. Moths have amazing hearing and olfactory sensitivity (§5.4.3.1), crabs can see stars (Doujak, 1985), cats are near optimal in low light and have exceptional smell and hearing. But man is the great all-rounder, possessing almost the best visual resolution (losing out to eagle), probably the best colour vision, but progressively less well equipped in hearing and olfaction compared to other mammals.

5 Hearing

Tones sound, and roar and storm about me until I have set them down in notes.
Ludwig van Beethoven

Vision is the sense with the greatest bandwidth, but our sense of hearing provides distinctive, equally vital information. Apart from communication through speech, tone voice conveys a full gamut of emotions. Moreover sound, with quite different properties to light, carries quite different environmental information, and is less directional than vision.

Human hearing is good, but by no means the most sensitive within the animal kingdom, or that with the greatest frequency range. But again, animals get very close to the limits imposed by physics. As computing power and miniaturisation have grown dramatically in the last decade, knowledge of hearing has become essential to determining the standards for the capture and storage of audio information from telephone conversations to games. In 2010 the directional information from sound is still not very well captured and utilised in artificial systems. Chapter 7 covers the processing of directional information.

Having got the theoretical building blocks behind us, sound is a good place to start in the study of the senses themselves. It embodies all the ideas of the chapters on information theory and Fourier Analysis, but, being one-dimensional, is somewhat simpler than vision. Moreover, the idea of frequency in sound is commonplace, and thus easier to discuss, whereas spatial frequency in vision is a less everyday concept.

But whereas the theoretical framework of Fourier Analysis is pivotal to understanding vision and hearing, the sonic transduction elements have a great deal in common with touch and we shall come back to them in Chapter 10. In fact these transduction elements are some of the oldest in evolution.

The physics and characteristics of sound, and in what ways and to what extent we are sensitive to it, are the starting point (§5.1). This gives us an idea of what sort of bandwidth hearing requires and we can move directly to ask how the recording and storage of sound fits in with these estimates.

The first thing to establish is the lowest and highest frequencies we can hear. As for all the senses, man turns in a pretty good performance but we are nowhere near the best. Cats and dogs easily surpass us, with cat sensitive out to 60 kHz (Sokolovski, 1974), over double human's maximum frequency. But even our furry friends can't compete with moths and crickets at the upper end of the frequency scale (§5.4.3.1).

But within these frequency bands, there are minimum and maximum sound levels we can detect or withstand. Beyond that there is the extent to which we can discriminate one frequency from another. The final psychophysical constraint we need to consider is *masking*, which plays a crucial role in the perceptual coders which drive MP3 and its forerunner minidisc.

We then move to the anatomy and physiology of how sound is processed. Some of the strategies now form part of audio compression techniques. Other aspects of hearing are yet to be exploited in artificial systems.

5.1   Physics of sound

Sound is a wave motion, supported by any of the states of matter, gases (air), liquids (e.g. water) or solids, but there are several differences from light. Firstly, it is a longitudinal rather than a tranverse wave. In other words the oscillating entities which make up the wave move in the direction of the wave itself, rather than bob up and down perpendicular to the wave direction (as in water and light). This distinction is not going to be esecially important here. Secondly, the speed of sound is very much smaller than light, about $335 \, \text{ms}^{-1}$. Frequency is much lower and wavelength much bigger too. One consequence of this is that we see the light from an event before we hear the sound, as in lightning and thunder. A second consequence is that sound diffracts around much bigger objects than light, the objects we see in the environment – it effectively bends around corners.

Animals have adapted to processing signals in sound and water, with some other adaptations for transmission of sound and vibration through the earth (§5.4.2). Our main concern will be transmission through air.

The natural world generates a huge range of different sound levels. We don't want to be deafened by a hurricane, an avalanche or a violent thunderclap. On the other hand it would definitely be to our advantage to hear a man-eating tiger padding up behind us. One of the loudest sounds we would encounter frequently in nature would in fact be thunder. Obviously, we wouldn't want our hearing to be damaged by thunderstorms, but the thunderclap is about at the maximum sound level we can tolerate. Again, a common theme in sensory systems is that they evolve just to the limits of selection pressure – in this case to be able to tolerate the loudest sounds commonly encountered.

The range of sound pressures we can tolerate in the short term without damage is $10 \log_{10}$ units, a massive 10^{10} vision has a similar huge range and uses the same mechanism, adaptation, to cope. Table 5.1 gives some examples of different sound levels.

The sound wave carries energy, and the basic physics of sound tells us that the energy delivered by the wave (the power) is the square of the amplitude. The unit of energy flow (or flux) is Wm^{-2} (watts). Since a lot of acoustics is concerned with the absorption of sound as well as its distribution patterns, the energy figure is frequently used and is what we use for the decibel scale, which we shall come to in §5.3.1.

Table 5.1 Sound pressure levels of common sounds (after Barlow and Mollon).

Sound pressure Nm^{-2} or Pa	Power (intensity) Wm^{-2}	Sound pressure level (dB SPL, i.e. referred to 20 μPa)	Examples and some effects (approx. only)
200	100	140	Jet engine; over-amplified rock group; threshold for pain
20	1	120	Damage to cochlear hair cells
6×10^{-1}	110	Threshold for discomfort	
2	10^{-2}	100	Motorcycle engine; orchestra
6×10^{-1}	10^{-3}	90	fff
2×10^{-1}	10^{-4}	80	mf; normal conversation
2×10^{-3}	10^{-8}	40	pp; quiet office
6×10^{-4}	10^{-9}	30	ppp; soft whisper
2×10^{-4}	10^{-10}	20	Country area at night
2×10^{-5}	10^{-12}	0	Threshold of hearing of young person at 1–5 kHz
6×10^{-6}	10^{-13}	−10	Threshold of cat's hearing (1–10 kHz)

The loudest sounds, those which are extremely painful and can cause hearing damage, have *much* more energy. In fact the ratio is as high as 10^{14}. Just stop for a moment and think how big that is. For example, it's bigger than the age of the universe in years, or about equal to the number of bytes on a 100 terabyte disc!

5.2   Anatomy and physiology

Auditory processing begins with the ear (§5.2.1), and, like the eye, there is some complicated preprocessing before the sound is transduced to a nerve impulse. Then, like vision, a nerve, the cochlear nerve (§5.2.2), takes the signal though several intermediate stages, to the thalamus and on to the auditory cortex (§5.2.3).

5.2.1   The ear

The ear itself has an amazingly complicated neuro-mechanical architecture. Figure 5.1 shows the anatomy, which we do not need to consider in detail. The ear has three components, outer, middle and inner, each with a distinct functional role:

- The ear has to couple sound waves in air to vibrations of a membrane, like the surface of a drum, to which neurosensors can be attached.
- The **outer ear** (§5.2.1.1) collects and focuses sound, adding some frequency tuning along the way.

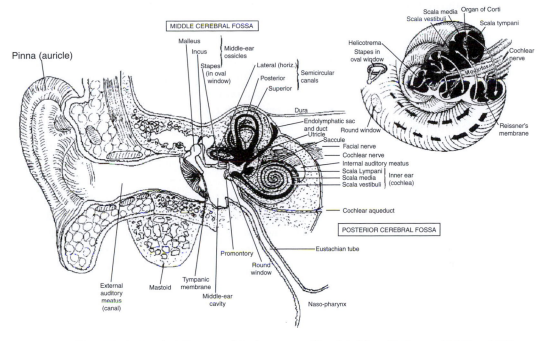

Figure 5.1 Anatomical diagram of the human ear. (Reprinted from Barlow and Mollon, 1982, p. 254.)

- The **middle ear** (§5.2.1.2) is a clever mechanical system which effectively couples the outer to inner ear, rather like the bridge on a violin; this is located in a cavity in the extremely hard temporal bone, but with vents, the *Eustachian tubes* to the outside air.
- The *transduction* occurs inside the cochlear, or **inner ear** (§5.2.1.3) on the *basilar membrane*.

In building robots or animats, the bio-mechanical system matters less to us than the signal processing strategies.

5.2.1.1   The outer ear

Horn-like systems are familiar to us in many devices. The acoustical gramophones of the early twentieth century had gigantic horns to amplify the sound without electricity, and some loudspeakers made by Klipsch Lowther and JBL amongst others may still be used as horns. Conversely, the first hearing aids were mechanical, the so-called ear trumpets, which were just large horns helping to focus more sound into the person's ear.

Ears in the animal kingdom frequently have much more extreme horn systems than ours – just think of the African elephant (see Plate 10.1) with ears as big as its head or basset hounds whose ears drag along the ground. Many animals such as cat, which we saw earlier has very sensitive hearing, can alter the direction of their ears at will from the perked-up alert mode, to pushed right back, to fight or flee. The pinna, the external 'flap' of the ear, then, plays a major role in sound collection and directional information (which we will come to in Chapter 7). Sound then travels down the ear canal to the middle ear. Just like the tuning of a tuba, the length, tapering and diameter of this tube filter the sound, emphasising some frequencies at the expense of others. In fact the speech frequency band of 2–5 kHz receives a boost of 10–20 dB. This is reflected in the minimal audible field (§5.3.2).

Anything which distorts the pinna will effect our subjective frequency response, and, as we shall see in Chapter 7, our appreciation of sound direction too. Headphones do this (with the exception of odd ear-embracing types such as those made by Stax and Jecklin-Float). But the recent trend in the ear headphones sold with many personal stereos are even worse. They actually impede the ear canal and thus effect its tuning – just like putting a mute on a trumpet. In-the-ear hearing aids potentially suffer from the same problem.

With digital signal processing, it is possible to make fine adjustments of frequency response. With a personal hearing aid fitted by an audiologist, or medical consultant, it is in principle feasible to customise the tuning – but the more complex the processing, the more power required, conflicting with the demand for light unobtrusive hearing aids.

With personal stereos at best we can have some filtering which will suit the average person, and thus not suit anybody perfectly. But any such preprocessing will mitigate other uses of the device or the use of alternative headphones. So, we basically have to live with the problems. But people are surprisingly tolerant and willing to adapt to all sorts of distortion in sound reproduction!

Since the outer ear is open to the world, this tube is air filled. Another way in which the ear canal's tuning may suffer errors is something many people experience at one time or another – the build-up of ear wax. This is in fact a defensive mechanism of the ear. Since the outer ear is open to the air, it is possible for dust, insects and other matter to get inside. Ultimately the canal could clog up completely. Small glands in the skin near the pinna end of the canal produce the wax. The function of the wax is to bind together this intrusive matter into a dry compact form, which just falls out of the ear in normal life. But some persistent irritation or infection in the ear will generate a defensive response by creating more and more wax until ultimately it is the wax itself that causes the blockage.

5.2.1.2   The middle ear

Stringed instruments such as the violin generate only a small volume of sound from the string itself. Most of the power comes from the resonant cavity of the case. Sound waves coming through the air from the string would, indeed, make the case resonate. But, in engineering parlance, there is a very *poor impedance match*. Thus the coupling is weak and not much would happen. For this reason stringed instruments have a mechanical coupling between string and sound box, the *bridge*. This greatly improves the impedance match and the effectiveness of the sound box.

Similar principles operate in the middle ear. As we shall see very shortly, the neural mechanisms deep inside the inner ear are extremely delicate and sensitive and operate inside a fluid-filled cavity. So, there is a need to create an effective coupling between the outer and inner ears.

The boundary between outer and middle ears is the *tympanic membrane*, which we often refer to as the ear drum. It vibrates in response to sound and triggers a series of bony couplings, analogous to the bridge on the violin, which pass these vibrations to the inner ear. Everything here is mechanical.

The three bones of the middle ear, the *ossicles*, are incredibly tiny, occupying a volume in man of less than an aspirin. There are three such bones, medically known as the *malleus, incus* and *stapes*, or in general based on what they look like as, respectively, *hammer, anvil* and *stirrup*. The stapes impact on the *oval window* of the inner ear. What this delicate bone linkage does is to transfer vibrations effectively from the larger tympanic membrane to the tiny oval window at the other end.

This tiny bone system is obviously vulnerable to direct damage, but also to any general degradation of bone. Thus a contributing factor to hearing loss in elderly people may be osteoporosis, the loss of bone density often experienced by women after menopause. But, like most of the joints in the body, the ossicular chain of bones becomes a bit stiffer, making it less able to transmit higher frequencies. This is another source of the gradual decline in hearing, *presbycusis*, experienced by most people with increasing age.

5.2.1.3 The inner ear

Finally the mechanical sound wave gets transduced into an electical neural impulse. This happens within the Organ of Corti[1], or inner ear. Like the pinna, it has something in common with brass instruments. The deepest instrument of a brass band is the euphonium, in which the player literally stands inside a long curled up tube. The low frequencies require a long tube, like a long organ pipe, and the only way to build a portable instrument is to curl it up.

The Organ of Corti is also a curled up tube, containing all the important basilar membrane where trandsuction occurs. The basilar membrane tapers from about 500 μm wide at one end to 200 μm at the other and is about 33 mm long in man (Stakhovskaya *et al.*, 2007). Since high frequencies are absorbed more readily than low in bodily tissues, the high frequency end is at the base, where the signal arrives. It has frequency tuned *hair cells* attached along its length. There are two types of hair cell, *inner* and *outer*, with differing roles. The inner ear is fluid filled. Movements in the fluid trigger movements of the hair cells, which in turn generate neural impulses.

Changes in the inner ear are probably the most common cause of presbycusis. The inner hair cells gradually wear out, making discrimination between sounds more difficult. Thus hearing sensitivity may be less affected than discrimination, leading to the unfortunate experience of being aware of a lot of conversation going on in a party, but being unable to clearly differentiate the words. It is interesting that hair cells are replaced in a range of other animals including sharks and chickens. Thus there may be the opportunity for gene therapy in the future to replace worn out hair cells (Brown, 1999). It turns out that complex genetic control mechanisms prevent hair cells from dividing, but that it might be possible to turn these control mechanisms off and Sage *et al.* (2005) already have a first demonstration in mice.

---

[1] Discovered in 1851 by the Marchese of Corti.

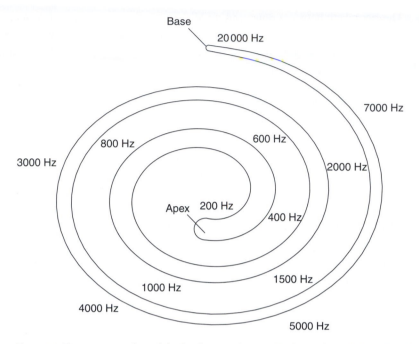

Figure 5.2 Frequency tuning of the basilar membrane. (Redrawn from Robert Port, http://www.cs. indiana.edu/port/teach/641/audition.for.linguists.Sept1.html.)

As a rule of thumb, the lowest frequency at which an object will vibrate is such that the wavelength is about equal to its size. Thus the size and sound transmission properties of the basilar membrane roughly determine the lowest frequency (about 20 Hz) and the highest (20 kHz falling with age). Cat, on the other hand, has a smaller basilar membrane and a highest frequency of around 60 kHz (Sokolovski, 1974).

The membrane varies not only in width, from 100 μ at the base to 500 μ at the apex, but also in stiffness going from the apex of the cochlear to the base, exaggerating the resonant frequency along the way. The net effect is logarithmic variation with frequency along the membrane in terms of hair cell tuning. Greenwood deduced the following formula, eqn 5.1 (Stakhovskaya *et al.*, 2007), for the frequency, F as a function of distance along the membrane, x:

$$F = A * 10^{\alpha x} - k \qquad (5.1)$$

with $A = 165.4$, $\alpha = 2.1$, x fractional distance from apex and $k = 145.4^2$.

But the mechanical properties are not quite so simple (Figure 5.2). Since pitch goes up in equal increments as frequency goes up exponentially (every octave doubles the pitch on a piano), pitch varies approximately linearly along the basliar membrane from 200 Hz.

---

[2] The Stakhovskaya paper seems to have omitted a multiplier of A to the constant k. The equation given here gives the correct frequency limits of 20 Hz and 21 kHz.

Helmholtz (1877) also contributed ideas to hearing and proposed the *resonance theory,* in which different parts of the basilar membrane vibrated at different frequencies. But it was Nobel Laureate, Georg von Békésy (1967), in the 1920s and 1930s who tested this experimentally. With some extremely delicate experiments he was able to observe directly the movement of the membrane. Sounds become travelling waves along the membrane, peaking at some point along the way. The peak of the travelling wave is roughly where one would expect the resonant frequency to occur in the Helmholtz model.

But as with many aspects of sensory information processing in this book, the broad concepts may be a century or more old but the fine details are still emerging. In March 2006 a theoretical paper in *Physical Review Letters* deepened our understanding of why the Organ of Corti and the basilar membrane are curved into a spiral (Manoussaki *et al.*, 2006). The conventional view was that the spiral saves space.

Naively one might expect low frequencies to be analysed at the outside of the spiral because of the reduced curvature (as in the difference between a trumpet and a tuba). Yet it is the opposite way round. Manoussaki *et al.* (2006) show the spiral has another function – it concentrates low frequency waves towards the outer edge akin to the whispering gallery of St Paul's cathedral in London. As the radius of curvature decreases, the amplification gets greater and for a factor of 10 change in curvature amplifies the sound by 20 dB. They also note that low frequency hearing threshold seems to be related to the number of spiral turns across the animal kingdom, which accords with their model.

The argument that thermal noise on the ear drum sets the sensitivity of hearing goes back to Sivian and White in the 1930s, but Harrison recently improved their estimate. The net effect of thermal excitation of the air and resonance in the ear canal are the dominant noise factors setting the threshold and the remainder of the auditory system matches this physically determined threshold.

5.2.1.4 Transduction on the basilar membrane

The basilar membrane is not merely small. Its vibrations in response to sound are *tiny.* At extremely high sound levels, 100 dBA, the vibrations are still only about 100 nm – well below the wavelength of blue light (about 400 nm).

The tiny hair cells, attached to the basilar membrane, convert these vibrations into electrical signals; neural signalling is primarily electrical and this is the stage at which the ear *transduces* mechanical vibrations to neural signals. The hair cells are exquisitely sensitive. A movement of a fraction of a nanometre, literally atomic scale movement, is enough to cause them to fire. There are four principal noise sources, a mixture of mechanical and transduction mechanisms, equivalent to a standard deviation of membrane fluctuation of a few namometres. This is almost an order of magnitude above the detection threshold of < 1 nm (Ricci *et al.*, 2005). They also respond very quickly, within microseconds. Herein lies a difficulty. Many of the transduction processes of the senses involve cascades of processes, as in the receptor to G-protein mechanisms in olfaction and taste. But cascades such as this take time. Thus one expects the hair cell transduction to be more direct, but we do not know the full details of how it works. But an exciting result appeared in a domain quite unrelated to humans or hearing. Walker *et al.* (Barinaga, 2000b; Walker *et al.*, 2000) discovered an ion channel mechanism in the bristles of the *Drosophila* fruit fly which has many of the characteristics of the hair cell channel. Having identified the genes which code for the transmitter proteins, the challenge will be to see if they can be exploited in human hearing pathology, another example of reuse of biotechnology across modalities.

As we saw above, the basilar membrane is tapered, sensitive to lower frequencies at the wide end and higher frequencies at the lower. In turn the hair cells, as we go along the membrane, are *tuned* to different frequencies. This is a *key idea*: the auditory signal is being broken up into different frequency bands before it is sent off to the brain.

This is quite different to a typical recording studio. A microphone generates a voltage which tracks the sound pressure incident upon it. It is one fluctuating voltage which is the superposition of all frequencies in the signal. If we send this to a digital recorder, the signal is just digitised as it comes in. It is sampled, but never split into separate frequency bands.

Frequency discrimination depends on two separate codes: the place code and the periodicity code. In the former the place on the basilar membrane is the indicator of frequency, which is fairly straightforward. There is also a mechanism of phase locking of cochlear fibres to stimulus frequency, but the details of this are not clear. This periodicity code dominates below 200 Hz.

Frequency is organised in an approximately logarithmic fashion on the basilar membrane, with about 2–3 mm corresponding to an octave with middle C located at around 6 mm along (Tramo, 2001) and around 10 octaves in total. An essential implication of this discussion is that the frequency selectivity is set at the periphery, i.e. in the ear and in the cochlear nerve. It is limited by mechanical and noise constraints in the ear itself. We shall see the same result appears in other senses, where light sensitivity etc. are set by the limitations of the eye rather than within the brain. This makes a lot of sense. Brain processing has adapted to make use of everything that comes in. The peripheral senses have adapted to be as sensitive as possible. An extreme example of this occurs in the enabling of trichromacy in mice by genetic engineering of photoreceptor pigments (§8.1).

However, we don't hear sounds at *all* frequencies. We coined the idea of *subjective bandwidth* (Snyder *et al.*, 1988) to capture the idea that a sense is interested in a particular range of stimuli appropriate to its lifestyle (Chapter 1).

There are countless examples of biological systems which have evolved to get close to theoretical optima. Rod photoreceptors can detect single photons. Blood vessels branch to new blood vessels such that the fluid flow is optimal. It should come as no surprise, then, that that the simple linear encoding of CD or DAT (digital audio tape) (§5.5.1)) is inefficient. Other more efficient systems code the signal in different frequency bands. Although the first to be widely discussed was Philip's digital compact cassette, shortly followed by Sony's minidisc, there were other schemes, routinely used before.

5.2.1.5   Ear asymmetry

Although the two eyes are functionally identical, subserving symmetrical binocular vision and stereopsis, the two ears may not actually be quite the same. Sininger and

Cone-Wesson (2004) showed that in human infants the left ear, which feeds the right hemisphere of the brain, is specialised for tones (possibly subserving music). The right ear, feeding the left hemisphere, is specialised for rapidly changing signals, typical of speech. It seems that this system with an innate bias for language is present at or soon after birth, well before the rest of the auditory cortex has matured. It would be interesting to know to what extent this asymmetry occurs in other animals.

5.2.2   The cochlear nerve

With the distinctive sensory organs, such as the ear and eye, the overall structure tends to be one of a vast number of detectors, often of several different kinds, which carry out the initial transduction. Then, maybe after some neural preprocessing, a set of cells carry the signal away to the brain. It is tempting to see this as like a telephone trunk cable, or a big optical fibre pipe under the ocean, in which the data is packed down tightly, for long-distance transfer. We shall see that in vision this is quite a good approximation.

In hearing there is not too much processing between the hair cells and the *cochlear nerve*[3]. In fact the ganglion cells of the cochlear nerve innervate the hair cells directly.

Most of the connections go to the inner hair cells, about 90% in fact. In humans there are about 33 000 cochlear neurons and about 3–4000 inner hair cells. Thus there is considerable *divergence*, each hair cell feeding ten or so cochlear neurons but each neuron receiving input from one and only one hair cell. The remaining 10% of fibres innervate the outer hair cells, of which there are about 12 000, and, here, we see *efferent* neurons going back the other way, from the brain to the hair cell. This is *radically different to the eye*, where there are neurons going back from V1 to the thalamus, but not from the thalamus to the eye[4]. The neurons tend to have a refractory period of around 1 ms and firing rates of up to about 500 Hz, putting them at the upper end of activity of nerve cells.

The function of the efferent fibres seems to be to change the tuning and dynamics of the basilar membrane. The outer cells on signals from the brain can bend the basilar membrane. This in turn biases the responses of the inner hair cells, thus tuning the signal they send to the brain.

In fact the outer hair cells can serve to protect the inner hair cells from damage by very loud sounds. Taranda *et al.* (2009) identified a particular protein, nAChR, which mediates outer hair cell change in sensitivity of the vulnerable inner hair cells. Mice genetically engineered to produce more of this protein were better able to resist loud noises without hearing damage. The possibility of taking this further to a protective drug is an exciting possibility.

5.2.2.1   Behaviour of individual cochlear nerve fibres

A concept which pervades sensory neuroscience is the concept of a *receptive field*. This is the range of stimuli to which the nerve cell is responsive. It could be the range of

---

[3] Nerves in the skull are given both functional names and earlier anatomical classification. Thus the cochlear nerve carrying auditory information is also known as the VIIIth cranial nerve.

[4] Or at least none have yet been identified, implying that they would have to be very few in number.

colours in an image, or in the case of hearing the range of sounds. Since, the cochlear neurons are arranged along the basilar membrane and tuned to different frequencies, we define the receptive fields in frequency space, i.e. as a function of frequency. It is essentially the filter set through which we hear the world.

Each cochlear fibre's receptive field has a frequency of peak sensitivity, referred to as the *characteristic frequency*, v_c. Its sensitivity either side falls off almost symmetrically, with a roughly Gaussian shape. We can specify the *effective bandwidth*, B, as the width of a rectangular filter with the same area as the response curve. The accuracy of the tuning may then be expressed with a parameter usually referred to as Q, or *sharpness*, given by

$$Q = \frac{v_c}{B} \tag{5.2}$$

The Q of individual fibres varies quite considerably, although across different species, say cat and man, the distribution may be quite similar. Q_x refers to the bandwidth taken at x below the peak.

The attempt to match tuning of cochlear fibres with the statistics of sounds has not been straightforward. Lewicki (2002) applied independent component analysis (§3.9.2) to three different classes of sound: environmental sounds (crunching of leaves, snapping of twigs, impulsive sounds with wide frequency spectrum); animal vocalisations, which have a more harmonic structure, especially birds; and human speech which is somewhere in-between the first two. Not surprisingly the filters predicted vary from Fourier-like for animal vocalisation to much more sharply temporally tuned filters for environmental sounds. The mix one would expect to find depends on the weights (and in effect the animal's lifestyle and priorities) with which these are mixed together. Lewicki (2002, Figure 7c) shows that the distribution of sharpness as a function of centre frequency is approximately a power law and matches experimental data from cat reasonably well.

Cochlear nerve fibres also display *lateral inhibition*. Excitation of fibres at one point on the basilar membrane (i.e. at one characteristic frequency) inhibits adjacent frequencies. Thus we get a narrowing of the effective frequency tuning.

Most fibres saturate at around 40–60 dB above threshold. About 10% of fibres show less background activity but extend out to 80 dB. But within their given dynamic range they behave in a roughly linear fashion.

5.2.2.2   Coding in the cochlear nerve

Coding in the cochlear nerve is still not fully understood, but we can make some general observations. There are 30 000 active fibres, of which about 95% assuming each carries 100 bps, our benchmark figure (§4.6.2) then the data rate would be 3 Mbps. Now if we compare this with the data rate from a CD, which remember was based on generous psychophysical parameters, we find an anomaly. The CD has a bandwidth of 0.7 Mbps. Thus we have an almost factor of three difference, but this does not allow for the sparse coding strategy common in animal senses. But the situation is even more complicated because it is possible to create a very realistic audio signal with considerably less than the bandwidth of a CD. Part of the reason for this seeming excess of fibres is the limited dynamic range of the fibres themselves.

Rieke *et al.* (1996) analyse the potential performance of auditory nerve fibres and compare them with precise psychophysical parameters, such as the threshold of frequency difference to tell two tones apart. The results are frustrating: we should be 10–100 times better than we actually are. But this assumes that all fibres tuned to the right frequency can contribute to the decision. Weakening this assumption leads to slightly better results, but the theory is still far from complete. There are still open research questions, even in the peripheral processing of sensory data.

5.2.2.3   Sampling properties

Chapter 3 looked at sampling a signal in time, and observed how this transformed the frequency spectrum. But the hair cells and cochlear neurons sample in both time and frequency at the same time. Each hair cell reacts to a sound stimulus and stops when the stimulus ceases if not before. But each is tuned to a fairly narrow band of frequencies. Thus the sampling strategy is akin to the Gabor (§3.8.1) or wavelet sampling models (§3.8.2).

Why should we go through this elaborate procedure to break the signal up into different frequencies and then have a lot of cells tuned to different characteristic frequencies? In hearing this is quite easy to grasp, but the same strategy which we shall come across in vision is less obvious.

In broad terms we want to do two things with sounds: we want to know where they come from and what they represent. What makes sound signals difficult to analyse is that they overlap one another (whereas in vision the situation is usually that one object occludes another). Directionality is a big help in unravelling several concurrent sound streams. Chapter 7 covers directional hearing. But directional resolution is not that good, and we do surprisingly well with monophonic sources (radio, older televisions) which have destroyed the directional information. The other strategy for unravelling complex sound sources is the time of onset, and the auditory system has a stream which processes on/off transients.

The nature of a sound and the information about what caused it is largely characterised by its timbre. The difference between the sound of a violin and a piano playing the same note is a function of the different mixtures of frequencies they contain. So analysing the sound into different frequencies is a *recognition strategy*. Electronic organ technicians will carefully adjust frequency balance to get the tone they require for a particular stop.

It is also a *compression strategy*. Different frequencies have different importance and our sensitivity to different frequencies varies. Thus we can assign more channel capacity to the important frequencies where we are most sensitive (§5.5.3). Section 5.2.5 discusses how both tonal and transient sensitivity vary in each ear according to the different functions of the two hemispheres of the brain.

5.2.3   Pathways into the brain

The cochlear nerve is the trunk line taking hearing information to the brain and as we shall see with most of the senses, the main exchange is the *thalamus*. But the cochlear nerve goes through a number of other stations along the way. Figure 5.3 shows a *system view* of the ear taken from Evans (1982). There are numerous different processing modules.

The first is the *cochlear nucleus*, divided into *dorsal* (DCN) and *ventral* (VCN) parts, located in the subcortical *pons-medulla* area. Here the cochlear nerve branches, one branch ascending to the *anteroventral* cochlear nucleus, the other descending to the posteroventral and dorsal cochlear nucleus. The DCN (dorso-cochlear nucleus) is concerned more with the *nature* of a signal, whereas the VCN (ventro-cochlear nucleus) is concerned with *location in space*. The split into what/where (§2.3.2) systems thus occurs very early on.

Just as we saw in vision in §3.8.1, the cells of the primary auditory cortex depend on external stimuli to develop their receptive fields. Thus if a rat experiences only particular tones during a critical period, it will grow up with a reduced sensitivity to everything else, just like the kittens reared with particular orientation of stripes (§6.7.2.3). But the critical period of development is not a fixed time scale, but is dependent upon appropriate organisation having taken place. If rats are reared with white noise (no correlations), then no correlated activity will develop in the cortical cells. In this case, the critical period is delayed (Chang & Merzenich, 2003). So, the critical period seems to be terminated by the formation of distinctive patterns in the the cortical cells, which occurs when they experience correlated inputs.

Cells in the nuclei are *tonotopically* organised, i.e. as we go deeper we effectively go along the basilar membrane from the apex to the base. This, again, is highly characteristic of sensory systems and not really very surprising. The early stages stay close to the organisation of the receptor system, in this case, its frequency tuning, with later stages reorganising the signal for pattern recognition.

However, the streaming of auditory information has already assumed a complex form in the cochlear nuclei as shown in Figure 5.4 (adapted from Evans, 1982).

The VCN has two principal types of cell: bushy and stellate, deriving their names from what they look like. Bushy cells respond to the *transients* in a signal, firing only at signal onset. Stellate cells have a so-called *chopper response*. They fire regular pules and to some extent smooth temporal irregularities in the input. The DCN has only stellate cells. This is consistent with the VCN assuming the locational role, where timing of arrival of signals is critically important (Evans, 1982). Thus the streaming of information here is according to signal character rather than signal function.

Three pathways lead out of the cochlear nucleus: the *dorsal acoustic stria*, the *intermediate acoustic stria* and the *trapezoidal body*, the latter being the most important. It projects to the *superior olivary nuclei*, still within, and on both sides of the brainstem. So we're still right at the very base (old) part of the brain. In here we find the *medial superior olive*, which codes interaural arrival times, and the *lateral superior olive*, which codes interaural intensity. These low level areas of the brainstem, part of the old brain, might be thought to have been fixed during evolution. But this is not the case: musicians show stronger activity in these areas, particularly in tasks that might be shared with speech, than non-musicians (Musacchia *et al.*, 2007).

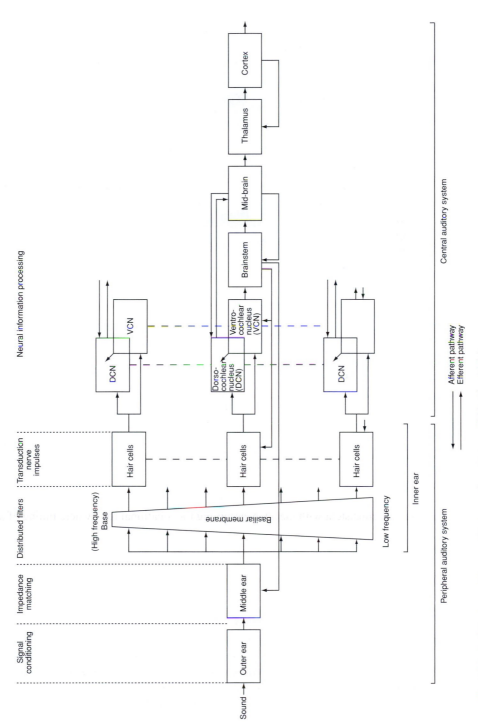

Figure 5.3 Architecture of the ear. (Redrawn from Evans, 1982, Fig. 14.1.)

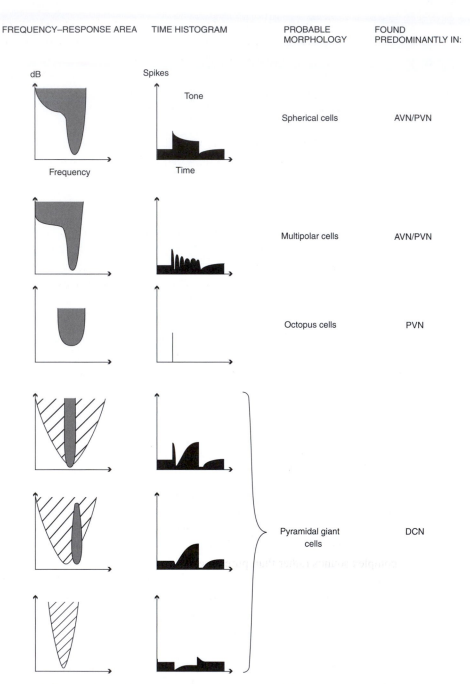

Figure 5.4 Cells of the cochlear nucleus (anterior ventral (AVN), posterior ventral (PVN) and dorsal (DCN)). The ventral cells are mainly excitatory, although not necessarily linear. So the octopus cells fire when a sound turns on. The dorsal cells are more complex with lateral inhibition effects analogous to the retina. (From Evans, 1982 – see text.)

The *lateral lemniscus* pathway takes us into the mid-brain to the *inferior colliculus* and eventually the *medial geniculate body* (MGB) of the thalamus. The MGB projects to the *primary auditory cortex* and the *superior temporal gyrus.* There seem to be two different types of synaptic connection, weak and strong. The weak synapses might code for sustained signals while the strong synapses code for transient (Atzori *et al.*, 2001). In rat barrel cortex (see Chapter 10) there may be even stronger synapses (Zador, 2001).

The auditory cortex preserves the tonotopic mapping with at least five areas containing six such maps (Talavage *et al.*, 2004). Considering that the basilar membrane is organised with frequency increasing from the apex to the base, one might think that pitch would be extracted really early on, maybe as early as the ear itself. But the pitch of complex sounds does not get extracted until the cortex itself, in a region just on the edge (anterolateral) of the primary auditory cortex (Bendor & Wang, 2005).

Within the auditory cortex itself, there seems to be further streaming of musical rhythmic information. The right hemisphere seems to handle variatons in note duration, whereas longer time pulse or metre seems to be a property of a different area of the auditory cortex (the so-called parabelt) and occurs on both sides.

Whenever we look at a complex system of this kind, particularly if we are interested in building a device or an animat to do the same sort of things, we have to bear in mind two questions. Firstly, these systems evolved, adding new functions on top of old, without the luxury of redesign, so we need to ask whether such a complex structure is an optimal one which we should copy. Secondly, having seen how the information splits up and is used for different purposes, we can ask whether we want all of that functionality. For example, our only requirement might be directional sensing without any interest in pattern recognition on the sounds themselves.

5.2.4   Specialisation in the auditory cortex

The auditory cortex exhibits distinct spatial segregation into two distinct information streams, which in turn project to different areas outside. In rhesus monkey the belt region around the core of the auditory cortex has three distinct lateral areas, the anterolateral (AL), middle lateral (ML) and caudolateral (CL). The belt area responds better to complex sounds rather than pure tones.

There are broadly two classes of cell[5]: those which respond best to higher contrast, which behave like linear filters; and those responding at lower contrast and not too high a contrast. In the marmoset monkey these latter cells are interesting: they are non-linear and give their strongest responses to wideband complex stimuli. They cannot be induced to give their maximum response to any pure tone (Barbour & Wang, 2003). These cells may be specialised for detection of particular complex signals, such as animal calls or human speech (Barbour & Wang, 2003).

Tian *et al.* examined cells in AL and CL using a complex sound natural to a monkey, its calls. AL is more sensitive to the type of call, CL to where the call is coming from. This

---

[5] Although it is convenient to describe two categories like this, the distribution is likely to be more of a continuum.

is further what/where streaming (§2.3.2) and entirely consistent with this streaming, CL projects to the dorsolateral prefrontal cortex whereas AL projects to ventral and orbito prefrontal cortex. ML has intermediate behaviour. Poremba *et al.* (2003) also map the auditory regions of the rhesus monkey and find them to be almost as numerous as those for vision. The same what/where dorsal/ventral organisation is apparent and the different streams are also preserved through areas where vision and hearing are integrated.

In humans the experimental data is not yet available, but the spatial topography is clearly differentiated. Seifritz *et al.* (2002) used fMRI to distinguish between temporal and sustained components of auditory signals. There is a marked transition going from the core to belt areas of sustained to transient. The what/where characterisation is only suggestive at this stage. But it does seem that the transient components are more likely to be associated with determining the spatial location of sounds. Further evidence from de Santis *et al.* (2007), and other evidence cited therein, shows strong evidence from evoked potentials that there is a distinct 'where' pathway selectively active in the right temopari-etal cortex. The existance of an *independently active* 'what' pathway remains in doubt.

5.2.5   Left–right specialisation in auditory processing

Because hearing is necessary for understanding spoken language, it would not be surprising to find evolution of some left–right specialisation in auditory processing. It does in fact seem to be the case that the right hemisphere (left ear) is more sensitive to tonal information with better temporal resolution. The right ear is more sensitive to speech-like signals. But recently Sininger and Cone-Weston (2004) have shown that even in infants there is specialisation in *the ear itself*, with the left ear more sensitive to tones and the right ear to clicks[6].

5.3   The range and sensitivity of hearing

In Australian cities such as Brisbane, Sydney and Melbourne, large colonies of bats occur in the botanical gardens and many of the trees. They make quite a lot of noise, but this is social communication. Many bats use echo location for navigation, but human beings can't actually hear the ultrasonic pulses, although we do hear the sounds bats use for communicating with each other. Like all animals our hearing cuts off at some upper frequency. But many mammals, such as cats, dogs and bats cut off considerably higher. Our hearing declines with age, even if we don't go to rock concerts, but young adults have a cut-off around 20 kHz, declining to below 15 kHz towards retirement age. One of the depressing signals of middle age is suddenly becoming aware that you can no longer hear the whistle from (cathode ray tube) televisions and computer monitors (about 16 kHz).

---

[6] An interesting observation has recently been made with the auditory hallucinations experienced by schizophrenics who are usually male. This is thought to arise because the female voice is higher pitched with a more complex spectrum, resulting in greater activation of the auditory cortex (Holden, 2005).

At the other end of the spectrum we can't hear the lowest frequencies: we cut off around 15 Hz. But to a large extent, we feel rather than hear the lowest frequencies. So, although high-quality headphones may have a perfectly flat frequency response to below audibility and may be capable of delivering high sound levels, we still do not get the same sense of bass energy that we get from subwoofer loudspeakers in a room – designers of virtual reality helmets take note. Section 5.3.2 delves further into the details of sensitivity with frequency.

There is another limitation of human hearing – intensity (§5.3.1). Some sounds are too soft, others so loud they cause pain and permanent hearing damage. At the low end, the limits are set by physics, something we shall see again for other senses. At the high end, we are limited by the strength and properties of the body – the Organ of Corti (§5.2.1.3).

5.3.1   The loudest and softest sounds

At the lowest level, what we call the *threshold of human hearing*, we have a sound pressure level of $20 \times 10^{-6} \, \text{Nm}^{-2}$, 20 micro-Pascals (µPa). This sound pressure is the amplitude of the sound wave. It translates directly to the force on our ears. It is tiny. The pressure of a human hair on a surface would be thousands of times greater. We measure pressure in SI units in Newtons (N) per square metre (Nm^{-2}), the Pascal. Cat, you will not be surprised to learn, has more sensitive hearing, at about 6 µPa.

No man-made detector comes remotely close to this range. The ear manages by using an *essentially logarithmic coding*. We say essentially because this is very dangerous ground to assume a simple law where there is much psychophysical sophistication in just what the mathematics of this coding are. So here we take a *simplistic* view, which is none the less adequate as an introduction for computing and engineering purposes.

In the nineteenth century, Weber and Fechner introduced the law named after them, sometimes called just Weber's law (Gregory, 1987). Basically, this states that the just noticeable difference in any stimulus is proportional to the strength of the stimulus, i.e. the minimum difference in sound level we can detect, ΔS is proportional to the sound level, S itself:

$$\Delta S \propto S \tag{5.3}$$

So if we say equal increments in loudness, ΔL, correspond to these changes in intensity, we have:

$$\Delta L = k \log_{10} \frac{S}{S_0} \tag{5.4}$$

with k a constant and S_0 a threshold intensity.

Thus the variation of sensitivity with sound level is approximately logarithmic. Unfortunately this is a bit of a simplification (Moore, 1997). But fortunately, the same principle holds pretty much true for other sensory domains, except perhaps smell and taste.

So we get to handle the huge dynamic range of hearing by this logarithmic law. The same also holds true for the huge range of light levels we encounter, from moonlight to the bright summer sun. It makes sense then to have logarithmic units. In sound, the unit is the *Bel*[7] which is a factor of 10 ratio in sound energy. That turns out to be rather a coarse unit in practice, so it is more common to use a tenth of this, the *decibel*, commonly abbreviated as *dB*. In other words two sounds differ in energy by a factor of 10 if they differ in energy by 10 decibels, by 100 if they differ by 20 decibels.

The situation for perceived loudness is we end up with a power law, found by Stanley Stevens (1975), to be common across most senses with varying exponents:

$$L = kS^{0.3} \tag{5.5}$$

given that a difference in sound level of 10 dB is required for an approximate doubling of loudness.

The decibel is a *ratio*. If we want an absolute value of sound energy we have to refer it to something. The obvious choice is the threshold of hearing, leading to the *dBA* unit. What does the A stand for? You can think of it as A for 'adjusted' for hearing sensitivity. It gives less weight to lower and higher frequencies where the ear is less sensitive. This brings us to our next topic, the variation of hearing sensitivity with frequency.

5.3.1.1   Pleasures of loud music

'Sock it to the sacculus' (2001) was the title of the Christmas editorial of *Noise and Vision*, a reference to recent work by Todd *et al.* (2000) showing the sacculus to be responsive to sounds over 90 dB with possible direct links to emotional pathways (Marks, 2000). The sacculus is part of the vestibular system (§10.7) but until Todd *et al.* (2000) was thought to have a hearing function only in fish. However, they found the peak sensitivity to be in the 300–350 Hz range and a bandwidth of 2–3 octaves. Interestingly, this bandwidth is above the thumping bass levels of much dance music!

5.3.2   Variation of sensitivity with frequency

It should come as no surprise that our sensitivity to sound varies with the frequency. Figure 5.5 shows the variation in minimal audible pressure (MAP). We are exceptionally sensitive to frequencies around those of a baby crying, but much less sensitive to very low or very high frequencies. We perceive sounds in the range of 0.5–5 kHz, the speech frequencies, more easily than lower or higher frequencies. What is more the relative sensitivity changes with loudness. So you won't hear really deep bass at low volumes. Thus at 50 Hz the sensitivity relative to 1 kHz is down 10 dB at 100 dbA but 35 dB at threshold. Maybe rock musicians know this.

Figure 5.5 shows the equal loudness curves from International Standards Organisation (ISO) ISO 226 (2003 update). These are based on curves originally derived by Fletcher and Munson (1933). The bottom, dashed curve is referred to as the Minimum Audible Field (which is actually calculated for both ears), an international standard ISO R226. A subjective measure, this is computed by averaging the judgements over many young

---

[7] After Alexander Graham Bell, inventor of the telephone.

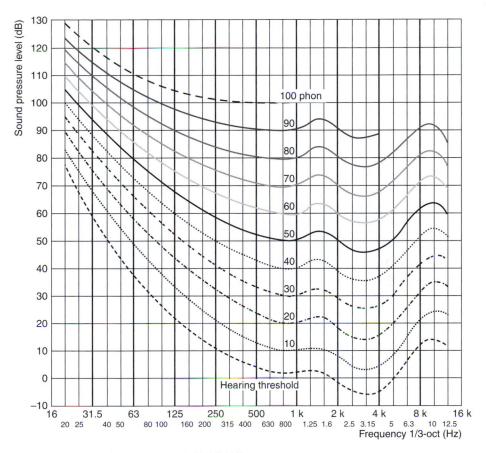

Figure 5.5 Equal loudness curves ISO 226 (2003).

people. Compared to the MAP there is a pronounced dip around 4 kHz, arising from two factors: diffraction around the head and resonance in the ear canal.

We can consider these curves as curves of equal loudness. Loudness is not a *physical quantity*, like the amplitude or energy of a wave. It is a *perceptual* quality. Hence we give it a different unit, the *phon*. People tend to be very bad at judging absolute values of any sensory signal, be it brightness, volume, colour balance or almost anything else. In general we tend to be much better at comparisons between stimuli than in absolute judgement of their magnitude. Hence we measure perception of loudness by getting an experimental subject to say when two sounds have the same loudness. Similarly we have another unit, the *sone*, measuring the subjective impression of loudness. There is nothing particularly special about these numbers. Like the foot or the furlong, they have a fuzzy history.

Another way to think of this is that if a sound at say 100 Hz is going to sound the same loudness as a sound at 1 kHz, then the lower frequency will have to have a higher absolute sound level. Thus the definition of the unit of loudness, the *phon* is the sound level of a 1 kHz tone which sounds equally loud. Because this is a subjective experience, the development of the phon had to be an average over many subjects. Not everybody would have the same individual phon scale. Figure 5.5 shows phon values.

To measure loudness in *phons* we compare the signal with a 1 kHz tone by adjusting the intensity of this tone until the loudness sounds the same. Obviously if the signal is a 1 kHz tone, then the intensity will be exactly the same, hence in Figure 5.5 all the curves go through the sound pressure level at the value corresponding to the phon value. We anchor the scale at 40 dB above the threshold of hearing. Thus a 1 kHz tone (the reference) at 40 dB above the threshold of hearing has a loudness of 40 phons. Because of the loss of sensitivity at lower frequencies, a 20 Hz tone has to have a much higher intensity to sound as loud. From Figure 5.5 it needs to have an approximate intensity *90 dB above the threshold of hearing.* There is a lesson to be learned here: we might experience very high bass energies without actually hearing them as very loud, creating the possibility of hearing or other damage as the rib cage rattles to the rock band. Headphones are a potential danger too. Because we do not get the high sound fields in the room, which make the rib cage rattle, we may not be aware of dangerously high levels.

The sound level in phons tells us about the perceived level. It tracks the logarithmic decibel scale. A slightly different quantity is how loud things sound. Thus, empirically, as given in eqn 5.5, when the *sound level increases by 10 dB, the subjective intensity doubles.* Hence we derive the *sone.* A tone of 1 phon creates a loudness of 1 sone, and as we can again see from Figure 5.5 the threshold of discomfort is reached at around 50 sones. Again the anchor point is 40 dB above the threshold of hearing and 1 kH. Note that unlike the phon scale, this scale goes up exponentially as a function of the intensity in decibels.

5.3.3   Masking

The idea of masking is very simple: a loud sound will hide a soft one. The sounds may occur together in *simultaneous masking*, but masking also occurs when the masking signal precedes (*pre-stimulus masking*), or follows (*post-stimulus masking*). As you might expect, the post-stimulus condition normally has a bigger effect than the pre-stimulus.

But the hearing system to a degree processes different frequency channels independently. Thus if two sounds are very close in frequency, one will easily mask the other. But if they are far apart, the masking is far less effective. Imagine how you might have difficulty talking over the top of a train's engine, whereas you easily hear the guard's whistle.

Fletcher, of the Fletcher–Munson curves (§5.3.2), proposed a simple conceptual model, which is right in principle, but mostly wrong in detail (Moore, 1995). The key assumption was that hearing operates independently in a number of channels, each of width equal to the *critical band*. If we take a signal within one band and mask it with broadband noise, then as we increase the noise bandwidth (centred about the signal), the masking will increase too. But as soon as the bandwidth of the noise gets bigger than the channel width, increasing it any further should have no effect. This is roughly what happens, but there are many complications.

The first complication is that the filters are not rectangular as implied. In fact the filter falls off approximately exponentially away from the centre, or *characteristic frequency*. We can, however, talk of the *equivalent rectangular bandwidth* (ERB), which is simply the unit height rectangular filter with the same area as the actual filter.

Given the logarithmic nature of frequency discrimination which we saw above, one might expect the width of the critical band to change with frequency. Indeed it does. The ERB increases inversely with the centre frequency, f, according to the following approximate equation (Moore, 1995, p. 175):

$$ERB = \frac{2.47}{4.37f + 1} \tag{5.6}$$

Masking is an important tool in audio compression (§5.5.2).

5.3.4 Discrimination of frequency

So far we have looked at our sensitivity to different frequencies, our tolerance of different sound levels and the way sounds mask each other most effectively if they are similar in frequency composition. But how easily can we tell sounds apart if they are of very similar frequency?

When do we know a musical instrument is out of tune? How big must the error in frequency be before we detect a difference in pitch? It transpires that the same Weber law (§5.3.1) holds for frequency discrimination too. As we increase the frequency, F, the difference in frequency, ΔF we can detect, the just noticeable difference (JND) increases in exact proportion, i.e.

$$\Delta F \propto F \tag{5.7}$$

The JND varies with frequency. As we might anticipate by now the best performance is in the speech range of 0.5 to 5 kHz. The behaviour outside this range is non-Weberian, falling rapidly with increasing or decreasing frequency.

Helmholtz, without the benefits of oscilloscopes and Fourier analysers as usual, found an innovative approach to measuring frequency discriminisation (Helmholtz, 1877). Using instruments with relatively pure tones, such as organ pipes, he looked for the detectability of *beats* between notes of similar pitch. Beats are low frequency patterns which arise when two waves are combined, akin to Moiré fringes with optical gratings.

5.3.5 Pitch

Frequency and pitch are closely related. After all a tuning fork, which vibrates at a particular frequency, provides us with a reference pitch. But with complex sounds, such as those from musical instruments, the perception of pitch is very much more complicated.

When several different frequencies are mixed together we may perceive them in one of two ways. In the first, *spectral pitch*, we hear two or more distinct tones. In the second, in *synthetic or virtual* pitch, we hear just one complex sound: as the components vary the quality, or *timbre* of this sound varies.

Listening to an orchestra, or even a piece of organ music on a small transistor radio, gives us a reasonable impression of the music. Yet, the bass response of the radio will not go anywhere near low enough to capture the fundamental frequencies of the double bass or even the lowest notes of the piano. In other words there can be a *missing fundamental* frequency, the frequency which we experience as the pitch.

So, why do we hear these low notes? For musical instruments such as these, each note has a rich collection of harmonics. The pattern of these harmonics is quite characteristic of the instrument. Thus the brain is able to fill in missing frequencies. It all happens without us being conscious of anything amiss.

In fact this pitch discrimination is so complex that there are pitch sensations created entirely by noise, such as Huggins pitch. In this, rather analogous to random dot stereograms, if white noise is presented to both ears, but in one ear a narrow band is phase-shifted, then a distinctive pitch is heard (Chait *et al.*, 2006). This is again a fundamental principle of sensory processing. Adjustments or assumptions are made in early processing to make our conscious life easier. *We are conscious of what we need to know* (Snyder & Mitchell, 1999). The penalty for this, mostly unimportant, is that sometimes our senses can be tricked.

5.3.5.1   Absolute pitch

One interesting variation in hearing amongst individuals is the perception of musical pitch. A small percentage of people are able to name notes directly and quickly, without reference, to hear C played on the piano and to immediately recognise it as such. This is an unusual ability in that it is transferring some exact, absolute value directly to consciousness. It is called *absolute pitch*.

Schlaug *et al.* (1999) show that in musicians there is a disproportionate increase in size in an area of the brain, the planum temporale, on the left-hand side in musicians who have absolute pitch. However, the processing of pitch itself seems to occur in the right hemisphere (Zatorre & Krumhansl, 2002). Although absolute pitch is rare, most of us have some degree of pitch differentiation: we can tell a high note from a low note. Deutsch argues that we have a pitch reference frame defined by the bandwidth of our speaking voices and make judgements relative to it. Thus people from different places, e.g. California versus England, may perceive musical pitch illusions differently, resulting from their different speech patterns (Deutsch, 1992).

But trying to locate musical pitch within the brain, Janata *et al.* (2002) found, in the rostral prefrontal cortex, in an area known as the superior frontal gyrus, a cycle of keys. In other words, the representation we need to be able to divorce the pitch relationships of a tune from the specific key in which it is being played. This is somewhat analogous to colour constancy (§8.4.2), where we correct for surface illumination.

There is another interesting parallel with vision. We shall see in §6.11 that there is some controversy about the nature of the face recognition areas of the brain, the extent to which they are dynamic or fixed and specialised. The Janata work shows that the tonality surfaces are dynamic, varying from session to session when they were measured.

5.4   Specialised hearing adaptations

5.4.1   Aspects of hearing in other animals

Hearing strategies in the animal world are very diverse. Whales and dolphins use echo navigation, still more sophisticated than any man-made system. Whales are thought to be capable of communicating in water over distances of hundreds of kilometres. Recently, very low frequency sound (below human audibility) has been discovered as a long-distance communication mechanism in elephants. As we discuss later, big animals have the potential to detect sounds of longer wavelength. Likewise they can tolerate higher intensities. It's difficult to win a shouting match with an elephant.

At the other extreme, small animals detect ultrasonic frequencies well in excess of man, from guinea pig at 25 kHz to moths very much higher. Nocturnal animals such as owls can hunt with high precision in total darkness (Helmuth, 2001) which we shall come back to in the stereo section (§7.3.4). So as with most of the senses, humans have good all round abilities, but are eclipsed by some animals which have developed specialist senses.

As we have seen the primary transduction mechanism is based around hair cells, effectively a development of a touch receptor. Thus it is not surprising to find a range of hearing mechanisms progressing from vibration sensitivity through to a fully fledged ear. We consider some vibrational mechanisms in Chapter 10.

5.4.2   Infrasound

Elephants, then, have an *infrasound*, i.e. a deep bass, long-range communication system, discovered only within the last few years (Payne, 1989). This infrasound is around 14 to 35 Hz, ranging from the low to down below the frequency limit for man. Elephant groups coordinate their behaviour over distances of over two miles by this means. One straightforward use is by females on heat for calling to males in the area. Somewhat more mysterious is the coordination of movement of separated groups. Although the African elephant has really large ears, the infrasound detection method is likely to involve the animal's versatile trunk described in §10.6.3.

Recently another terrestrial animal with at least one long dimension, the giraffe, has also turned out to use infrasound communication (Mason, 2002). One of the world's largest birds, the cassowary, has a booming sound which travels well through the dense jungles of New Guinea. They have a monstrous headpiece, which may be more than just decoration. It is probably a sensor for infrasound (Holden, 2003a).

Less obvious is the tiger. As one would expect because tiger is bigger than domestic cat, the peak energy of vocalisations is lower, in fact peaking around 500 Hz. But there is also considerable ultra low frequency, while the size and structure of the inner ear suggests that tigers do indeed respond to this infrasound (Walsh *et al.*, 2003). Tigers have large territories and these long-range acoustic signals presumably act as some sort of territorial markers.

Whales also have huge skeletons and are capable of processing very low frequency information. As yet little is known about how and what they use such communication for. One particular whale, the narwhal, actually has a specialised device to tune it to the calls of its own species. This Pinocchio of the sea has a huge bony extension to its nose, about 15 feet long! This resonant device filters out background noise, a very useful feature in these days of increasing ocean subsonic noise from shipping. But it seems our understanding of the precise function of the narwhal's sensor is still growing. Recent research at Harvard has shown that it has over 10 million nerves (i.e. similar to the size of the optic nerve in humans) and serves a variety of sensory functions including in-water temperature, pressure and salinity (Choic, 2006).

One interesting example of low frequency animal sound was resolved quite recently in November 2002. Back in the 1950s, US Navy sonar had picked up a strange low frequency groan, unidentifiable at the time. Using sophisticated computer software the sound was tracked by a research ship, which eventually homed in on a minke whale! The sound is thought to be a love call.

But concensus is growing that the sound of shipping is not good for whales. The mass beachings which occur all too frequently may result from confusion brought about by ship-bourne interference. Even worse a pod of 14 whales was found to have died from cranial haemorrhaging, probably caused by intense sound waves from sonar (Williams, 2001). At present the US Navy is testing new low frequency sonar which produces prodigious sound levels. Hopefully ecological sense will prevail. Apart from the effect on navigation, shipping noise is just simply stressful, as evidenced by the rise in the stress hormone, cortisol. However, there is an interesting adaptive effect here: for high intensity shipping noise there is a cortisol secretion increase in some fresh water fish. But the same intensity level with a Gaussian spectrum has less effect (Wysocki *et al.*, 2006). Other studies show that temporal resolution in hearing of fish is impaired by loud noise, taking several days to recover (Wysocki & Ladich, 2005).

There have been all sorts of anecdotal suggestions that humans are sensitive to infrasound and some recent experiments confirm that this is indeed the case. At a recent London concert, sound at 17 Hz, just below audibility, was injected into the concert hall. Audience members reported distinctly different (although not entirely consistent) emotional responses when the infrasound was active (Holden, 2003b). There is anecdotal evidence that this is also relevant to the reproduction of music in the home (§5.5.4.1).

5.4.3   Use of very high frequencies

At the opposite extreme to elephants we have some examples of ultra-high frequency sensitivity in insects. Bats eat moths and they find them using echo location. These ultrasonic pulses are up in the 200–250 kHz region, over ten times the cut-off frequency of man (§5.4.3.1). So, it is of immediate survival value to a moth to be able to hear the pulses from a bat in the vicinity. Bats are discussed further in §11.4.1.

5.4.3.1   **Moths and crickets**
Cricket, *Grylluss*, face a similar problem. In this case wasps are the predators. They do not use echo location, but their wings beat at a characteristic frequency, around 150 kHz, to which the cricket needs to be especially sensitive.

Crickets do not have ears as such, but two appendages at the rear, known as *cerci,* which are vibration sensitive. The cells innervating the cercus are each connected to a thousand or so hairs of three types. Two are of no concern to us (one touch, the other gravity sensitive), but the third are thin filiform hairs which are sensitive to air flow. These cells then code hair movements as a spike train.

One of the prime, life-saving functions, of this system is to enable the cricket to fly off in the other direction when it hears a wasp approaching. The wasp's wings beat at about 150 Hz and when the wasp is a couple of feet away, the cricket is in what we call the *near field.* At this range, the direction of air flow, as well as the changes in pressure of the sound wave, can be determined and this the cricket does by having hairs set to move in different directions.

Precise measures of the information transmitted by the spike train (Rieke *et al.*, 1996, §3.3) yield a modest SNR of about 1 over a bandwidth of 300 Hz. This equates to about 300 bits per second or 3.2 bits per spike. The temporal resolution seems to be about 0.4 ms. Measuring the spike timing to an accuracy greater than this does not contribute significantly to the information.

For a given average firing rate, it is possible to ask what temporal spike distribution would give the maximum possible entropy. From this maximum, one can then determine the coding efficiency relative to this theoretical limit using any given (in this case measured) function for transforming the spike train back to the stimulus. Cricket cerci cells seem to transmit at least 50% of the maximum.

The moth's 'hearing' is even more primitive but has a similar main function to cricket: detecting ultrasonic pulses from bats (predator) and taking evasive action. Moth[8] has *just one or two* neurons on each side of the head to do this job (Rieke *et al.*, 1996, §4.1.1).

5.4.4 Inducing infidelity

The sweetness of bird sound in the country sometimes hides some nasty goings on. Bird song is often a competition for mates, an avian Meister Singer. Recent experiments by Mennill *et al.* (2002) report how female chickadees eavesdrop on the performance of their mates in the song world. Chickadees are monogamous – but their nests often contain eggs of dubious parentage. Aggressive males compete by matching the frequency band of their songs to other males (and sing at the same time too). Submissive males choose different frequencies. Unfortunately this is just the excuse the females need to play around!

5.4.4.1 Masking internally generated sounds

Circadias can make a really deafening noise, upwards of 100 dBA close up. Dolphins generate sonar signals over 200 dB above background. One might wonder why animals are not confused or even injured by their own signals. The corollary discharge principle (§2.4.4) comes into play: the sensory systems receive a message which instructs

---

[8] Strictly speaking the moth genus *Noctuidae.*

them to ignore the signal the animal is itself about to produce. In cricket, Poulet and Hedwig (2006) elucidated the neural circuitry crickets need to do this.

5.4.5   The frog sacculus

The countryside at night can be amazingly quiet, but next to a river one particular animal holds forth in endless amphibian partying, the frog. But frogs do not only have finely tuned calls to one another, they have an additional vibration sensitive organ, the *sacculus*. It uses the same sort of hair cell-like mechanisms found in the ear but is exquisitely sensitive to vibrations in the ground (Rieke *et al.*, 1996, §3.3.2).

Compared to the cricket, the bandwidth is narrower with a lower maximum frequency approximately 40–70 Hz. But the SNR is higher at about 3–4. The information rate is 155 bits per second working out at around 3.2 bits per spike and a coding efficiency of 50–60%. Again these measurements are made using noise stimuli.

Something really interesting turns up in frog hearing. Measured using random signals, the information rate is about 46 bits per second or about 1.4 bits per spike. However, the frogs' calls have a distinct frequency spectrum. If the measurements are made with signals with this spectrum, the information leaps to 133 bits per second and 7.8 bits per spike. Thus, as the *signals approach the natural signals to which the animal is adapted, the information rate goes up and approaches the theoretical maximum.*

Unfortunately the sounds of the frog are sometimes picked up by mosquitoes and used to home in on their victim. It used to be thought that some female mosquitoes were deaf, but the familiar buzz is now known to be used as a mating signal. The sexes interact by matching the frequencies to each other. In the Dengue mosquito matching is to a harmonic at 1200 Hz of the basic female buzz frequency of 400 Hz (Cator *et al.*, 2009). This latter study revised the upper limit of mosquito hearing to 2 kHz, whereas before it had been thought to be tuned to 300–800 Hz.

5.5   Audio encoding

Even though the Sony Walkman, the first personal stereo, appeared three decades ago, the sheer ubiquity of personal music, where every commuter, jogger, office worker inhabits their own private world, has been driven in part by miniaturisation. Advances in computer hardware and battery power have gone alongside more compact ways of storing audio data. Section 5.5.1 looks at the first digital recording of data, in which the goal was to cover what seemed to be the entire gamut of human hearing without regard to data compression. Section 5.5.2 takes up compression, and §5.5.3 the most common implementation. Section 5.5.5 looks at artificial hearing.

5.5.1   Linear encoding

When the CD was developed, considerable thought went into the parameters of sampling frequency and sample accuracy. Audiophiles would say that the designers got it wrong, but the sonic errors may have more been related to the recording and reproduction technology. In any case, the CD is not far from indistinguishable from the original, *monaurally.* The issue of directionality and multiple channels is far more complicated and we shall consider this in Chapter 7.

The first parameter to set for the CD was the sampling frequency. For the vast majority of people, hearing cuts off at 20 kHz. However, the analogue filters of the day could not be made to cut off abruptly at some specific frequency[9]. So, some margin for error was allowed with a maximum frequency of 22 kHz at a sampling rate of 44.1 kHz.

Now each sample has to be *quantised*, i.e. represented as a binary number. The simplest quantisation is linear. We just divide up the intensity range into equal steps. If we use 16 bits, that gives us a ratio of 65 536 voltage levels, or a power range of 96 dB (since power is the square of amplitude). From our earlier discussions this should be enough[10]. In fact orchestras do go louder than this. But for rock music this is considered a safe level. In fact some local councils set this as the limit for rock concerts to avoid hearing damage to young fans. Thus the data rate for a CD is 16×44.1, around 0.7 Mbps (megabits per second).

The CD exploits just two features of human hearing: the cut-off frequency at the upper end and the maximum sound level normally occurring in music. But it's already clear that this is extravagant. Since the human auditory system is less sensitive to low and high frequencies, all frequency components do not need to be stored to the same precision. The beauty of digital signal processing is that many transformations of the signal are feasible before storing or on retrieval. In fact the CD's linear encoding stood for ten years and a tape format DAT (digital audio tape) with a similar linear encoding (but 48 kHz cut-off) is still the standard in recording studios.

The first attempt to exploit human perceptual limits in a more sophisticated way was Philip's digital compact cassette, a great idea but a commercial failure. The first real success was Sony's minidisc, now superseded by a vast array of MP3 and other *codecs* and the appearance of radically new business models for selling music such as Apple iTunes. Before studying MP3 there is yet more to learn about the information processing restrictions of the auditory system.

5.5.2 Compression

Compression is all about noise, keeping noise at bay. Dynamic range is also inextricably linked to noise. As we saw in §5.1, the human ear can take over 100 dB of sound levels before discomfort sets in. A compact cassette had at best 60 dB before the tape saturates. But if there was no noise on the tape, then this would not matter. We could just shrink the input signal on recording and stretch it back again on replay. But this does not work

---

[9] The problem arises because the sharper the filter, the greater the ripples introduced lower down in the (audible) frequency spectrum. At the time of writing three new 96 kHz sampling frequencies are competing: DVD movies, DVD audio and an audio-only format SACD introduced by Sony. Audiophiles say the difference is clear.

[10] In fact the story is a bit more complicated, owing to the introduction of so-called dither noise. This increases the signal to noise ratio to over 100 dB.

because of noise: shrinkage can push two similar sound levels so close that they are no longer distinguishable.

Noise is the enemy and it confronts in apparently two different ways. The first is the obvious one: we can hear it. It's the hiss on a tape or the cracks and pops on a vinyl record. But it also may just make the original information hard to retrieve. The sound may be distorted, fuzzy, lacking in detail or things which were audible on the original signal are just no longer there.

Fortunately for the audio world, human hearing is not uniform throughout its range and it's possible to exploit two properties:

- The varying sensitivity with frequency (§5.3), meaning that there is effectively less information at very low and very high frequencies and we can allocate less bandwidth to these frequencies.
- The masking of soft sounds by louder ones (§5.3.3).

Although various analogue compression systems were around for a long time, such as those used by the BBC in delivery of FM radio, two companies established themselves as household names in audio compression: *dbx* and *Dolby*. When digital arrived things got a whole lot more complicated as we shall see. In particular, we have an extra noise source to deal with – quantisation noise.

The principle behind dbx was compression/expansion, or *compansion*. The idea was that to put the signal through a non-linear compression reduces peak loudnesses and brings up softer passages to store the audio information. This was necessitated by the limited dynamic range of the storage media of the time, principally magnetic tape, compared to that of human hearing. Since according to Weber's law our discrimination of intensity changes falls with frequency, this should work: we are effectively stretching the loudness scale at higher intensities.

On replay, the inverse of the compression was applied to expand the audio signal again, and dbx worked fairly well in practice. The problem with analogue compansion systems was that the expander has to recognise when to expand and there are no perfect algorithms for doing this. So sometimes a distinct 'breathing' would occur as the expander was continually trying to catch up, particularly on percussive music such as piano.

Dolby followed similar compansion practices but adopted a somewhat different strategy. The idea was to do nothing unless the noise was audible. Loud sounds are not troubled by noise. They mask it. Softer sounds are so troubled and they are brought up in level to increase the signal to noise ratio. Recall (§5.3.3) that masking is frequency dependent – high frequency sounds are not masked by low and vice versa.

Thus Dolby A, the professional version, operates by splitting the audio signal into four bands. Within each band, loud sounds are left untouched, as the tape noise will be masked by the music itself. Softer sounds are amplified to increase the effective signal to noise ratio. Dolby encoded tapes played back without Dolby thus sound bright, because of effective high frequency emphasis applied at the encoding stage.

Variations on the original Dolby principle were used for consumer items. Dolby B and Dolby C were the most common variants.

5.5.3   MP3 encoding

Despite initial audiophile oppostion, digital audio opened up many opportunities for sound reproduction, offering an almost complete absence of background noise and in many systems a much clearer, distortion free sound. The first mainstream consumer digital format was of course CD, which uses very little insight from human hearing (§5.5.1).

But soon *perceptual encoders* appeared, which gave subjectively similar quality using far less storage space. In recent years, the format which has come rapidly to the fore, mainly due to the internet, is MP3. Whereas CD makes only minimal use of human perception and does not use statistical encoding, MP3 does both.

We will find the basic idea of MP3 very similar to JPEG compression, which we consider in §7.6.2. The principle is to split the music signal up into bands and encode them according to their importance to the human auditory system. Each band still has plenty of statistical redundancy in it and this is reduced using Huffman coding (§4.2.6).

MP3 is actually part of the MPEG standard (§5.5.4) which covers both moving pictures and audio. With video there is a natural chopping up of the signal into frames, the audio is also divided into frames for encoding purposes. At which point things get rather complicated, since there are different versions of the standard and there is considerable freedom for implementers to introduce their own encoding and decoding procedures with some for proprietary.

5.5.4   MPEG audio

MPEG stands for *Motion Picture Experts Group*, a subcommittee of the ISO. This first standard of the mid-1990s had five parts, covering both video and audio. Part 1 covers the integration of visual and audio streams, Part 2 video, Part 3 audio, Part 4 testing and conformance and Part 5 a technical report and software implementation. The notation MP3 comes from this third part concerning audio.

However, MPEG-1 did not last long and was soon replaced by MPEG-2. The new specification originally had nine parts (but Part 8 was subsequently dropped) of which the first five retained roughly their original meanings. The new parts addressed specific computing issues, video and audio now having moved firmly into the computer arena. In this book, the only one of these which might concern us is Part 7, an all singing all dancing multi-channel specification with up to 48 channels and sampling rates up to 96 kHz!

MPEG-4 is on the way with further computer applications and bells and whistles and exceeds 2000 pages of description! MPEG-1 audio had three levels, carried over into MPEG-2. Going from Level 1 to Level 3, the bit rates go down and the coding sophistication goes up:

1. Level 1 is for coding at above 128 kHz. Philips digital compact cassette, one of the first consumer perceptual encoders, used 192 kbs. The frames are 512 samples long. It uses a psycho-acoustic model referred to as Model 1.

2. Level 2 operates at around 128 kbs. The frames increase by a factor of 3 to 1152 samples. Model 1 is still used.
3. Level 3 operates below 128 kbs using more powerful psycho-acoustics, Model 2.

Despite being an international standard, MPEG is not free of propietary control. In fact one of the core algorithms was developed by the Fraunhofer Institutfür Integrierte Schaltungen in conjunction with the University of Erlangen. Creating standards such as this is a long and often difficult process. In fact the DVD and HDTV in the USA ended up adopting Dolby AC3 (Dolby Digital) rather than MPEG!

5.5.4.1   The Fraunhofer algorithm

The principles of masking and frequency importance need considerable research and experimentation to implement successfully. The first thing we need is a set of frequency bands and the perceptual noise thresholds in each. The narrower we make the bands, the more precisely we can set the masking thresholds, but the more one band can potentially mask an adjacent one. MP3 was not the first standard to tackle these issues. Dolby Digital had already done this in the analogue domain.

In division up into frequency bands there is a trade-off between simplicity of the filter system and the variations of sensitivity and frequency discrimination across the spectrum of human hearing. In fact 32 sub-bands were chosen, each with a constant bandwidth of 625 Hz, making a range of 20 kHz. Since the first two bands accomplish a lot in human hearing and the first half dozen encompass almost all the bandwidth of speech, the bands vary considerably in subjective importance. The variation of human perception *within* the band is more significant for some bands than for others.

The MP3 coding is quite complicated, however. Having split the bands up, we now proceed to encode each band. This is quite a CPU intensive process compared to the subsequent replay. It considers two nested iterations.

The inner loop is a lossless Huffman encoding of the quantised signal. The quantisation is a power law encoding with an overall gain setting. The signal is quantised and Huffman encoded. But the frame has only a certain amount of bandwidth allocated, so this code might run out of bits. If it does, the overall quantisation gain is changed so that the signal fits into a smaller number of bits. The encoding is repeated and the process iterates until the code fits the available space.

But all is not yet over! This quantisation may have introduced a level of quantisation noise which is above the perceptual noise threshold. In this case we have to start again with a new scale factor for the band and repeat the inner loop. This goes on, until the Huffman encoded band fits in the available space and the encoded signal does not exceed the perceptual noise threshold.

The MPEG-2 Layer 3 standard has two different extensions over MPEG-1. One goes up to 1 Mbps. The other has half-sampling rate options of 16, 22.05 and 24 kHz with bit rates of 32–256 for Layer 1 and 8–160 kbs for Layers 2 and 3. The multi-channel specification allows for a 6th channel (the 0.1 of 5.1) which goes up to 100 Hz only. This is the subwoofer channel, or to give its conventional name, the low frequency enhancement (LFE) channel.

Subwoofers have proliferated following the advent of home cinema. But prior to that, they had a mediocre reputation, since integration with other loudspeakers was non-trivial. Richard E. Lord, founder of REL, took a different perspective, that very deep sub-bass was *essential* to a sense of realism, and started a successful company to implement such ideas. Anecdotal reports of the success of such designs is consistent with the emotional effects of injecting low frequency noise (§5.4.2) and the mysterious properties of the sacculus (§5.3.1.1).

5.5.5 Artificial hearing

Hearing was the first sense in which an artificial implant, the *cochlear implant*, replaced the ear. With tens of thousands of successful operations, the cochlear implant is a major success story, allowing previously deaf people to hear. As with vision, restoring hearing to somebody who has lost it through injury or disease is easier than to somebody who is congenitally deaf. Early results suggest, however, that if the patient receives an implant sufficiently early in life the auditory cortex will develop normally (Rauschecker & Shannon, 2002).

The cochlear implant injects signals directly into the auditory nerve. If the nerve is damaged, then an implant directly into the brain is necessary. This is far from easy and the hundred or so experiments so far have not enabled patients to achieve satisfactory speech recognition. The difficulties lie in the way frequency is organised – implants need to penetrate deep into areas such as the cochlear nucleus to provide a reasonable range of frequencies. The technology to do this, microelectrode arrays, is improving rapidly and hopefully will eventually restore hearing to such patients.

6   Basic strategies of vision

If only we could pull out our brain and use only our eyes.
Pablo Picasso

6.1   Overview

Vision in the animal world is unbelievably diverse. The compound eyes of insects have a resolution of a few cycles per degree (cpd), whereas eagle resolution is almost a hundred times greater at almost 200 cpd. Insects frequently have colour vision and some have ultra-violet sensitivity, whereas many mammals have weak colour vision and few are trichromats like man. Most mammals are dichromats (Ahnelt & Kolb, 2000). Stereopsis is present in mammals, but not all animals. On the other hand the flicker fusion frequency (§7.5), the frequency at which images blend together as in the cinema, is about three times greater in flies than in man.

In the face of such diversity, what are the useful information processing principles in common? They are as discussed in Chapters 1 and 2: the nature of the physical stimulus and its fundamental limitations; transduction, the conversion of light to an electrical signal; noise and information; and information streaming. These principles are what the designer of robots or virtual reality systems needs to take away. They find use in human computer interaction and the compression of sensory data for optimal trade-offs between bandwidth and perceptual quality. This chapter deals with the overall architecture, later chapters look at colour and object properties, movement and binocular vision. So the plan is to start right at the optics of the eye and track upwards as visual data is streamed into the brain. Needless to say our tour of this immensely complex computational structure will be a whistle-stop tour at best. With vision conferences, such as that of the American Research in Vision and Ophthalmology, alone attracting tens of thousands of delegates every year, the knowledge and growth of knowledge are truly vast. We will trace a very thin line through key ideas and the way they relate to man-made systems using these cardinal principles of information content and streaming.

Section 6.2 deals with the first principle, just what information is carried by light in the visible spectrum, in other words what information enters the eye. Turning to the visual system itself, §6.3 sketches briefly the overall architecture before going a little deeper in subsequent sections. §6.4 covers the preprocessing stage, before transduction. The functional elements are the pupil, which allows light into the eye and the optical system, which focuses an image on the light sensitive surface, the retina (§6.3). The emphasis is on the optics of the vertebrate eye (actually quite distinct from the compound eye of insects, and looks at how the retinal image is constrained by physics. The next stage, transduction, discussed in §6.5 is accomplished by the photoreceptors, which sample the retinal image. This sampling stage defines what goes in and constrains all subsequent processing. There is no wavelength sensitivity at this stage and there will be no colour vision. From the receptors on, information starts to be segregated into different streams and §6.6 discusses the nature of this streaming process, why and how it evolved, and lays the groundwork for later chapters. As with all the senses, the first stop after the sensory organ, the eye in this case, is the thalamus (§2.5.1). Thereafter information streams to a variety of cortical areas. This will complete the snapshot of the physiology, how the visual system is put together, although it is but the briefest of discussions. But now armed with this knowledge, it is possible to ask how these anatomical and physiological features constrain perception. Do the physical limits and biological constraints match what we get at a perceptual level? Section 6.7 considers psychophysical correlates in resolution, sensitivity and information flow. Other elements of perception appear in later chapters, notably colour (§8.5), texture (§8.8), motion (§7.5) and stereopsis (§7.2.2).

Although the emphasis is on mammalian, particularly human, eyes, it is worth a brief mention of other designs (§6.8). Robots do not have to be anthropomorphic – look at the robot vacuum cleaner developed at MIT; more flying saucer than humanoid. So, seeing how unusual eyes have evolved to fit particularly environmental niches provides inspiration for robotics. But it also provides insight and inspiration for the designers of animats, the virtual robots of virtual worlds and computer games. The range of possible eye types is immense, and this section gives no more than a mere soupçon to whet the appetite. Land and colleagues have studied many such different kinds of eyes and written numerous highly readable articles and books thereon (Land & Nilsson, 2002), while §6.10.2 discusses briefly some of the progress in visual implants, where biological and machine vision fuse.

6.2   The physical limits

The optical imaging process sets two fundamental limits on the information which can get to the brain:

1. *Photon noise* arising from the *quantum nature* of light.
2. Resolution imposed by *diffraction*, a property of the *wave nature* of light.

In a strange irony of the development of scientific knowledge, Newton, who amongst his many achievements had considerable insights into optics, firmly believed in a particle theory of light. After fairly acrimonious debate, he lost out to Thomas Young, advocate of a wave theory. In the early twentieth century, the rise of quantum mechanics saw the theory of light take on a dual character as both wave and particle and the properties of *both* limit vision. The visual system of most animals is sensitive to just a small range of the electromagnetic spectrum, from about 700 nm (red) to 400 nm (blue). Why should we be restricted to this range? Why have we not evolved infrared detectors to operate at night or X-ray eyes to see through solid objects beloved of science fiction writers?

At the infrared, low energy end some animals have indeed evolved detectors. We discuss the rattlesnake pit in §11.2.1. But the rattlesnake is cold blooded. In warm-blooded animals, such as mammals, the interference from our own body heat makes infrared very difficult to use. No mammals have any significant infrared vision[1].

At the high end, the ultra-violet, insects and birds do go out further than we do. But it gets progressively more difficult to use these high energies, because they damage so much organic tissue. Ultraviolet light is in fact very damaging to mammalian eyes (see §6.2.1 for an avian solution).

But such arguments are always fragile. Evolution finds clever solutions when there is a need. However, both our visual bandwidth and the absorption band for cholorphyll, the photosynthetic pigment of plants, is actually the peak band of energy transmission from the sun. For photosynthesis in plants, this obviously is a good idea: they want to capture as much solar energy in the photosynthetic pigment, chlorophyll, as they can. For animals, using sunlight for vision, the more light energy we use, the greater the signal to noise ratio, and the better their vision.

6.2.1   Nature of the illumination

The illumination conditions, $I(\lambda)$, of the natural world vary according to time of day and environment. Every object radiates electromagnetic energy. How much and at what wavelength depends upon its absolute temperature. Physicists idealise a radiant object as a *black body*, which has known radiant energy distribution. In the quantum theory of light, the energy is discretised into packets, *photons*, of which the energy is:

$$E = h\nu = \frac{hc}{\lambda} \tag{6.1}$$

where h is Planck's constant, c, the speed of light in vacuo and ν the *frequency* of light. Frequency tends to be a rather unfamiliar concept for light outside of spectroscopy. It is exactly equivalent to frequency of sound, but in vision and optics we tend to think of light more in terms of wavelength, λ, related to frequency through the speed of light, c, i.e. $c = \nu\lambda$.

The light from the sun during the middle of the day may be taken as a black body, the radiation, in photons, of which is given by

$$n(\lambda) = \frac{2c}{\lambda^4 (e^{\frac{hc}{\lambda kT}} - 1)} \tag{6.2}$$

where λ is the wavelength, k, Boltzmann's constant and T the absolute temperature (degrees Kelvin) (RCA, 1968). $n(\lambda)d\lambda$ gives the number of photons emitted per unit

---

[1] There is an interesting parallel to hearing here: we are not sensitive to very low frequencies, below around 20 Hz, because they would be subject to a lot of noise from the mechanical operations within the body.

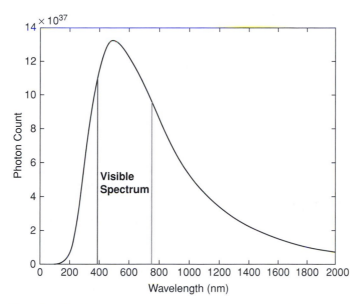

Figure 6.1 Black body radiation from the sun. The vertical lines show the approximate limits of the visible spectrum.

surface area per unit solid angle in a wavelength interval. At wavelength λ the corresponding radiant energy is given by:

$$E(\lambda) = \frac{n(\lambda)hc}{\lambda} \tag{6.3}$$

The denominator in eqn 6.2 contains a dimensionless Boltzmann ratio of energy of photon over kT. Since exponentials go to infinity faster than any polynomial, the denominator of this expression goes rapidly to infinity as λ goes to zero, so the number of photons becomes very small as the wavelength becomes very small.

The temperature of the sun is around 5800 K (Pidwirny, 2010), giving the curve in Figure 6.1. At dawn and twilight, sunlight has a much redder quality, since the pathlength through the atmosphere is longer and there is more scattering of blue light out of the light path. On the other hand in lush vegetation, much of the ambient reflected light is green. Figure 6.2 shows the illumination from different sources within a forest (Endler, 1993), obtained by Chiao *et al.* (2000) with pronounced emphasis in the green for much of the reflected light. The brain copes with these variations remarkably well with a mechanism known as colour constancy (§8.4.2).

For comparision figure 11.1 shows the black body radiation from a human being with body temperature 37 °C or 310 K. The heat from our own body would interfere with attempting to use this part of the spectrum. Rattlesnakes, however, do make use of the infrared (§11.2.1), but exploit it for finding warm-blooded prey.

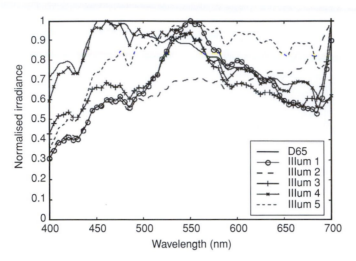

Figure 6.2 Irradiance inside an Australian forest. Illuminations 2 and 5 are from the gaps in the canpopy (direct sunlight) and are roughly similar to sunlight itself. One and 3 are forest and 4, woodland shade respectively. These have a prominent peak around the reflectance of chlorophyll. D65 is the standard CIE white illuminant (§8.7.1). (From Chiao *et al.*, 2000 – see text.)

Ultra-violet light gets more and more damaging to animal tissue as the wavelength drops below 400 nm. Hence mammals have introduced pigmentation mechanisms to soak up ultra-violet before it can cause damage. Birds, on the other hand, have other filtering methods. In §8.3.1.2 we see how bird photoreceptors contain oil droplets which have very steep cut-offs at low wavelength. Thus they can control ultra-violet via oil droplets, which gives them the possibility of ultra-violet sensitive pigments[2].

Budgerigars have found an alternative use for ultra-violet, however. The crown of their head fluorescence under ultra-violet illumination, producing a strong yellow signal in the range 500–600 nm, ideally placed for chromatic vision of the main cones. In fact this fluorescent display has become sexually selected (Arnold *et al.*, 2002).

As for X-rays, this will always be the stuff of science fiction. Apart from the huge energy (and damage potential) for X-rays, they are hard to capture[3]. But, the intensity of X-rays from the sun is quite low anyway.

6.2.1.1   Measures of light

There are several different measures of light flux from a surface or emitted by a source, but they fall into two different categories. *Radiometric* units give energies (e.g. Joules in SI units) and are independent of human vision. *Photometric* units give energies filtered through the wavelength sensitivity of the human visual system (Figure 6.3).

---

[2] I am indebted to Daniel Osorio for pointing out this distinction between mammals and birds.

[3] The high energy means that atoms of high atomic weight are necessary to have the required energy levels. Medical X-ray intensifying screens, for example, use materials based on elements such as tungsten and the rare earths such as lanthanum.

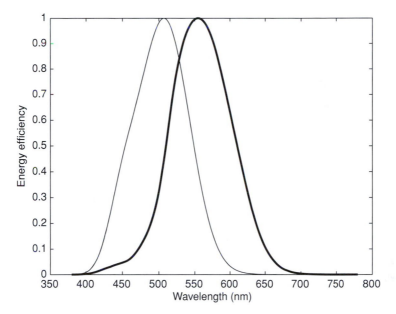

Figure 6.3 Sensitivity of the human visual system in bright light, *photopic*, thick curve, and low light, *scotopic*, thin curve.

Then, again, different units describe different features, e.g. total emission from a source of electromagnetic radiation or the energy incident upon a surface. There are corresponding quantities for the radiometric and photometric cases. Table 6.1 summarises these units.

The candela, the SI unit, began life as a candle[4]. It was then given a more precise definition in terms of black body radiation, the black body in this case being a lump of platinum at its melting temperature. But for ease of measurement a further definition was introduced in 1979: the light from a monochromatic source at 540 terahertz, or approximately 555 nm (green, approximately peak sensitivity of the eye) with an intensity of $\frac{1}{683}$ watts sr^{-1}[5]. This odd figure was chosen to get close to the original candle definition.

A further refinement, often used in the visual psychophysics literature, is the eponymous *Troland*, named after a Leonard T. Troland, which is a unit of retinal illumination, found by multiplying the luminous flux incident on the eye by the area of the pupil.

But in this book, where the focus is on information, the most important quantity is the photon count, rather than the energy, for two reasons. Firstly, the signal from a photoreceptor depends upon the number of photons captured. Once a photon has activated a receptor, its precise energy becomes irrelevant (§8.2.1). Secondly, the information is limited by noise, and the noise arises from the discrete (photon) nature of light (§6.2.2).

---

[4] Whale watchers will be pleased to know that this definition has been superseded. It referred to the light from a candle weighing one-sixth of a pound burning 120 grains per hour according to the 1860 Metropolitan Gas Act in London. The grain was a crystal of *spermacetti* extracted from the head of the sperm whale!

[5] Solid angle is measured in *steradians*. There are 4 π steradians to the surface of a sphere.

Table 6.1 SI (Système Internationale) photometric and radiometric units. The candela (which replaces the candle) is one of the six fundamental units (along with mass etc.).

Description	Radiometric	Photometric
Light emitted by a point source	(Radiant intensity) Watt sr^{-1}	Candela
Brightness of a surface	Radiance	Luminance (lumen m^{-2}) Watt m^{-2}
Light incident upon a surface	Irradiance	Illuminance
Retinal illumination	Scotopic Troland	Phototopic Troland
Flux	Radiant flux (Watt)	Luminous flux (lumen)

An important concept, familiar to many people these days given the ubiquity of digital cameras, is the **f/number**. Imagine pointing a camera at a bright surface. The amount of light received by each pixel (or each photoreceptor in the eye) depends on the brightness of the surface and the size of the aperture (pupil, hence the troland). It also depends on the focal length of the lens. If the lens is a telescopic lens it will see only a small part of the surface and capture less light than a wide angle lens. Thus the amount of light captured depends on the ratio of the aperture, D, to the focal length, f, known as the **f/number**, F,

$$F = f/D \qquad (6.4)$$

Note that because D occurs on the denominator, small f-numbers correspond to allowing *more* light in.

6.2.1.2   Range of light intensities

In the natural world the range of light intensities is huge. At one extreme there is starlight, in which some animals, such as leopard, can still hunt. At the other there is bright summer sunlight in the middle of the day, where the pupil closes right down to restrict the amount of light entering. In humans the ratio of maximum to minimum pupil size is around three, so the intensity range the pupil can control is an order of magnitude.

6.2.2   Photon noise

As the light level falls, human vision degrades smoothly. Thus we find it very difficult to read in moonlight, but are not consciously aware of why. In fact the signal noise has become too great to reliably discriminate shapes so small, yet the visual system has filtered out this noise (and with it a lot of the image) without conscious awareness (Snyder & Srinivasan, 1979). The effect of photon noise is easy to understand. A given amount of light energy will correspond to a particular number of quanta. If the light level is low, we have few quanta and the image takes on a pointilistic form, like a Seurat image. In medical X-ray imaging we have the same effect, but this time because the quanta are big, corresponding to the much higher energy of X-rays. Since dose has to be minimised to avoid health risks from the X-rays themselves, radiographs often have a mottled look, arising from precisely this problem.

At high light levels, such as bright sunlight, quantum noise is not a problem. But the visual system has other sources of intrinsic noise (§6.5.1).

6.2.3 Diffraction

The wave nature of light imposes a different restriction. When objects are of the same order of size as the wavelength[6], light seems to bend around them and no longer travel in straight lines. This is the process of *diffraction*. Photoreceptors in the retina are packed together very tightly and are comparable in size to the wavelengths of visible light. Hence the imaging of light and its capture on the retina are strongly influenced by diffraction phenomena.

In fact, what happens is that the entrance pupil itself limits the maximum spatial frequency which can pass through. Thus no matter how small we make the receptors, there is a maximum resolution possible set by diffraction. This limit, v_{co}, is given by

$$v_{co} = \frac{D}{\lambda} \tag{6.5}$$

where D is the pupil diameter and λ is again the wavelength (Snyder *et al.*, 1988).

Thus one might expect that the best human resolution would occur at large pupil sizes but the opposite is the case. The best resolution gets close to the physical (diffraction) limit *but occurs at the smallest pupil sizes.* At larger pupil sizes aberrations in the lens (§6.4.2.2) of the eye reduce the resolution well below the theoretical limit (Campbell & Gubisch, 1966). Just as camera lenses tend to lose quality as one goes towards maximum aperture, it is more difficult in engineering terms to get rid of aberrations for big lenses. In fact light coming in at the edge of the lens, and impacting the cones obliquely, has less of an effect than light which enters centrally, partly as a result of optical waveguide effects (Snyder, 1973). This is the Stiles–Crawford effect (Stiles & Crawford, 1933). There is also a Stiles–Crawford effect of the second kind in which rays show a slight colour shift as they move from the centre (Westheimer, 2008).

But a solution to this problem has emerged during evolution. Eagles are approximately diffraction limited with pupil sizes of 6 mm, about the size of our maximum pupil.

It's a different but widespread mechanism at work. Evolution exerts pressure for practical advantages, not for excellence per se. The pupil of the eye opens up in dark conditions to allow in more light. In such circumstances, the resolution becomes limited by photon noise. Thus there is no selection pressure for optimal lens performance. There is such a pressure in bright sunlight when the pupil is small (Snyder *et al.*, 1988).

There is a further possibility deriving from the wave nature of light. Light can leak out from photoreceptors, giving them a fundamental minimum useful diameter (Snyder, 1973).

---

[6] Light at the red end of the visible spectrum is about 0.7 µ; human cones are about 2 µ but some avian cones are as small as 1.5 µ (Snyder, n.d.).

6.3   Overview of the architecture of the visual systems

Vision is the most complex of the senses, processing the largest information flow and taking the most brain resources. In man about 10% of the entire cortex is devoted to vision, and a large part of this is to the early stages of processing.

The eye is often thought of as the zenith of evolutionary complexity, so much so that it was for a while held to be an argument against evolution. It even spurned a title for Dawkins' book on evolution, *The Blind Watchmaker!*

Animals have two quite different types of eye:

1. Some insects have so-called *compound eyes* consisting of numerous facets, each with an independent aperture and lens. Such eyes are of lower resolution than the simple eye of most vertebrate eyes. Ironically much work on information theory in vision, the early work on scene statistics and the formative ideas of neural coding from Laughlin, Snyder and others owes much to insect (compound) eyes (Laughlin, 1981; Snyder *et al.*, 1977a; Snyder *et al.*, 1977b).

   The compound eye trades off resolution for a very wide field of view. Artificial compound eyes are just about to become a reality for robotics and other applications, using novel polymer growth and assembly techiques (Jeong *et al.*, 2006).

2. All vertebrates have simple eyes, which have an optical system much more similar to that of a camera. They have a single lens which forms an image at the back of the eye. In the camera this image is recorded by film or, in a digital camera, a CCD or CMOS array of photosensitive cells on a microchip. In the eye, the image is captured and transmitted to the brain also by a set of photosensitive cells, the photoreceptors, but this time of biological construction. Following transduction by the photoreceptors, a mass of neural tissue preprocesses the visual signal before sending it to the brain along the optic nerve in the *ganglion cells*. This assembly of receptors and neural processing is the *retina*.

The mythical Cyclops with just one eye has no visual counterpart in the animal world. Most classes of animal have at least two eyes, although some arthropods have more; some spiders, for example, having as many as eight. Multiple eyes can achieve a greater field of view, something which many insects achieve through compound eyes.

Although the principles of image formation on the retina were well understood by the sixteenth century, the happenings from there on out remained a mystery. Kepler, who gave the first correct description (Wade, 1998, p. 26), expresses his ignorance in a neat homuncular[7] perspective in 1604:

I leave it to the natural philosophers to discuss the way in which this image or picture is put together by the spiritual principles of vision in the retina and in the nerves, and whether it is made to appear before the soul or tribunal of the faculty of vision by a spirit within the cerebral cavities, or the faculty of vision, like a magistrate sent by the soul, goes out from the council chamber of the brain to meet the image in the optic nerves and retina, as it were descending to a lower court.

---

[7] The ancient, outmoded, idea of some sort of consciousness, or *little man* inside the brain looking at data as it comes in. This sort of idea and various manifestations of it is essentially an infinite regress.

The huge variation in eyes across the animal kingdom does not allow us very many generalisations in what information travels from the eye to the brain. We look at a few interesting variations in eyes in §6.8, but our discussion will be somewhat vertebrate, especially mammal, sensitive.

Firstly, there are two pathways: one (§6.3.1) deals with image analysis, vision as we normally think of it; the second (§6.3.2) deals with a range of other things, such as eye movements.

6.3.1 The cortical pathway

The cortical areas to which we can clearly attribute specific function are the visual areas known as V1–V5, in the *occipital lobe of the brain* at the back of the head. V1 and V2 are the early stages of processing, where the input from the eye arrives, and handle all aspects of the visual image. These take up a lot of space; this early visual processing is computationally very expensive.

It used to be thought that this was a feed-forward information flow, but extensive feedback occurs. Kosslyn and König describe how extensive experiments reveal that mental imagery activates V1 in just the same way that actual visual stimulus does. More recently Li *et al.* (2004) have shown that the information carried by V1 neurons is increased by context.

From V1 and V2, visual information flows along the ventral (what) (areas V3, V4) and dorsal (where) (areas V3, V5). V3 has a strong role in processing dynamic form. V4 is the colour processing centre, where we see cells responsive to colours as opposed to wavelength of light, something which §8.4.2 takes up in more detail. V5 processes moving stimuli (§7.5). As techniques improve subdivisions of these areas have become apparent, V3 and V4 have subsidiary areas V3A and V4A, primarily fed by V3 and V4 respectively, the combined areas sometimes referred to as the V3 and V4 complexes.

We discuss this in more detail in §6.6.2. Part of this complex modular structure is an accident of evolution. But there are ways in which it has evolved to be computationally efficient. (See colour plate section.)

6.3.2 The subcortical pathway

The second pathway proceeds to the *superior colliculus*, a subcortical unit underneath the thalamus (§2.5.1). Some information now travels on to a different part of the thalamus, the *pulvinar* and hence on to the *amygdala*, an area associated with emotional responses such as fear. Section 6.7.1.2 discusses a particular example of this pathway in the perception of the white part of the human eye. Other areas subserved by this pathway are eye movements (§12.1) and eye control. Separate pathways control the pupil size (§6.4.2) and the focusing and accommodation of the eye (§6.4.2.2).

6.3.2.1 The simple eye

The mammalian eye is an extremely complex biological entitiy, assembled from many different types of cells and biological materials.

The primary features for visual information processing are:

- The pupil, or iris, which controls the amount of light entering the eye, equivalent to the aperture of a camera. We saw in §6.2.3 that this controls the optical information entering the visual system in terms of signal noise and resolution.
- The lens which focuses an image onto the retina at the back of the eye. The lens separates two compartments of the eye, both fluid filled by the *aqueous* and *vitreous humours* respectively. Since both of these have a different refractive index to that of air, this makes the optical system somewhat different to that of a traditional camera.
- The photoreceptor mosaic where light is transduced to an electrical (neural signal). Unlike amateur digital cameras, this is not a uniform mosaic, but has highly variable topography. In humans there is an area of very high density in the centre (the fovea), but in other animals the topography can be quite different.
- The blind spot, where the optic nerve exits the eye, an area of about 4×6 degrees, 15 degrees medial (towards the nose) from the fovea (Tong & Engel, 2001). It is somewhat surprising to find all the blood vessels (and asssociated neural tissue) *in front* of the photoreceptors, so light passes through them before capture[8].
- The optic nerve which carries visual information to the brain.

The human eye lacks an anatomical component found in many animals which operate in low light conditions. The reason a cat's eyes look so striking in dim light is because they have a reflective mirror at the back of the eye, *the tapetum*. This mirror helps them to capture just that little bit more light when they are out and about at night.

6.4   The optics

Just as with a camera, the lens and formation of the optical image on the light detecting surface are not just the first stage of processing but also underpin everything that happens thereafter. There are numerous similarities to the camera and quite a few differences. Some variations derive from the biological implementation, but others are more fundamental. In general terms, almost everything seen in the man-made optical imaging systems has already appeared in the animal kingdom. The one exception so far is a negative refractive index material, predicted by Veselago (1968), and realised experimentally three decades later. Resolution beyond the diffraction limit becomes possible.

---

[8] The blood vessels are not completely transparent, since blood, at least, is red. So photoreceptors in the shadow of blood vessels receive less light. In squirrel monkey the shadows leave traces in the sensitivities of visual cortical cells (Adams & Horton, 2002) although these do not seem to occur in man. The fovea is free of large blood vessels.

6.4.1   The cornea and refraction

Just as a modern camera lens contains numerous elements, the eye is not just a single, simple lens. In fact there are four refractive surfaces, which arise because of the variation in the refractive index alluded to above. They are the front and rear (anterior and posterior) surfaces of the cornea and the front and rear surfaces of the lens. The discussion here follows that of Millodot (1982).

The functional importance of these refractive elements is in their *dioptrics*, the refraction of light through to the retina at the back of the eye. But there is a slight loss of information as some light is also reflected at each of the surfaces, the so-called *catoptrics*[9]. There are four possible reflected images from the eye, the *Purkinje images* I, II, III and IV.

The first of these images is the brightest, the direct reflection from the cornea, something we see readily when we look into somebody's eyes. The other images are less easy to see, but III and IV are particularly useful to optometrists, since they show the process of change in shape of the lens as the eye focuses, the process of accommodation (§6.4.2.2).

There is a third category of image in the eye, the *entopic images*, which are formed by detritus within the eye. The most common example are the *muscae volitantes*, or visual floaters, bits of dead cellular material floating around in the fluids of the eye. Needless to say these images are not particularly wanted or useful.

The refractive power of a surface, P, determines the degree to which it bends light incident at an angle to the perpendicular (normal) to the surface. It depends on the radius of curvature, r, and the difference of refractive indices, μ_1 and μ_2 according to

$$P = \frac{(\mu_1 - \mu_2)}{r} \tag{6.6}$$

Power is usually measured in *dioptres* (D) as the reciprocal of the focal length in metres. Table 6.2 shows the typical refractive power of each of the four surfaces. The cornea dominates, because the air/water refractive index difference is far greater than the differences within the biological tissue of the lens and the aqueous and vitreous humours. With such importance to the performance of the eye, the cornea has one of the most rapid cellular repair mechanisms in the body, with a 24 hour turnaround on corneal tissue.

The total refractive power of the eye is not the simple addition of the powers of the individual surfaces, but is somewhat less. For two surfaces the total power is:

$$P = P_1 + P_2 - (d/\mu)P_1 P_2 \tag{6.7}$$

where d is the separation and μ the refractive index separating the surfaces (Born & Wolf, 1980, p. 157).

The total power of the eye is around 60 D, a bit less than the total in Table 6.2. We can approximate this by just a single spherical surface to give us the so-called *reduced eye*.

---

[9] Dioptrics technically refers to situations in which when the object moves, its image moves in the same direction. In catoptric systems, as with a simple mirror, the reverse is the case (Born & Wolf, 1980, p. 155).

Table 6.2 Refractive power of the surfaces of the eye. Refractive index of air taken as 1.0. The lens surface is taken in the relaxed (infinity) position. μ = refractive index; r = radius. (Recast from Millodot, 1982, p. 50 – see text.)

Medium	μ	Refractive surface	r (mm)	Power
Cornea	1.376	Anterior surface of cornea	7.8	48.2
Aqueous humour	1.336	Posterior surface of cornea	6.8	−5.9
Lens	1.41	Anterior surface of lens	10.0	8.4
Vitreous humour	1.336	Posterior surface of lens	−6.0	14.0
Total (additive)	**64.7**			
Total power (actual)	**60**			

Using eqn 6.6, we calculate the radius of curvature of the reduced eye refractive surface as about 5.5 mm. The anterior focal length, f_a, is given by $f_a = -1.0/60 = 17$ mm and the posterior by $f_p = 1.33/60 = 22$ mm. The actual eye is about 24 mm putting the reduced eye surface 1.66 mm behind the cornea.

The *optical axis* passes through the centres of curvature of all four surfaces, including the *nodal point*, the centre of curvature of the reduced eye effective surface. The *visual axis* passes through the *fixation point* of the eye (out in the real world), through the nodal point to the *fovea*, the area of acute vision. The angle between visual and optical axes is about five degrees, with the optical axis towards the nose.

6.4.2   The lens and pupil

The pupil, or iris, controls the amount of light entering the eye. In man the range of diameters is from about 2 mm to a maximum of 8 mm, but this varies quite a lot from person to person. This fourfold difference equates to a factor of 16 in light capture[10]. However, the eye has to contend with a light level range of 10^{10}, so the pupil could not be the only source of intensity control.

The pupil is somewhat akin to the muscular control in the middle ear, which immediately shuts down the ear to loud sounds. The pupil can change with light level much more quickly than the biochemical mechanisms of adaptation which cover the entire range. Its size is also affected by emotional changes, widening with fear (Whalen *et al.*, 2004). It may also be dilated by drugs such as atropine, commonly used by optometrists when carrying out a retinal examination.

The pupil size controls the depth of field of the eye, just as with a camera. High f/numbers (§6.4) give greater depth of field, i.e. the range of distances for which objects are in focus. It seems unlikely that the visual system ever directly controls depth of field. In bright light, the pupil is smallest, resolution highest and depth of field also at its maximum. Two separate pathways control the pupil via the superior colliculus (§6.3.2). One via the Edinger–Westphal nucleus controls the pupilary sphincter muscle

---

[10] There is a slight qualification to this factor arising from the Stiles–Crawford effect (§6.2.3).

for contracting the pupil in bright light. The other via the *cilio-spinal nucleus* controls the pupillary dilator muscle, which widens the pupil in dark conditions. Aspects of the control of pupil size have appeared in just the last couple of years, an evolving story we take up in §6.9 (Erichsen & May, 2002).

Unlike the camera, the focusing mechanism of the lens is different. In a camera, the lens moves backwards and forwards to achieve focus. Using the lens equation:

$$\frac{1}{f} = \frac{1}{u} + \frac{1}{v}$$

where f, u, v are focal length, distance of lens plane to object and distance of lens plane to image respectively. In the camera the focal length remains constant during focusing while u and v change (with the proportional change in v being much larger). In the eye u and v are constant but the muscles of the lens cause it to contract, thereby changing the focal length. This process of *accommodation* becomes less and less easy with age reducing the ability to focus close in – *presbyopia.* In some animals, such as chameleons (Reptilia) the cornea, too, can change shape and accommodate the eye (Pettigrew *et al.*, 1999). Scientists interested in the history of physics, particularly optics and vision, often have the highest regard for the nineteenth-century pioneer Hermann von Helmholtz. His work culminated in a treatise on physiological optics (Helmholtz, 1896). In 1855 he identified the muscles in the eye (the ciliary body) which control accommodation through making the lens fatter amongst other things (§6.4.2.2).

6.4.2.1   Aberrations

Optical lenses are rarely perfect. They suffer from a variety of imaging defects known as *aberrations.* A spherical refractive surface, such as we use in the reduced eye, does not focus all parallel rays incident upon it to a single focal point. Rays off the optical axis come to a slightly different focus and the locus of focal points forms a cusp. This is *spherical aberration.*

But the spherical lens model is a convenient mathematical idealisation. Astrononical reflecting telescopes, for example, use a parabolic surface, which does bring incident parallel rays to a point focus. In practice the shape of the cornea is not perfectly spherical and the aberrations are less than might be expected. But for reasons we shall come to later when we consider sampling, we do not necessarily want perfect optics at the larger apertures. We want aberration-free imaging for *small apertures* in bright light and this we pretty much have.

Chromatic aberration is also a problem, but again the sampling and post-processing make these difficulties fairly small. Shorter wavelength blue light comes to a focus in front of the retina, red light slightly behind. The difference in power between extremes of blue and red is about 1.5 D. Thus blue and red will not simultaneously be in focus.

An example of how *not* to exploit this occurs with lights against a dark background. So a bright neon sign at night with red and blue lettering might be difficult to see as a single sign mixed with many other signs.

6.4.2.2   Accommodation

Schiener in 1619 in his tract *Opticus* (Wade, 1998) was the first to observe accommodation. He showed that double vision could occur for objects some way from the point of fixation. The eye has a complex muscular control system, integrated with movement of the head and partly controlled by the vestibular system (§10.7). Apart from specifically optical issues, we leave the control of eye movement until later (§12.1).

Whereas in a camera the lens (or elements within it) move backwards and forwards, in the human eye the lens changes shape. In other animals accommodation mechanisms are thought to also include lens movement (fish) and change of corneal shape (birds).

The lens of the eye in a young person is flexible. At rest it is held flat by tension in the suspending ligaments, the *Zonule of Zinn*, a network of collagen fibres. A flat lens corresponds to minimum refractive power and thus focuses at infinity. As the ciliary muscles contract, they release tension on the zonule and relax the lens.

As the lens relaxes it becomes fatter, so increasing its refractive power and moving the focal point closer in. The lens is held within a membrane, *the capsule*, which is thicker at the edges than in the middle on the optical axis, determining the shape the lens assumes as the muscles contract. This mechanism requires the lens to be flexible enough to change shape, something which becomes less easy with age (Heron & Charman, 2004, p. 139).

Two reference points are used to define accommodation:

1. The *punctum remotum* defines the greatest distance, r, of clear focus, or *far point*, from the cornea.
2. The *punctum proximum* defines the nearest distance of distinct vision, p, i.e. the state of maximum accommodation.

The range of accommodation is simply the distance $r - p$. The amplitude of accommodation is the range of power achieved by the lens, i.e. $(\frac{1}{r} - \frac{1}{p})$ dioptres with distances in metres (Millodot & Newton, 1981). The amplitude for young adults is around 7–8 D. Accommodation occurs pretty quickly in around 300 ms. It seems to operate via a range of mechanisms (Campbell & Westheimer, 1959):

- Chromatic aberration, as we have seen, brings red and blue components of the image into focus at different depths, hence this is an accommodation cue. It cannot be the only one of course, because not all images have strongly different colours and colour blindness is quite common (§8.3.1.4).
- There are also some variations in spherical aberration with accommodation, but again this seems like a minor phenomenon in bright light.
- The eye is continually adjusting accommodation by tiny amounts of around 0.1 D at a frequency of around 0.5 Hz optimising contrast like a digital camera.

Accommodation is not perfect. At night or in the air there may not be sufficient image detail on which to focus, resulting in errors.

6.4.2.3 Refractive errors

Not all eyes are perfect! In principle, a distant object should be brought into focus on the retina in a correctly functioning, or *emmetropic*, eye. In practice there are several common abnormalities:

- Distant objects may be blurred and the distance from the eye at which objects are first in focus, the far point, is much closer. This is *myopia* and normally arises because the eyeball is too big and the distance to the retina is too great, meaning that the in-focus image falls short. About 20% of the western world is thought to be myopic.
- Distant objects may be clearly in focus and the effective far point is virtual, actually behind the retina. This is *hyperopia or hypermetropia* and, although it provides the best overall vision, the constraint strain of accommodation may lead to headaches.
- *Presbyopia* is a failing of accommodation which starts to appear in middle age as a result of increasing crystallinity and lack of flexibility in the lens.
- *Astygmatism*[11] is usually a failing of the cornea, where the radii of curvature in the horizontal and vertical planes are different; thus the focal point of a line depends upon its orientation.

Only the last of these specifically involves the cornea, but laser corrective surgery adjusts the corneal surface (power) to correct the refractive power of the whole eye. This is usually feasible since the cornea has the largest refractive power of the eye's refractive surfaces (§6.2).

6.4.2.4 The ophthalmoscope

The optical system of the eye can be turned back on itself, to view the back of the eye or *fundus*. Charles Babbage, inventor of the first mechanical computer, was probably the first to view the fundus in this way, but the development of a suitable instrument is, yet again, accredited to Helmholtz in 1851.

Light is shone into the eye via a semi-silvered mirror and the observer views the reflected light. If the eye is emmetropic, the reflected light will be focused at infinity and will be viewed without optical correction. We shall describe later that we can see much of the inside of the eye *but not the photoreceptors in man* (§6.5.2), although they are visible in some animals such as the garter snake (Land & Snyder, 1985).

6.4.2.5 Comparison with the camera

The first concepts of camera-like imaging go back a long way, with the first pinhole camera-like devices appearing in the fifth century BC in China with emprical studies by Ibn al-Haytham as far back as *c.* 1040 and da Vinci *c.* 1500 (Wade, 1998, pp. 27–28). However, Porta in 1589 is often accredited (incorrectly) with the discovery of the camera obscura, and certainly popularised the link between the camera and the eye (Wade, 1998, p. 28):

---

[11] Greek '*a-*', without, and '*stigma*', point.

If you put a small centicular Crystal gall to the hole, you shall presently see all things clearer... Hence it may appear to Philosophers and those that study Optics, how vision is made... The image is let in by the pupil, as by the hole in the window; and that part of the sphere that is set in the middle of the eye, stands instead of the Crystal Table.

Porta was confused about the position of the lens, however. The first correct diagrams seem to be those of Scheiner in 1619, who also had a grasp of why the retinal image is inverted. The correct description of image formation comes from Kepler in 1604.

Although the first cameras had a simple lens, the modern camera lens is a much more sophisticated device with numerous elements, not all of which may move as the lens focuses. A very recent innovation in commercial camera equipment from Canon and Nikon is the use of vibration reduction or image stabilisation relative to camera movement. We shall see later that human eye movements have extremely clever adjustments of this kind (§12.1).

Cameras have two generally used auto-focusing mechanisms. Infrared or ultrasonic systems are *active*: they send out a signal and record its reflections to estimate the distance to the object to be focused, rather like echo location in a bat. These mechanisms are fast and effective at short range but are vulnerable to interference by neighbouring objects, intervening glass etc. They generally are used on older fixed lens compact cameras.

Most SLR cameras use *passive systems*, in which the image itself is used for focusing. The lens is moved backwards and forwards until the contrast is maximised. Such systems have advanced a great deal in the last few years, especially in terms of focusing on moving objects, and now abound on digital cameras. In speed and accuracy they now compete in speed with human accommodation.

6.5   Transduction: photoreceptors and sampling

6.5.1   The photoreceptors

Light lands on the back of the eye, the retina, much as light arrives on the film at the back of a camera. In conventional film, light energy causes the reduction of silver halide to silver. The silver halide crystals are scattered randomly through the film emulsion. In a digital camera, however, the light sensitive cells, or pixels, form a regular, usually square mosaic. Light incident on these cells leads to small voltages inside charge coupled devices (CCDs) or complementary metal oxide semiconductor (CMOS) elements, two different technologies for converting light energy into electrical signals.

In a biological eye, we get a similar photovoltaic process. A visual pigment, or *opsin*[12] such as *rhodopsin* aborbs light and moves to a higher energy state[13]. It then

---

[12]   There are five opsins in the vertebrate retina, but fewer, usually two, are found in most mammals (Ahnelt & Kolb, 2000).

[13]   Rhodopsin type pigments are found throughout living systems, including bacteria. Some cyanobacteria, for example, adapt their photosynthetic absorption to the spectral composition of the incident light, and it is possible that a single rhodopsin pigment with two different photoactive states can act as the controlling sensor (Vodeley *et al.*, 2004).

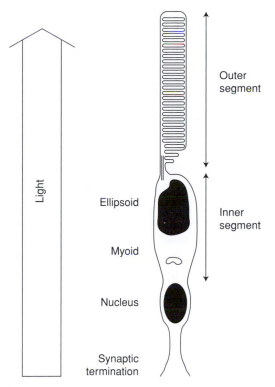

Figure 6.4 Basic components of a photoreceptor. The discs which contain the photopigment are visible in the outer segment. Light passes through the inner segment first. (Adapted from Rodieck, 1973 – see text.)

drops back down to the original state producing an electrical signal in the process. This operation takes place within a photoreceptor, which transduces light to tiny electrical signals. Rhodopsin has several states, some of which can lead to thermal isomerisation, introducing noise into the visual system (Barlow *et al.*, 1993).

In a vertebrate eye there are two types of receptor, rods and cones, named by Schultze (1866) and shown by Müller (1872) to be different types (Ahnelt & Kolb, 2000). The receptor in Figure 6.4 has two main parts, the inner and outer segments. The outer segment actually faces backwards and the image is formed underneath the mass of transparent neural material which forms the retina. The outer segment contains the photopigment where the light is absorbed. The inner segment contains many mitochondria, the powerhouse of the cell. It is here energy is produced to create the electrical signal and generate the output of the receptor[14].

A photoreceptor looks something like a champagne flute. Just as the flute traps bubbles, so the receptor traps photons through total internal reflection and optical waveguide

---

[14] However, mitochondria may also have a role in the waveguide properties of the receptor in the periphery (Hoang *et al.*, 2002).

properties (Snyder, 1973). But capture is most likely when photons travel in along the axis. Photons which come in at an oblique angle are more likely to travel straight through.

6.5.1.1   Rods

The rods are long and thin and sensitive at low light levels. They are so sensitive that they will generate a detectable voltage from the absorption of just a single photon! There are many more rods in the human eye than cones, about 100 million rods compared to around 7–8 million cones. Yet it is the cones that support high resolution vision.

The low resolution of the rod system arises functionally because it operates at low light levels. It arises anatomically because there is huge *convergence* in the rod system: many rods feed into a signal neuron. So, why go to the trouble of having many rods, why not just one big one?

The answer seems to lie in the properties of the photopigment. It spontaneously flips creating spurious signals at random. If we want to detect a single photon, then that photon can flip only one photopigment molecule. So we want the probability that a flip actually arose from a photon and not from noise. Thus to keep the noise as low as possible, we need the smallest amount of pigment in each rod. Animals have, too, involved various mechanisms for making the thermal noise as low as possible (obviously an advantage for cold-blooded animals with good nocturnal vision) which can allow them to adjust the noise according to time of day (Barlow *et al.*, 1993).

The ratio of rods to cones varies widely. In cat, which hunts at night, the ratio of rods to cones is 20:1 even in the *area centralis,* compared to the fovea in man which is rod-free (Ahnelt & Kolb, 2000).

6.5.1.2   Cones

Cone photoreceptors on the other hand tend to be somewhat fatter, containing more photopigment and thus having a greater dynamic range, but a higher level of intrinsic, thermal noise (Sampath & Baylor, 2002). Whether this arises because the cone photopigment is noisy, i.e. isomerises spontaneously at a much higher rate than rhodopsin, or whether the noise in the cone channel arises from transduction noise further upstream has been the cause of some discussion. Fu *et al.* (2008) used genetic techniques to get red cone pigment into a mouse rod, where the amplification is much higher, and thus show that it is *not* the photopigment that is the primary source of noise. Partly as a result of increased noise, they are less sensitive and operate in bright light conditions. It is the cones that provide us with high resolution *and* colour vision.

Cone sizes vary by an order of magnitude across species from around $15\,\mu$ in garter snake to around $2\,\mu$ in humans. Around $1.5\,\mu$ seems to be as small as cones ever go. Whether this relates to anatomical or physical constraints, or whether it arises from the progressive tendency of light to leak out from the cone as its size gets close to the wavelength of light, is still an open question.

They not only vary in size and shape but also exhibit more radical differences. They may occur in conjoined pairs as *double cones,* common in reptiles, birds and monotremes and marsupials amongst mammals. There is speculation that double cones subserve special functions such as motion detection (Ahnelt & Kolb, 2000), but they have disappeared in placental mammals. The cones of these animals also frequently contain oil droplets (§8.3.1.2).

Humans like all mammals have a short wavelength (S) cone system and it is quite different to the long (L0) – medium (M)) wavelength system. Only about 10% of the cones are S and they form an independent regular mosaic, which is somewhat less regular around the fovea (Ahnelt & Kolb, 2000). The density of S-cones varies little between individuals, but the L/M system shows considerable variability, but usually with more L than M (Hofer *et al.*, 2005).

6.5.1.3   Photoreceptor topography

The human retina, especially in the foveal area is somewhat like a digital camera with a regular mosaic. Whereas in cameras this mosaic is usually rectangular[15], in the mammalian eye this mosaic is hexagonal, with slight deviations from perfect symmetry (Curcio & Sloan, 1992). Other regular mosaics do exist with rectangular mosaics occurring in teleosts, and a high degree of regularity in sub-mosaics in avian retinae (Kram *et al.*, 2010).

At the opposite extreme to the almost crystalline regularity of the human fovea is the American garter snake, the retina of which is more or less completely disordered. The large-scale organisation differs even more strongly. It has a huge concentration of cone receptors in the centre of the eye, the fovea, where there are no rods at all. As we move away from the centre, the cones start to get bigger and move apart with rods appearing in-between. This absence of rods in the fovea accounts for slightly greater sensitivity at night of the axis. It is tempting to assume that all animals have a fovea like ourselves. But this is far from the case and in mammals the fovea is essentially confined to primates. For example, cat has a less specialised high resolution area, the *area centralis*; rabbit has a visual streak along the horizon; while some birds have two foveas, one straight ahead and one out to the side.

A final interesting feature of receptor mosaics is whether we can see the receptors by looking in through the eye with an ophthalmoscope. This links to the functional properties of the photoreceptor mosaic sampling of the image (§6.5.2, §6.5.3).

6.5.1.4   Adaptation to different light levels

A good photographic film has a latitude, or dynamic range, of around 4 log units[16] while current digital cameras respond accurately to a rather low range of light levels (§6.10.1). Yet the cones of the human eye operate over at least 10 log units. The change in pupil size is worth no more than 1 log unit and this remarkable range results from the ability of the eye to adapt to the average luminance level. We are all familiar with this adaptation process going from light to dark conditions.

---

[15] But there are exceptions; Fuji has an octagonal sensor.

[16] Using (by convention) logs to base 10. So the ratio of minimum detectable intensity to saturation level is 10^4.

Adaptation is a very widespread mechanism of animal senses. In vision it starts right at the photoreceptor level, but occurs also at stages further up such as in the magnocellular pathway (§6.6) (Lee *et al.*, 1999) with a total range of 6 log units. A receptor measures light level and adapts its response to the average level. As we go higher up the visual system, where neurons respond to complex patterns, pattern adaptation also occurs. Section 6.10.1 discusses how the limitations of low dynamic range in displays and hard copy can be overcome by digital image processing.

6.5.2   Sampling

Although frequency is a natural concept for sound, it is also extremely useful in optics and vision. In this case the frequency is spatial: sinusoidally varying patterns of light, grey level, photographic density of a film or intensity on a computer monitor. Section 6.2.3 showed that diffraction limits the maximum spatial frequency which gets through to the image on the retina. Thus from the sampling theorem (§3.6), following the sampling arguments, we can see that we do not need an arbitrarily fine receptor mosaic. There is a receptor density which will be sufficient to exactly capture the image (in the absence of noise).

For a rectangular sampling grid, we just need a density of receptors corresponding to the sampling frequency in both directions. So, using eqn 6.5 above, we can work out the density we would need. For a hexagonal matrix the same principle applies, but we do not need the centres of the receptors to be quite so close together. We can get by with about 10% fewer receptors.

What happens in practice is really interesting. It turns out that aliasing problems are not so serious for many natural images, because they lack huge amounts of high frequency energy (Huck *et al.*, 1983). Aliasing usually just ends up as another form of noise, along with photon noise and noise generated within the receptor. However, if we improve the contrast, by increasing the cut-off frequency we can increase the signal/noise ratio. Such an argument suggests that animals should *undersample* and, indeed, in most cases they do (Hughes, 1996; Snyder *et al.*, 1986; Snyder *et al.*, 1988). It is also interesting that the design of digital cameras has independently evolved the same strategy. Digital cameras frequently undersample too (at the time of writing, but evolution of such cameras is occurring very rapidly).

Data for topography of the photoreceptor mosaic is hard to obtain and thus rather few cases are known precisely. But one which is quite well documented is man. Man does not seem to undersample to any significant extent in the fovea. The best explanation for this lies in the multiple spectral types, thus we postpone further discussion until the section on colour vision (§8.3.2).

If the sampling is matched, then the maximum frequency visible applies both ways. So, we cannot see the photoreceptors through the lens of the eye with an ophthalmoscope. Land and Snyder (1985) were able to see the receptor mosaic in the garter snake, because this animal undersamples severely.

6.5.3   Wavefront correction ophthalmoscopy

The image quality of the lens rapidly deteriorates as the pupil opens. Thus the best resolution is at the smallest pupil size, the one used in bright light when the signal to noise ratio is highest. But a solution to the problem of being able to see the photoreceptors is to use a larger pupil size but correct the lens aberrations.

The technology to do this already existed in astronomy, where atmospheric distortions cause image degradation, fixed by *deformable mirrors*. If a mirror is not optically flat, its reflection is distorted, which is how the distorting mirrors in a fairground work. It is possible with computer control to build a mirror out of many little movable segments. Thus the distortion of the mirror is variable and can be made exactly equal and opposite to the distortion caused by lens aberrations.

A number of labs began development of these systems, notably Jagger and Hughes at the National Vision Research Institute in Melbourne (Hughes, 1996). This technique can now resolve photoreceptors and it was used to establish that there are in fact two types of colour blindness: one (the familiar one) where the pigments are identical and the other (the new one) where one class of cones is defective (Mackenzie, 2003). It has also been used to address issues of the uniformity of distribution of L- and M-cones (§8.3.2).

6.6   Transmission to the brain

The retina is a highly complex multi-layered neural system, preprocessing visual information before sending it on to the brain. Part of the preprocessing is about compression reducing the amount of information to be sent, making transfer faster and requiring less energy. But the retina also begins the streaming process by which visual data is sorted into different categories to be used for different things (He *et al.*, 2003). The retina has a range of different cell types, notably *horizontal, amacrine, bipolar* and *ganglion*. For our purposes we can think of the first three as doing the preprocessing and the last, the ganglion cell transmitting information along the optic nerve to the brain.

Figure 6.5 shows schematically the retinal wiring, with the principal, but not all, interconnections. There are several parallel pathways, the on and off systems which are distinct from the bipolars to the ganglion cells. Another pervasive parallel subsystem is the transient and sustained pathway. Bipolar cells exhibit both categories, but the ganglion cells may preserve one or the other or integrate both to give wideband temporal properties. There are at least 55 distinct cell types across the above categories (Masland, 2001). The horizontal cells feed back onto the photoreceptors and play a role in controlling the overall gain of the retina. Bipolar cells, onto which the photoreceptors also have synapses, connect both to amacrine and ganglion cells. The rods and cones have their own bipolar cells, but the rod system is evolutionary younger and is thought to have become grafted onto the existing cone system. There are 29 types of amacrine cells which carry out a range of preprocessing tasks (Masland, 2001).

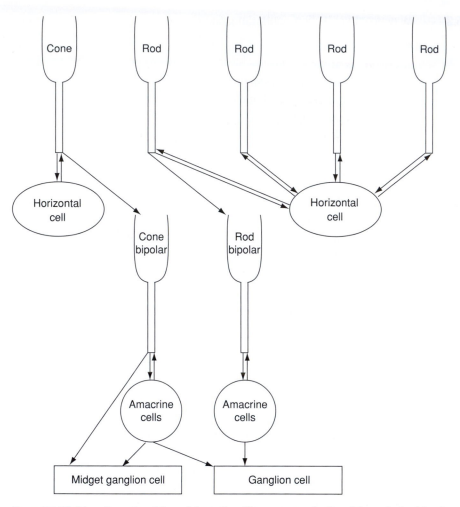

Figure 6.5 Highly schematic wiring of the retina. The vast complexity of the retinal wiring is still only partially understood at a functional level. The three classes of internal cells, horizontal, bipolar and amacrine have numerous subclasses with different properties, wiring and function, tailoring the image data to go to the brain along the ganglion cells. A full discussion appears in Masland (2001) – see text.

In many animals a number of specialist cells appear in the retina, such as cells tuned to edges or movement (e.g. in rabbit). But, as we shall see later, for animals with stereoscopic vision the cells subserving binocularity tend to be simple centre-surround cells. The essence of a centre surround cell is that the background is subtracted from the signal in the centre. The surround is that bigger and takes a sort of local average. So the effect of the surround is to take out the average light level and signal variations against the average. If the surround was infinite and uniform this is exactly what it would do. Thus the huge range of light levels we operate in are removed at the retinal level. Very little information about absolute light intensity goes to the brain.

A second feature of the surround is that it takes out the lowest spatial frequencies. These are very gradual intensity gradients, which do not usually contribute anything useful about the scene.

There are least 13 subtypes of ganglion overall (Levick, 1986) with huge variations across the animal kingdom in general (Hughes, 1977). However, as Sincich and Horton (2005) remark, we know a huge amount but major issues still remain unresolved. Primates have a further specialisation associated with colour vision and high resolution form vision. In primates there are three distinct pathways to the brain, known simply as the M, P and K (Konio) cell pathways. These labels derive from the pathways out from the thalamus. From the retina there are (in humans):

- The M or *parasol* cells which project to the *magno* layers. These feature in most mammals.
- Midget/bistratified: these become the P and K cells and carry spectral and high resolution information and are found primarily in primates.
- Sincich and Horton (2005) suggest that the consensus is that the blue–yellow bistratified ganglion cells with blue (S-cone) centres have their own distinct pathway to the K cells in the LGN. Some debate has centred around the pathways for the S- (short wavelength) cones. They contribute to the surrounds of several pathways (allowing, for example, blue–yellow opponency), but they appear in the centre of the cells of only the K pathway.

The M-cell pathway has a conduction speed of around $15\,\text{ms}^{-1}$ compared to $6\,\text{ms}^{-1}$ for the P cells. It also has a higher contrast sensitivity, greater than 60 supersecs compared to less than 20 supersecs for the P cells. The M-cell pathway is essentially the entire set of cells in animals such as cat which are largely achromatic. Its cells subserve movement detection, large-scale form recognition and binocularity. Although they respond more rapidly than the P cells, they have lower resolution. The centre of each M cell is fed by several cones and there is no wavelength discrimination in the cell.

Compared to the M-cell pathway, the P cells tend to have lower sensitivity and lower temporal resolution, but outnumber the M cells by about four to one (Lennie, 2000). The greater density of P cells would make them more suitable for carrying high *spatial frequency* information. This they do, in the sense that they have smaller receptive fields. But some of the high resolution tasks of vision such as stereopsis are mediated by the M cells. It is very easy to confuse resolution and high spatial frequency. By resolution we usually mean the smallest things, or the smallest gaps between things which we can see. It is very tempting to assume that this implies the need for high spatial frequency. But in fact these resolution tasks often depend as much on high contrast at low frequencies. It is *fine-grained, repetitive* patterns which demand high frequency sensitivity. It is interesting that the evolution of the P pathway may have been a prerequisite for trichromacy (§8.6).

There is little rod input to the P pathway, if any, which is consistent with its use in bright light. But there are a lot more P cells, approximately one per cone in the fovea. Thus they have the most accurate sampling of the retinal pathways.

Compared to the M cell, the P-cell pathway appears to be a separate additional evolution in primates. It samples the retinal image more accurately, with one ganglion cell per cone. Because the cones have wavelength sensitivity, it follows that the P cells must have distinguishing spectral properties and they are the pathway which supports colour vision. The cells tend to be wavelength antagonistic. So, if the centre is red, the surround will be green in what is referred to as an *empcolour* opponent cell.

Streaming is a fundamental feature of animal visual systems. Processing modularity, with functional modules and brain areas for movement, colour, stereopsis and further areas which integrate the senses. It's also a particularly useful idea for engineering. Since many systems are limited in computer power or bandwidth, selecting out the useful and only the useful information is very efficacious.

6.6.1   Compression

The image that lands on the retina contains information, some of which is not essential for pattern recognition, object identification and location, or navigation. Thus this information gets stripped away before forwarding the useful information onwards. Some image compression happens within the retina itself:

- The rods are pooled with the signal from a larger number feeding one ganglion cell. There is also some pooling of cone photoreceptors away from the fovea.
- The absolute light level is removed and replaced by a contrast relative to the background.
- The lowest spatial frequencies are removed.

In fact the last two operations implement a compression algorithm reinvented in the engineering world and known as predictive coding.

6.6.1.1   Predictive coding

Srinivasan *et al.* (1982) proposed a coding strategy for retinal ganglion cells which relies upon adjacent intensity values being correlated with one another. As we have seen, correlation is introduced by the optics, but the very existence of structure in images implies correlation in the scene itself.

Ganglion cell receptive fields are cylindrically symmetric, within an inner circle and outer, annular surround. The centre is excitatory, the surround inhibitory. The net effect is to have no response to a uniform field, and to subtract out the average intensity and just transmit scene contrast.

Srinivasan and collleagues were able to show that the centre and surround have a precise mathematical relationship, corresponding to the engineeering technique of *predictive coding*. The surround of the receptive field may be used to predict the value at the centre. Of course this prediction will contain errors, but the error signal will be smaller and less costly to transmit than the full centre value.

The predictive coding idea proved to be very influential and numerous extensions followed. Derrico and Buchsbaum (1991) introduced colour, showing that a single high resolution achromatic channel and a low resolution colour channel sufficed for the image databases around at that time. Since then many studies have used higher resolution data in space and wavelength and found similar subdivisions.

6.6.2   From retina to brain

Visual information leaves the eye in the optic nerve and heads towards the *thalamus*. As we saw in the introduction, the thalamus is a subcortical unit in the diencephalon and acts as a relay station for all the senses. The visual pathways go into a thalamic area referred to as the *lateral geniculate nucleus (LGN)*.

From the LGN the pathway continues to the back of the head, to the *visual cortex* in the *occipital lobe*. There are numerous distinct visual areas. In the much-studied macaque monkey there are at least 30, but in man there are thought to be as many as a hundred. In the discussion that follows we tend to describe the pathways in the direction from the retina to more and more specialised processing as a sort of information flow. But the actual wiring is highly reciprocal. Often as many wires go back as go forwards. Handshaking and top-down iterative refinement are extensive and remain active areas of research.

As we move deeper into the cortex, the functional nature of distinct areas becomes less clear. But at the time of writing we are seeing explosive growth in our understanding of the modular structure of visual computation as a result of two experimental techniques. The M and P cells get their names from the areas of the LGN to which they project (M for magno and P for parvo referring to large and small cell sizes). More recently a third pathway has been diffferentiated from the parvo pathway, the Konio (K) cells. There is still uncertainty as to the exact pathways for different cone types, but there is good evidence that an *on* pathway for the short wavelength cones feeds into the K pathway (Szmajda *et al.*, 2008).

Like the cortex, the LGN has six layers: the top four have the smaller cells to which the P ganglion cells project; the bottom two are larger and receive the M-cell input[17]. People with various sorts of brain damage may have no conscious sight but still have *blindsight.* They are not aware of being able to see something, but are nevertheless able to make correct decisions based on visual data.

From the thalamus *optic radiation* takes us into the first visual area, V1, Brodmann's area 17, the biggest area, where the input from the LGN arrives and the first extensive signal processing occurs. From V1 in the main we progress to V2 where cells with more complex behaviour are found, such as cells responsive to illusory contours, of which Figure 6.6 shows an example (Heydt *et al.*, 1984).

There are three other distinctive visual areas before we leave the occipital lobe. V5 handles movement, V3 dynamic form and V4 colour. Distinctively visual areas occur also in the temporal cortex, where we find various sorts of object recognition and the parietal cortex which in very general terms seems to handle spatial relationships.

Hubel and Wiesel won the Nobel Prize for their definitive classification of orientation-selective cells in V1. This area has a very distinctive stripey appearance when stained for cell bodies (its cytoarchitecture) and is thus known as the *striate cortex.* The area in front, which includes many of the other visual areas, is referred to as the prestriate

---

[17] There is another much less important pathway from the retina to the superior colliculus and also to another thalamic area, the pulvinar. Phenomena such as blindsight (Kaas, 1995) arise from these pathways.

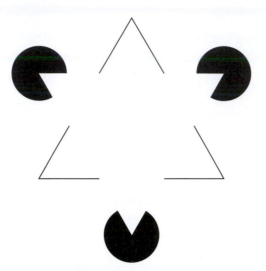

Figure 6.6 The Kanizsa triangle: a well-known illusory contour in which there appears to be a distinct triangular figure.

cortex, from its physical location in the brain – *not* because information goes there before the striate cortex!

For a couple of decades, the Hubel and Wiesel maps of orientation columns and hypercolumns dominated thinking about V1 (Hubel & Wiesel, 1962). A variety of information theoretic models, which we discuss in §6.7.1, accounted well for the structure and topography, with one small difficulty. Any model arguing for information theory as a metric for cell organisation tends to imply that the number of cells is optimal. V1 has far too many cells, almost an order of magnitude too many, for these models.

In the 1980s, further work by Livingstone and Hubel (1988) elucidated further cell classes, with distinct wavelength dependency. Using cytochrome oxidase (§2.6.1) staining, two distinct categories of cells emerge in V1. The M-cell pathways enter in layer 4B. The P cells entering layers 2 and 3 generate two distinct staining patterns. Since observation of the patterns usually precedes functional understanding, these categories were referred to then as the *blobs* and *interblobs* – and the names stuck. The blobs essentially extract *spectral information*, the interblobs *form information*. Beyond V1 the pathways get a little complicated. Most of the connections out from V1 go to V2. But some signals go directly to V3 from the interblobs and V4 from the blobs.

V2 is the second large area of cortex which has a general role. It receives the majority of output from V1. The cytochrome oxidase architecture is again distinctive. A repetitive cycle consisting of thick stripe, interstripe region (also known as pale stripes), thin stripe, interstripe region repeats over a distance of about 4.5 mm of cortex. *Each one of these regions forms a distinct topographic map of the visual field.* However, these topographic maps are not uniform in space. Just as the most accurate vision is in the fovea, with resolution decreasing with distance out from it, so there is a *cortical magnification factor*. More cells are devoted to areas with greatest ganglion cell density, but this is not surprising in information terms.

The blobs of V1 project to the thick stripes. As you would now expect, this is essentially a colour pathway and the thick stripes project out to V4. But V4 processes spatial information too. These broad generalisations are not perfect!

The nature of the wiring from say the blobs in V1 to V2, V3 and V5 looks as if it is part of a general strategy. These multiple projections may occur through cell branching, but it seems that the blobs themselves have regular structure.

However, the fine details are still controversial. Sincich and Horton (2002) cast doubt on the tripartite pathway from V1 to V2, finding, again using cytochrome oxidase staining, that there are really just two subsystems, essentially the dorsal (MT) and ventral (V4) pathways for what and where respectively. The blobs project to the thin stripes while the interblobs project to both thick and pale stripes. Thick stripes in V2 then project on to MT while pale and thin project to V4.

This section is pretty complicated as it stands, so we have left out one whole big aspect of these pathways – the binocular one, to which we shall return later.

6.6.2.1   Illusions from motion

It might seem from the above discussion that there is a definite wiring path from retina to primary visual cortex. A point in the visual field activates a particular point in the retina, stimulates particular ganglion cells and ultimately ends up in a predefined place in the cortex. Recent studies suggest otherwise!

Motion in the visual field produces many illusions. But how is an illusion represented in the cortex? Earlier work shows that some illusionary lines (Figure 6.6) show up in V1, others not until V2. Conversely, when an illusion makes a physical stimulus active on the retina seemingly disappear, the activity at the corresponding location in the cortex disappears too (Eysel, 2003). A similar phenomenon occurs with touch (§10.4.9).

But the story is even more complicated (and still unravelling at the time of writing). Whitney *et al.* (2003) showed that moving patterns within a stationary window appear subjectively in a different place in the visual field to the corresponding location of the activity they generate in V1. So the location in V1 is neither where the stimulus is physically on the retina nor where it appears to be subjectively! The explanation seems to be along the lines that motion is not yet being interpreted in V1. Motion analysis occurs in areas such as MT and MST, which feed back to the primary area V1, and influence where and how the representation occurs.

6.6.3   Dorsal and ventral streams

Once into the visual cortex the streaming is partly determined by cell type. The M cells input into layer 4Cα of V1 while the P cells input into 4Cβ. The ventral stream is clearly identifiable via the blobs and interblobs. Layer 4B is more complicated. It has two distinct cell types: pyramidal cells which feed the thick stripes in V2 and V3; spiny-stellate cells which feed area MT. V2 and V3 also project to MT, thus providing an indirect pathway via pyramidal cells. The spiny-stellate cells receive little if any input from $4C\beta$ and thus very little P input. The receptive fields in $4B$ are large and achromatic and are directional, orientation specific with binocular disparity. Early models would have suggested that there would be no P-cell input, but more recent studies cast doubt on this, perhaps indicating a P role in low luminance or high spatial frequency conditions (Yabuta *et al.*, 2001). The different cell types suggest, however, that the P- and M-cell contributions to MT might subserve different functions. There is speculation that these different cell types might have different levels of feedback from higher levels, notably the spiny-stellate direct pathway being primarily feed-forward (Levitt, 2001).

The M-cell pathway may also have a role in the top-down processing of visual stimuli. Using magneto encephalography, Bar *et al.* (2006) were able to show that the orbitofrontal cortex (OFC) is active about 50 ms *before* the object recognition areas in the temporal cortex. Thus the OFC is thought to have a role in rapid detection and prediction of visual content, i.e. facilitating object recognition. Moreover, this activity in the OFC is driven by *low* spatial frequencies and is thought to arise from the M pathway.

6.7   Psychophysical limits

6.7.1   Spatial frequency channels

One of the most intriguing and in many ways controversial aspects of information streaming was the idea of spatial frequency channels. This is an extremely useful and important concept, underpinning compression systems such as JPEG. In hearing we accept frequency, and we saw earlier that transduction on the basilar membrane leads naturally to the cells of the auditory nerve coding sound in separate frequency bands, just as when we write notes on the stave in a musical manuscript we are encoding the music as a series of frequencies. If we were to take this really seriously, however, these cells are not actually encoding a pure frequency, because a frequency in the Fourier sense stretches from the beginning to the end of time. What we have is more like a pulse of frequencies, or in engineering parlance a windowed frequency band. Instead of a wave stretching from the beginning to end of time, we have a wave packet (§3.8.1).

Experiments in the 1960s showed that we process visual information in separate bands, but in fact far fewer bands than auditory information. Estimates vary around four. The sort of way we can see this is that we can psychophysically adapt or change one channel without influencing the others (Blakemore & Campbell, 1969; Woodhouse & Barlow, 1982). In Figure 6.7 there are several different frequency patterns. They vary in frequency and orientation. In the centre of the figure we have another frequency at a much lower contrast. In the experimental setting it is just visible. But if you stare for about a minute at the top left grating of the same frequency and orientation something interesting happens. Just as we adapt to bright light or a bad smell, so we adapt to frequency. Thus the grating in the middle becomes invisible. It will reappear after 10 seconds or so when the adaptation has died away.

Figure 6.7 Independent adaptation to different spatial frequency channels. (Redrawn from Woodhouse & Barlow, 1982 – see text.)

But could this be relevant to real images? Let's return to the square wave and its harmonic components (§3.3). If we adapt to the higher frequency components then the square wave will start to look like the sine wave of the same frequency.

This effect appears in a famous illustration created by Harmon and Julesz (1973). In a coarsely pixelated picture of Abraham Lincoln the face is very hard to see, presumably because of interference from the sharp edges of the squares. Blurring the image brings out the face (e.g. by moving further away), as the high frequency components are removed. But there are other more subtle effects, shown by Pelli (1999) to occur in paintings of Chuck Close.

Again the picture appears as one moves away to go through a striking transition from a flat, multi-coloured surface to a three-dimensional face at a sufficient distance. However,

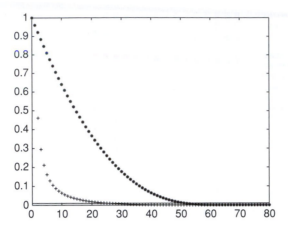

Figure 6.8 The asterisks show the modulation transfer function (§3.5.3) (squared) of the lens aperture arising from diffraction. The + signs the net spectral energy on the retina after the fall of the input information with frequency is taken into account. The line parallel to the x-axis shows a S/N ratio of 10 (low light). At this S/N most of the high frequency power is below the noise level.

this transition seems to occur not at a fixed number of markers per face, but rather when the marks fall below a particular size, around 0.3 degrees. This is about the maximum size in which a three by three grid would fit in the central one degree of the fovea and suggests some special size-related processing. That size has some preferential treatment appears from another direction – size is preattentive (Stuart *et al.*, 1993).

Controversy has centred around whether this was really Fourier Analysis or whether these cells were actually just tuned to edges and bars of different orientations. The answer seems to lie somewhere in-between. These spatial frequency properties of perception mirror the Gabor-like cells found in the V1 area (§3.8.1), the first stage of visual processing in the brain.

Why go to this trouble of separating out spatial frequency information in this way? Partly because there is different information in the different bands and so the low frequencies give us the broad outline of objects, but we need the high frequencies to process the fine texture of surfaces (§8.8). Since we need a much lower bandwidth for the low frequencies, it can be efficient to do the preliminary segmentation of a scene into objects using this low frequency channel. Broadly speaking, this is what happens.

Another reason is that noise in the photoreceptors is independent of spatial frequency; it's approximately white noise. Yet the signal after passing through the optics falls rapidly with frequency. Thus the signal to noise ratio falls too. So it makes sense to split up the information and assign less channel capacity to the higher frequencies. This is the strategy used in JPEG compression too. Figure 6.8 shows how in fact sensitivity falls with spatial frequency, but also how it depends on light level. Here we plot the spectral *power* using a common model for the fall of signal power with frequency of $\frac{1}{f}$. This power spectrum is then multiplied by the square of the modular transfer function (MTF) of the lens. The noise is constant, independent of spatial frequency, since the stochastic catch of each photoreceptor is independent of its neighbours. The brighter the light, the better the signal to noise ratio and the further we can push the resolution.

6.7.1.1   Special rapid action pathways

One recently observed facet of spatial frequency channels is in special purpose pathways designed for rapid response. When we come across something dangerous, fear is a natural consequence. Emotions, and particularly fear, evoke strong activity in the amygdala. Obviously fear benefits from rapid detection and response mechanisms, and it appears that the amygdala (§2.5.1) have a special, low frequency pathway to get at key visual information as soon as possible, before detailed pattern analysis has occurred (LeDoux, 1998). In such a situation passing all the visual information to the amygdala would just generate overload. The information which is selected turns out to be just the low frequency band.

Whalen *et al.* (2004) using fMRI imaging showed activity in the left ventral amygdala in response to fearful eye whites (i.e. larger than normal). This response seemed to be based just on the size of the white patches, independent of other facial details, and independent of the shape or contour of the eye. In other words this is essentially a low resolution only response[18].

6.7.1.2   Hyperacuity

At first one might think that it would be impossible to see detail smaller than the size of a receptor, but the sampling theorem tells us that this is not the case. If the noise level is low enough it is possible to interpolate to arbitrary accuracy between receptor centres, providing that the optics and sampling array are matched, as they are in the human fovea. In fact an example of subreceptor interpolation has been known for a long time – vernier acuity. We can see that a line is broken as opposed to continuous when the break is almost ten times smaller than the receptor (cone) diameter (Westheimer, 1981).

Investigations of hyperacuity revealed many other stimuli and configurations, such as the width of bars, which are also discriminable on such a small scale. Stereoacuity also has this sort of resolution in humans (§7.2.2) and in the cat vernier and stereo acuity are lower relative to the cone size but still of similar order (Bellevill & Wilkinson, 1986).

Various sorts of hyperacuity involve time and space. When measured with a spatial component, such as when two lines close together are presented in quick succession, time intervals as low as 2 ms may be perceived (Westheimer & McKee, 1977). But if the stimulus is designed to work purely on time, then the thresholds are much higher around the temporal sampling interval (25 ms for a flicker fusion frequency of 40 Hz (§7.5)) (Zanker & Harris, 2002).

Why should we need to resolve below the size of a cone in one eye, as hyperacuity, apart from this being a byproduct of stereopsis? One intriging possibility is that it can

---

[18] Precisely to provide the high contrast for communication and gaze, Tomasello argues that humans evolved eye whites, unique amongst primates, precisely for communication and gaze direction detection (Bickham, 2008; Tomasello *et al.*, 2007).

help us with very precise estimates of eye direction and eye contact. Baron-Cohen, in his analysis of autism, invokes an Eye Direction Detector model as a crucial feature of social interaction, absent in autistics (Baron-Cohen, 1997).

Suppose somebody is a distance D away and we want to determine if they are making eye contact, or looking just to the side of the head, say h away. Their eyes have to rotate an angle $\frac{h}{D}$, so the edge of the eye moves by an angle $\frac{rh}{D}$ where r is the radius of the eyeball. This distance now subtends an angle of rh over D^2 at the observer. Thus very small changes in eye direction for people relatively close are possible through hyperacuity (Land & Tatler, 2009).

6.7.2 Preattentive vision

Somewhat related to it, but probably using different pathways, is preattentive vision, studied extensively by Julesz (1975). In visual displays or patterns a feature may pop out immediately, within a few milliseconds, so fast in fact that it cannot be the result of sequential processing of features within the image. A cross in a pattern of circles will pop out. A red circle in a field of blue circles also pops out immediately. Julesz studied the characteristics of texture features which make them pop out in this way against a background. It has been suggested that one possible advantage of, or need for, preattentive judgements is for breaking camouflage, i.e. for rapidly spotting predator or prey (§8.8).

6.7.2.1 The iceberg effect and dithering

Noise does have its advantages, though, in a phenomenon known as stochastic resonance, now demonstrated in cells of the visual cortex of the cat. As discussed in §4.6.4 a small amount of noise can push low signals over threshold. Volgushev and Eysel (2000) and Anderson *et al.* (2000) show how orientation tuning is maintained independent of contrast.

As the orientation of a line deviates from the optimal orientating tuning of a cell, the sensitivity falls. So low contrasts should disappear as soon as the line rotates, meaning that the tuning is narrower for low contrast signals. But this is not the case. Anderson *et al.* (2000) explain this neatly by invoking stochastic resonance. Noise pushes the low contrast signals over threshold some of the time, giving on average a continuous response down to zero.

In a neat example of convergent evolution, a similar process is used in digital sound recording, where it is referred to as *dithering*. In the early days of the CD many people complained that the sound quality at low levels was terrible. After much acrimonious argument, the consensus emerged that the sound was indeed less than perfect because there was an abrupt cut-off at the lowest level of the first bit. The solution was to add a small amount of noise, of average amplitude, a fraction of the lowest bit. Thus the lowest level signals were never just on or off, but a fluctuating quantity which averaged out over time to values below the value of the lowest bit. It worked.

6.7.2.2   Gabor functions and localised Fourier Transforms

This compromise between the Fourier and space or time viewpoints awakened interest in a discovery by Denis Gabor. He invented the electron microscope, for which he received a Nobel Prize before receiving substantial local recognition. He had proposed a set of functions which were Gaussian windowed sinusoids (which is how the packets in Figure 3.6 were actually drawn). Stepan Marcelja spotted their application to vision, creating a cottage industry in visual neuroscience and computer pattern recognition. In one dimension the Gabor function is given by:

$$g(x) = trig(2 * \pi * \nu x)e^{-b(x-x_0)^2}$$

where *trig* is either cos or sin (giving even or odd functions about the centre), x_0 is the centre of the function, ν the frequency and b the degree of damping of the Gaussian component. If we take the Fourier Transform of a Gabor function, we get two peaks, in positive and negative frequency, centred at the frequency of the cosine wave. Thus filtering with a Gabor function slices out a piece of the Fourier spectrum (rather than a single frequency).

The way these orientation selective receptive fields are created within the brain relies on inhibition from other cells. As we age inhibition weakens, and our orientation sensitivity weakens. But there is hope! Recent experiments on monkeys (Leventhal *et al.*, 2003) demonstrate that GABA can be infused back into the inhibitory cells in V1 restoring the original selectivity. Monkeys in the human equivalent age group of 78–96 had function restored equivalent to a 21–30 age group. Benzodiazepine tranquilisers, such as Valium, are already know to enhance GABA activity, so drugs may exist which can ameliorate this age-related decline (Miller, 2003).

6.7.2.3   Hebbian learning and Fourier components

Gabor functions serve as a good general purpose compression strategy. The process of Hebbian learning itself promotes the formation of principal components and these turn out to look very much like Gabor functions. So, in other words, if we take a layer of untrained neurons, and expose them to pink noise, then, through the process of Hebbian learning, they will achieve oriented wave-packet receptive fields.

The Gabor decomposition of an image does not increase the number of functions – n pixels leads to n Gabor functions, where the intervals sampled in space are reduced as the number of orientations and spatial frequencies of the functions goes up. This is true for any linear transformation between complete basis sets, including principal component analysis. But this does not fit the primary visual cortex. V1 has far more cells. Stevens (2001) advanced an argument for this based on the way orientation is encoded. He found in over 23 homologous primate species (haplorhines) a scaling law of the form:

$$N = 161n^{\frac{3}{2}} \tag{6.8}$$

with N the number (in millions) of cells in V1 and n the number of cells (in millions) in the LGN. He argues that to preserve angular resolution the number of cells in a pinwheel must go up as $n^{\frac{1}{2}}$ to preserve angular resolution. This is a sparse coding-like argument, where each cell is used to signal a particular orientation. It is *not* a requirement of the Gabor or wavelet transforms, and would be unlikely to be used in a man-made computational system.

So, what happens if the input is not pink noise? For natural scences, the Fourier spectrum and second order correlations are very similar, so we end up with the Hubel and Wiesel organisation. But Blakemore showed that if kittens are reared in a restricted environment, in which only some components are present, they develop only a restricted set of orientation sensitivities. This same sort of behaviour exists much more generally and it comes up again in §7.2.1 on ocular dominance. We discuss the auditory cortex in rat in §5.2.3.

At birth, in humans especially, visual acuity is very low but increases rapidly during the infant's first year of life. But it requires visual input to drive it. Maurer *et al.* (1999) studied babies with congenital cateracts, which seriously blur the visual image. Until the operation to remove the cataract, there is little improvement in acuity. But in as little as one hour after surgery, acuity is on the rise! Thus self-organisation in response to visual input to the cortex is extremely rapid (although it still takes months to years to reach full adult acuity).

Gabor functions were, with the computers of the time, tricky to use. But not long afterwards a similar way of doing the same thing, but with much greater computational efficiency sprang up: the wavelet (Strang & Nguyen, 1996). Where Gabor functions remained a specialist interest, wavelets became big business. They now are used in many image compression and pattern recognition techniques and form the basis for the latest JPEG specification[19].

6.7.3   Scene statistics and independent component analysis

The assumption of pink noise captures the gross characteristics of natural spectra, but does not look very much like any actual scene at all. As with many areas of science, rough heuristics give way to more precise measures as technology develops. In this case the primary technology was more computing power and storage space, but also, to some extent, the advent of digital cameras and the ease of collecting data from the natural world. (Hoyer and Hyvärinen (2002) used 100 000 natural image patches in their computer model.)

In early work computing the information theoretic constraints on eye design, Snyder *et al.* (1986) used an exponential correlation function for natural scenes, giving a power spectral density varying asymptotically as $\frac{1}{f^2}$. Extensive studies (e.g. Field, 1987; Ruderman & Bialek, 1994; Simoncelli & Olshausen, 2001) show a power spectrum of approximately this form, giving *scale-free* or scale-invariant properties of natural scenes. In an attempt to explain such properties, people have looked at distributions of object sizes. Ruderman found that an object size distribution varying as $\frac{1}{r^3}$ gave a similar power spectrum. Refinements by Lee and Mumford (1999) take into account object occlusion.

---

[19]   Mathematical software packages now often come with wavelets built in. MATLAB, for example, has a wavelet toolbox.

So just by taking a model of circular objects of varying sizes, allowing them to overlap will lead to the sort of power spectrum observed in natural scenes.

6.7.4 Non-linear behaviour

There are many examples of non-linear behaviour in sensory coding, from contrast adaptation to spatial properties, such as *end-stopping* where cells cease to fire when stimuli extend beyond their borders. Schwartz and Simoncelli (2001), for example, in which the variance of cell activity is made proportional to the summed squared responses of the immediate neighbourhood, capturing measured properties of visual and auditory neurons. These non-linear aspects are an important growth area in understanding the finer details of sensory coding.

The Hebbian, self-organisation models of receptive fields discussed above are essentially linear. But economy of neural activity can be achieved with sparse coding, in which non-linear receptive fields inhibit activity within classical linear receptive fields (Vinje & Gallant, 2000). The synaptic energy efficiency also seems to be a consideration in retinal processing (Vincent & Baddeley, 2003).

6.8 Evolution and diversity of eyes

We remarked at the beginning of this chapter that there is an entire (vast) range of animals with different sorts of eye to that of man. Land (1990) points out that the animal eye typical of vertebrates from fish to man has evolved numerous times across many orders.

It seems to be fairly easy to evolve a camera obscura type of eye – just a gradual increase in curved surface with a few receptors. Some primitive animals still have an eye of this sort, such as the primite mollusc, the Nautilus. Such an eye is not very efficient, allowing a very low light level onto the receptors.

Land also describes other unusual optical structures. Scallops have a mirror rather than lens, rather like a Newtonian reflective telescope. The shrimp eye also uses mirrors, but instead it uses rather a lot of them. The array of tiny mirrors directs light onto a receptor array in a similar manner to a conventional lens.

6.8.1 Extreme performance

The largest eye of the animal kingom belongs to the giant squid. It is a simple eye with a lens aperture of a mind-boggling 10 inches across. It has adapted to life at the bottom of the ocean, whereby it sees a photon now and again, a bit like the task of an astronomical telescope.

It's not hard to guess who has the best resolution. If you fly at great heights, looking for small prey on the ground, you need excellent resolution in bright light. The eagle has it and tops the table for spatial resolution.

The prize for the best colour vision is hard to award. Amongst land dwellers, it would certainly be man. But some birds have four as opposed to three colour pigments, but how this translates to visual performance is not well known. A quick snorkel in a coral reef, surrounded by fish of all imaginable coloured patterns, suggests fish are pretty good too. Some fish also have four receptor types.

Birds also come in for another prize – the smallest photoreceptors, close to the limit set by diffraction (§6.2.3).

Box jellyfish are one of the terrors of the idyllic beaches of Northern Queensland in Australia. They are one of the most dangerous of all poisonous creatures, and make swimming in the ocean at the wrong time of year highly risky. They also have unusual eyes. Most jellyfish have simple light sensitive pits, with no image forming capability. Box jellyfish have 24 eyes in four clusters of six. Each cluster has four simple pits and two simple (human-like) eyes (Lawton, 2003).

6.9   Non-visual radiance detection

A few years ago a magic drug for curing jet lag appeared – melatonin. Jet lag is costly to businesses and holiday makers, and anything which reduces its ill effects belongs in a traveller's luggage. Melatonin occurs naturally within the body and its levels go up and down with bright light such as sunlight. Strong light resets the body's clock to a new time zone, a process referred to as *photic entrainment*. The question is how is the light detected and used for managing the body's clock, or *circadian rhythms*?

Many vertebrates, such as birds, amphibians and reptiles have light sensors within the brain itself (Menaker, 2003). But in mammals, the light sensing is done within the eye. Only in the last five years have the mechanisms come to light, and there are still significant unanswered questions.

The first surprise came when the light sensitive body clock was found to continue to operate in mice, even when all the rods and cones had been disabled (Barinaga, 2002a). So, there is something else besides the known photoreceptors which responds to light.

These light sensitive cells in the mammalian eye are, surprisingly, specialised retinal ganglion cells. There are only a few of them, about 1–2%, maybe less than 1000 in the whole retina and they project mainly to two parts of the older brain: the supra chiasmatic nucleus (SCN), which maintains the body's circadian clock, and the olivary pretectal nucleus (OPN), which plays a big role in control of the pupil response.

Using backward tracing, the ganglion cells which project to the SCN were found to respond vigorously to light even when the rods and cones had been disabled. These cells have huge receptive fields, allowing a small number of them to cover the whole retina (Barinaga, 2002a).

The challenge was then to identify the light sensitive pigment. The rod/cone pigments were nowhere to be found. The first candidate to emerge was menanopsin, a pigment with the opsin component found in other photopigments. Hattar *et al.* (2002) developed antibodies for melanopsin and were thus able to show that it was present in the ganglion cells which project to the SCN and OPN. Berson *et al.* (2002) showed around the same time that these cells which project to the SCN are light sensitive and that their spectral sensitivity and kinetics matched the characteristics of the body clock's photic entrainment.

Investigations of mice without melanopsin show considerably reduced entrainment by light of their body clock – but there is still some residual effect. There are other light sensitive pigments in the retina too, *cryptochromes*, which are known to play a role in entrainment in insects (Menaker, 2003). If mice lack cryptochromes, then they lose circadian rhythms and show a much weaker pupillary response.

Finally, it now seems that the mechanism by which melanopsin operates is actually different to the photopigment operations within the retina. It behaves at a biochemical level more like invertebrate photoreceptors where the regeneration cycle after exposure to light follows a different biochemical pathway (Panda *et al.*, 2003; Panda *et al.*, 2005).

Melanopsin has another likely role in the control of the pupil contraction in high light level through the retinal ganglion cells which project to the OPN. Lucas *et al.* (2003) showed that mice which do not have the melanopsin gene lose pupil contraction at high light levels. But the melanopsin mechanism seems to operate in conjunction with the rods and cones at lower light levels. The full story is not yet fully worked out.

Although this chapter has observations on the eye and vision going back hundreds of years, we find that early vision is still springing surprises upon us.

6.10   Man-made visual devices

Visual displays have got steadily better over the last few decades. With the advent of LCD and related technologies, they have also got smaller and lighter. The gamut of colours has increased, although the blacks from LCD systems were still in 2011 only just approaching the quality of the best cathode ray tubes (CRTs). Two relatively new advances into relativley new areas are high dynamic range lighting (§6.10.1) and retinal implants (§6.10.2).

6.10.1   High dynamic range images

The human eye has an astonishing dynamic range, able to operate over some 12 log units (to base 10) from moonlight to the brightest sunlight. It manages this through adaptation to the average light level, but it still manages at least 4 log units at a given adaptation level (Ward, 1994, 1998). In a bright sunlit scene we are hardly aware of this, until, that is, we try to capture and reproduce such a scene.

A matte photograph has a maximum reflection density of around 1.4, thus falls well short of what the eye can take in. Yet the French Impressionist painters still manage to capture convincingly the bright light of the Mediterranean summer countryside. Equally striking painters, such as Arthur Streeton, were similarly able to capture the very intense Australian sunlight on canvas, in a way a camera cannot straightforwardly do.

Camera films often had considerable *latitude* with a non-linear compression in the foot (low light level, shadow areas) and the shoulder (high light level) of the characteristic curve of film density versus exposure, thus giving some tolerance to a wide range of brightness. Ansel Adams, one of the pioneers of landscape photography, formulated a procedure, the *zone* system for optimising exposure and print density (Adams, 1981). He used ten zones, each a stop apart (giving about 3 log units). The photographic print was then created to map particular exposure zones to particular grey levels. For the computer graphics in this paper image capture is relevant, but the zone system forms the framework of *tone mapping*, the mapping of scene light levels to pixel values on the display device.

Digital cameras do not have this compressive behaviour for shadows and highlights, but have a sharp cut-off at saturation and a hard noise floor. Debevec and Malik (1997) developed a method of circumventing this by taking multiple photographs at different exposure levels and essentially carrying out tone mapping using the low exposures for detail in the very bright areas and the high exposures for the shadow detail. This is usually referred to as *high dynamic range* radiance or lighting. A variety of tone mapping algorithms have subsequently been formulated over the last decade. They may be classified into:

1. *Pixel-based* or region-based transforms, where the transform may be applied to a pixel or where some operation may be carried out using a collection of pixels.
2. *Local* or *global* where the treatment of each pixel may vary locally or may be spatially invariant.

Algorithms from Pattanaik *et al.* (2000) and Drago *et al.* (2003) use spatially invariant pixel transforms. Ward-Larson *et al.* (1997) used a spatially invariant histogram approach where the pixel brightness levels were allocated according to their probability of occurrence.

The approach used by photographers for mapping a high range negative to the much lower reflection density range of a print is spatially variant, effectively a pixel-based method of what photographers call dodging and burning in. Essentially what this does is to decrease and increase the exposure of the print in highlights and shadows respectively. Reinhard *et al.* (2002) developed an algorithm from this procedure.

Yoshida *et al.* (2005) compare these and other approaches over two images of 5.5 and 4.4 log units dynamic range using five criteria: brightness, contrast, detail in dark and bright regions and overall naturalness of the result. No one method stands out as superior across all criteria. The Drago method was one of the most robust, although it is spatially invariant.

Some specialist hardware is now available. The SpheroCam camera which made its debut in the premier computer graphics conference, SIGGRAPH 2010, has 26 f-stops of exposure latitude (nearly 8 log units) (SpheroCam, n.d.).

6.10.1.1   Application to games

The requirements of tone mapping for games are different to either static images or special effects in film – images have to be mapped in real time. From DirectX 9 launched at the end of 2002 high dynamic range (HDR) has been available on graphics cards through the Shader software. In successive Shader versions the number of bits per pixel has risen, reaching 128 bits in Shader 4 in DirectX 10. But the appropriate transform is (user) software based, although carried out on the graphics processor. The major competitor to DirectX (OpenGL, not owned by Microsoft) now also supports high dynamic range rendering (HDRR).

Early adopters of HDR were Valve Corporation in an expansion of *The Lost Coast* for the popular game *Half Life 2*. The Electronic Artsgame *Need for Speed: Most Wanted* used HDR extensively, producing very striking effects; for example, from wet surfaces in bright light (Millar, 2006).

6.10.2   Retinal implants

Over 30 years ago, experiments to stimulate the visual cortex directly with arrays of microelectrodes were carried out. But this cyberspace scenario still offers very little in the way of real vision (Zrenner, 2002).

But there is a need for ways of handling blindness, particularly in elderly people. Zrenner (2002) estimates that 17 000 people go blind in Germany every year. In half the cases the problems are with the retina and, frequently, the loss of the outer surface where the rods and cones are located. Hence there is strong motivation to develop implants which can replace the signals in the optic nerve to the brain.

The epiretinal implant injects electrical pulses directly into the ganglion cells. It does not contain the light sensitive elements themselves. These are embedded elsewhere in the eye, or even in some external camera.

The subretinal implant replaces the rods and cones. It consists of light sensitive photodiodes or polyimide film which injects signals into the retinal neural apparatus in a similar manner to the photoreceptors themselves (Sachs *et al.*, 2005).

Both technologies are at an early stage of development (compared to similar implants for hearing) with low resolution and many medical problems still to overcome (Cohen, 2002).

Nomad (Scott, 2001) is a system from Microvision for writing directly onto the retina using a micro-scanning mirror and red light emitting diode (LED). It can be used to write text information for pilots with very many possible uses in the pipeline. Being head mounted it is much less sensitive to external vibration than a screen.

6.11   Controversies over image recognition

At the time of writing some of the ideas on where and how pattern recognition takes place are undergoing some change. Almost 20 years ago, Rolls, Perrett and others discovered cells in a brain area referred to as the inferotemporal cortex which were particularly sensitive to faces. It is the last area of the visual system before integration with the other senses takes place. At first the idea that cells might be so specific was greeted with considerable scepticism. Were these cells really responsive to just oval blobs, brush-like objects or other quite generic things? Gradually these other possibilities were eliminated and face cells were accepted.

However, the last few years have seen the growth in powerful new technologies of brain investigation. We are literally in the throws of the brain imaging resolution. With techniques such as fMRI, we can see specific parts of the brain light up as we carry out different tasks. In reading, for example, we can see all the visual mechanisms, the motor systems that control eye movements and the language centres processing the text. These new techniques have identified an area heavily involved in face processing, the *facial fusiform area* (FFA), yet at the same time thrown doubt on the face cell model. It seems that these same cells, which are responsive to faces, are actually responsive to any patterns which we know really well, any patterns in which we are a special expert. So a bird watcher might find activity in this area brought on by birds, a car enthusiast by cars.

To really pin down this distinction between specifically faces or an area for catching familiar things or making fine discriminations, Gauthier and colleagues created some artificial creatures, greebles, which look nothing like any object or animal on the planet. Individuals soon learn to recognise and classify greebles – and they use the face area to do it (Helmuth, 2001b). Using event-related potentials, Gauthier *et al.* (2003) showed that there is some degree of holistic processing of faces, but that this also occurs for other areas of expertise, and the two interfere (e.g. cars and faces). Thus there seems to be a common mechanism (and possibly a common area) for holistic processing of images associated with expertise.

Recently, Haxby *et al.* (2001) used fMRI to look at recognition of various objects including faces. They found that particular categories of object produce distinct patterns of activity in various cortical areas. The pattern could be used to predict successfully what sort of object was in the field of view of the subject.

The FFA does indeed respond to faces. But the patterns of activity there could also be successfully used to predict other objects in addition to faces. So, this area is not exclusive to faces. Another intriguing dimension is that cells in this area are also responsive to images where a face would be anticipated, e.g. an image where the face has been blurred out, as news broadcasts do with victims or criminal suspects (Cox *et al.*, 2004). Likewise other areas give patterns indicative of a face. But the debate is not over! Tsao *et al.* (2006) have just found an area in macaque monkey in the same general region as the FFA in which 97% of the cells respond uniquely to faces. Nevertheless this high specificity of a particular area to a particular type of object may be restricted to faces or a very limited group of other objects (Kanwisher, 2006).

At the same time Downing *et al.* (2001) have shown that one particular area of visual cortex in the lateral occipital lobe has a very specific sensitivity to images of the human body or parts thereof. This is reminiscent of the discovery of the face recognition cells some years before. As Cohen and Tong (2001) discuss in the same issue, the jury is still out. Other recent work identifies seven different category regions in the occipitotemporal cortex: two face, three object-related and two building-related regions (Hasson *et al.*, 2004). Finally an old idea, the grandmother cell, is back with cells in the inferotemporal cortex responsive to the concept of, say, Bill Clinton (§4.3.1). *Plus ça change . . .*

7 The correspondence problem: stereoscopic vision, binaural hearing and movement

> The fundamental object of the invention is to provide a sound recording and reproducing system whereby a true directional impression may be conveyed to a listener thus improving the illusion that the sound is coming, and is only coming, from the artist or other sound source presented to the eye.
>
> Alan Blumlein, inventor of stereo recording, British patent 394325

7.1 Introduction and overview

Vertebrates have two eyes[1], ears and nostrils, in fact they are pretty much bilaterally symmetrical (although we have just one liver, for example). We can think of several reasons why this might be the case. Animals fight and get injured, or they get injured in other ways, so having two eyes or two ears (vets do a lot of business sewing cats' ears back up) provides some degree of redundancy. Having an eye on each side of the head makes it possible to see a large portion of the world, a more panoramic view. But in the sensory domain two channels enable the extraction of directional information. In the case of olfaction, the information, the differential arrival of odours at the two nostrils gives some indication of the direction of a source, but it remains imprecise. But for hearing and vision it is very precise indeed.

This chapter covers an important theoretical idea, the *correspondence problem* which underlies several aspects of sensory processing. When the information is collected by two spatially separate detectors, the signals coming in to each do not necessarily match perfectly. The differences can be used to infer something about the spatial properties of the signal. The two primary sonic differences are a variation in time of arrival and a difference in intensity arising because the signals have travelled slightly different routes.

Similarly when something moves in an image, its different location in two successive image frames has to be matched. In each case there is a common underlying strategy in biological systems in which the correspondence between small tokens is measured. The correspondence problem is *solved very early in the signal processing*, before object or sound identification occurs.

Stereopsis, the extraction of depth from binocular vision, evolved quite a long time ago and seems to be present in all mammals and in the species from which mammals

---

[1] Invertebrates such as spiders may have considerably more.

evolved. Since it restricts panoramic vision, it must be useful. By carefully looking at the wiring diagram of stereopsis pathways, it seems very unlikely that there could have been any common ancestor for the bird and mammal architectures (Pettigrew, 1986). In other words it is practically certain that stereopsis evolved independently at least twice. Less substantial evidence suggests it evolved a third time in sharks. It must be useful!

Binaural hearing similarly provides directional information. But we can hear a lot of useful sounds with just one ear. Diffraction allows us to hear, say human speech, from the other side of the head. So, the argument for covering more of the world is not so strong. Nevertheless most animals have two ears and binaural hearing is extremely useful too.

Both vision and hearing have additional strategies for getting height information. In the case of hearing this results largely from the peculiar shape of the pinna of the ear (§7.3.2). In vision it's a complex computational process.

Stereopsis and binaural hearing both extract azimuthal information, i.e. information in the horizontal plane. Both have a common computational problem, the need to match tokens: image features for fragments in vision; sound fragments in hearing. But the information has different purposes. In vision, the direction from which light is coming is known for each eye. The azimuthal information enables us to determine distance away from the eyes, or depth, in the horizontal plane. In hearing, this information is not available directly and it has to be computed. Thus binaural hearing is used to calculate the angle of a sound source relative to the direction straight ahead. Olfaction is similar.

Both have a clear evolutionary advantage but stereoscopic vision also engenders trade-offs in the field of view (§7.2).

This chapter differs a little from other chapters in that the receptor systems have already been covered. The focus is more on biocomputational issues of how sensory data is exploited. Section 7.2 covers biological stereo vision and §7.2.3 discusses computational aspects. Section 7.3 describes binaural hearing and the height information we get from the distinctive shape of the pinna (§7.3.2). A brief look at artificial systems follows (§7.4) for vision (§7.4.1) and panoramic sound (§7.4.3). Section 7.5 covers the detection of movement, which chapter concludes with a look at compression in images and video (§7.6).

7.2   Stereopsis

Although directional and depth information are really immmportant in evolutionary terms, we can often function quite well when our stereo vision or hearing is deficient. If we have a blocked ear, through a cold or infection, the world does not suddenly lose any sense of three dimensionality – other cues and vision fill in the gap without our conscious awareness of the problem. We have many cues for visual depth and may not realise that something is adrift with stereo until we come across a very specific task – rather like the way people may be unaware that they are colour blind.

There are a number of ways in which we can extract depth information:

- **Accommodation**, the processing of focusing on an object provides muscular feedback on the distance away; the closer the distance, the greater the tension in the ciliary muscles; this is a relatively crude measure.

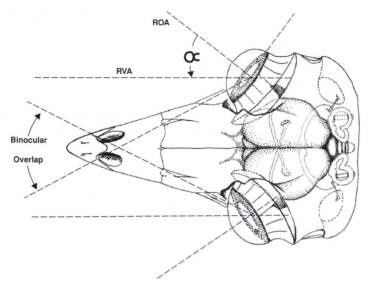

Figure 7.1 Overlap of visual fields from the two eyes. In such an avian retina, there are two foveas, a central one and a second temporal (towards the nose), binocular fovea. The angle between the optical axis through the centre (ROA) and the binocular visual axis (RVA) may be large. (After Pettigrew, 1986.)

- **Judgement about objects which we know to be of a particular size**; such judgements can let us down as in the moon illusion[2].
- **Motion parallax**, which is in some ways similar to stereopsis; by moving the head we can get several perspectives on the object.
- **Parallax from one eye**. Sandlance (*teleostei*) and chameleon (*reptiloa*) exhibit converging evolution with a one-eyed method for judging depth. The eye rotates and gets parallax. In sandlance, the eye drifts back to the central position, slowly, waiting for the next object anywhere in the visual field. The advantage to this strategy is in remaining difficult to detect by prey, as opposed to moving the head like a pigeon (Pettigrew *et al.*, 1999).
- By inferring three-dimensional (3D) shape from object structure and shading; this too can let us down as shown.

Stereopsis is more accurate than all of these and completely independent of our knowledge of the world and its objects. It tends to operate though at relatively close distances. For remote objects we are left with cues such as the actual size of objects themselves.

The first of two requirements of stereopsis is that the visual fields overlap. So images of the same things occur in both eyes. It is not essential that the optical axes of the eyes point in the same direction. In some animals such as the kingfisher, the axes of the two eyes may make an angle as high as 50 °C as shown in Figure 7.1 (Pettigrew, 1986). In

---

[2] The moon appears larger when it is close to the horizon, compared to when it is high in the sky. One common explanation is that this arises because looking out horizontally we have cues from trees, buildings, mountains which enable us to estimate its size, which we do incorrectly through being unable to estimate just how far away it really is.

this area of binocular overlap we perceive a single image; we fuse the images from the two eyes. The images must satisfy some constraints on how different they are for this fusion to be possible. In fact both must lie within Panum's fusion area, about ten minutes of arc in the central fovea. Outside this area, we get *binocular rivalry* in which the image in one eye dominates[3] or we get diplopia or double vision. As we move out from the central fovea the size of Panum's area increases quite rapidly.

One of the consequences of binocular overlap is that we lose some degree of panoramic vision, for a given focal length and aperture of lens. So there is a trade-off which obviously favours some level of stereopsis. However, some animals have the best of both worlds. One species of owl exhibits *voluntary strabismus* in which one eye can drift out of alignment. Thus in a defensive mode, the owl can increase the width of the visual field. The second requirement is that some procedure exists to find the correspondence between the images in the eye, an important issue which we shall come to shortly. However, firstly we need to examine the actual algorithm, based on disparities (§7.2.2).

7.2.1   Ocular dominance

Where binocular overlap exists, as is necessary for stereopsis, parts of the topographically organised visual cortex are going to have input coming in from different eyes. Just as different orientations are organised in columns, so there is an additional columnar arrangement, the ocular dominance columns, corresponding to each eye.

Just as in the maturation of orientation selectivity discussed in §6.7.2.3, ocular dominance columns have a critical period for their formation. Once this period has elapsed, no further change in ocular dominance occurs, even if one eye is deprived of input. New genetic techniques have given new insights into how the critical period works. At least two factors are at work (McGee *et al.*, 2005). Myelination seems to terminate the critical period at the same time that GABA neurons mature (Fagiolini *et al.*, 2004; Hensch & Stryker, 2004). Such inhibitory neurons, as in orientation selectivity (§6.7.2.2), drive an inhibitory mechanism for ocular dominance. There are some indications that blocking the GABA neurons in the visual cortex (in mouse) might partially return ocular dominance plasticity (McGee *et al.*, 2005).

7.2.2   Disparites and depth horopters

Figure 7.2 shows the image geometry of two points at different distances from the eyes. Corresponding points are different distances apart in the two eyes. The difference, usually given in angular terms, is the *disparity.* There are two types of disparity: crossed and uncrossed as shown. Simple geometry then relates the relative distance of the two points to the disparity angle.

---

[3] Rivalry is driven by interocular competition rather than competition between different patterns (Tong & Engel, 2001).

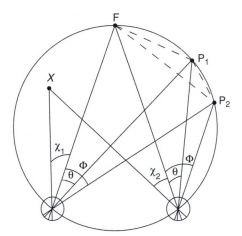

Figure 7.2 All the points, **P**, on the larger circle (the horopter) have the same disparity even though the images fall on different parts of the retina. This is true for any chord (dashed line) on the circle which goes through the fixation point, **F**, and the nodal points of the two eyes. The point X has disparity $\chi_1 - \chi_2$.

However, a little thought will show that the depth result is not unique. An entire surface, or *horopter of points,* will create the same disparity as shown in Figure 7.2.

Stereopsis is *extremely sensitive* with disparities as low as a few seconds of arc, considerably *smaller than a photoreceptor* detectable in optimal conditions[4].

At the other extreme, stereopsis gives way to other mechanisms when the disparity in the fovea exceeds about 1 degree. Outside the fovea the disparity gets larger, but in practice little use if made of such information.

This mechanism is extremely sensitive. Two seconds of arc corresponds to about 0.05 mm at 50 cm. So we can effectively see that a hair on a surface at 50 cm is a hair and not a line on the surface, by the variation from flatness. It's about at the level possible through noise in the visual system. As we saw above, we resolve down below the level of a cone. The sampling theorem would in principle allow us to dig down to arbitrary resolution. The limit to *stereoacuity* then is set by noise.

There are two other visual properties which have the same sort of resolution as stereopsis and are almost certainly limited by the same fundamental constraint. Just as we can see a few seconds of arc as a difference in depth, so we can see a break in a line when the break is similarly a few seconds of arc. This is known as *hyperacuity* (§6.7.1.2). But somehow stereoacuity and hyperacuity interfere: if the two parts of the line are at different depths, then vernier acuity is significantly impaired (Heinrich *et al.*, 2005).

---

[4] This might seem counterintuitive. If the optics were perfect, which would mean a high level of undersampling by the receptor array, a point source would be a point on the retina. So we could never locate it to better than the diameter of the cone. We would know which cone captured it, but nothing more. However, if the optics are matched to the receptor array, a point source is spread out into a blob (the Airy disc) which covers a number of receptors. It is then possible to infer from these several receptor values where the peak value precisely is.

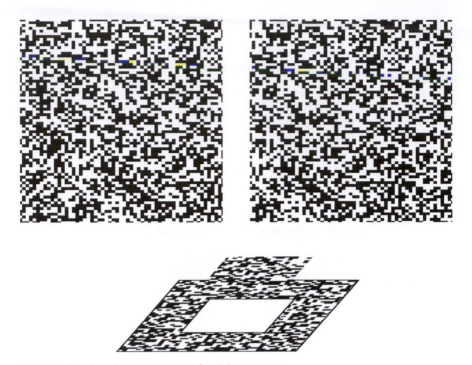

Figure 7.3 Random dot stereogram. After Julesz, see text.

Similarly, although the flicker fusion frequency is quite low, corresponding to a period of about 20 ms (50 Hz), we can detect much smaller time intervals as errors in the regularity of a series of bars presented to the visual system. In the same way a musician is extremely sensitive to the smoothness of a series of notes.

7.2.3   Computational strategies

When we look at stereoscopic disparities, we assume that we know which dot corresponds to which on the retina. Then everything is straightforward. But in real life in real scenes, there may be many possible *correspondences* between points on the two images. Computational stereo algorithms have been incredibly hard to develop. But random dot stereograms illustrate how good human algorithms are. Figure 7.3 shows two patterns, one presented to each eye, random dot stereograms as introduced by Bele Julesz (1971). An optical system of some kind is used to ensure that each eye sees just one of the patterns. The patterns consist of random dots with just one change. There is a square in the middle in which the pattern, though generated at random, is the same in each eye. What is more, it is shifted slightly between left and right eyes. The result is that it appears to the visual system as a square floating above the background.

Since these patterns are nothing but black and white dots, this seems like a very difficult thing to do. Somehow we've got to check every dot and find out if there is a matching dot somewhere in the other field and then assemble them all together. One thing that is immediately obvious is that stereopsis is not dependent on extracting known objects. There are no objects in either image alone. There are not even any significant micro-patterns such as lines or edges. The system is also very robust. We get depth from flashed stimuli (so useful as an early warning mechanism) and the match is very robust to noise and slight image distortions. So, dirty spectacles, glaucoma or other slight mismatches in the eye, do not stop us from getting a depth match.

The strategy of matching very primitive tokens, rather than objects or fragments of objects, is a strategy which operates through all known animal visual systems, regardless of which evolutionary path they have followed. The ganglion cells of the optic nerve can be sensitive to a wide variety of shapes, patterns and movements across the animal kingdom. But the cells which subserve stereopsis all have a simple, cylindrically symmetrical profile, which effectively measures local contrast at a point. Thus it seems that there is a fundamental principle at work here. Stereomatching *has* to be done before object recognition or feature extraction (Pettigrew, 1986).

7.2.4 Shape from shading

Another cue to get depth information is from shading. This is again a cue which is not dependent on specific objects but on the gradients of intensity in the retinal image. If we imagine light shining onto a sphere, just as the sun illuminates the earth, there will be a gradual change in illumination as the surface recedes away from the light. The angle which the surface makes with the direction of the light falls off from 90 degrees when the illumination is maximal to zero when it is parallel, in fact following a $\cos^2$ distribution.

Shape from shading is a non-conscious clue. We do not conscioulsy know when and how we process this information and where it occurs in the brain is not yet known. However, Lehky and Sejnowski (1986) put forward an interesting perspective from artificial neural networks thinking. They trained a feed-forward neural network to determine shape from shading and then examined the hidden units. They found receptive fields very much like the frequency and orientation-tuned cells of the primary visual cortex. We know (§3.8.1) that these cells decorrelate to second order, thus it is interesting that they are optimal for other tasks as well.

7.2.5 Depth from texture

Texture also provides depth cues. But it differs from shading in that the visual system must first make some judgements about the individual elements, the *texels,* which make up the surface. As a surface gets further away, the texels get smaller and closer together (Figure 7.4).

For a long time discussion raged in the visual world about whether textures were discriminated on the basis of statistical properties or whether the actual shapes were important. Julesz of the random dot stereograms (§7.2.3) pioneered this topic, establishing a range of patterns, where a texture boundary would 'pop out' rapidly. Where a

Figure 7.4 Three-dimensional shape from texture. We see a bump on the surface, so are implicitly assuming that all the texels (the black circles) are the same and the distortion results from perspective.

stimulus is presented very briefly and then turned off or replaced by something else, the percept is said to be *preattentive*, something considered for texture in §8.8.

The preattentive texture story turns out to be very complicated and it will take us too far from our main themes. From a game designer's point of view it does allow us some possible optimisations. For textures which are going to be presented only very briefly, we may not need to reload a new texture, or we might be able to make a simpler version to economise on texture map memory in a games console. One of the driving forces for understanding texture discrimination is camouflage breaking, and, again, the applications in games are plentiful.

Texture gradients can be used for inferring surface orientation. Tsutsui *et al.* (2002) found in the *caudal intraparietal sulcus* (CIP), in macaque, cells which respond identically to binocular disparities *and* texture gradients of sloping surfaces, with some cells responsive to just one of the depth cues. Previous work (Connor, 2002) had found cells responsive to *both* binocularity and perspective cues. The cells seem to respond to a particular surface orientation regardless of what form the information takes. CIP is part of the dorsal (where) pathway as we might expect.

7.3   Binaural hearing

Binaural hearing is also a masterpiece of biocomputation exhibiting three features very similar to vision:

- The correspondence problem is solved.
- The sensitivity is extremely high, at the maximum resolution of the auditory system; a time interval of arrival of sound at the two ears can be detected as low as 10 μs.

- Just as stereopsis allows us to separate objects out from one another based just on their distance away, so binaural hearing allows us to separate sounds coming from different directions, the so-called *cocktail party phenomenon.*

Hearing requires two mechanisms for sorting out directions, the *duplex theory* enunciated by Lord Raleigh in 1907 (Stern & Trahiotis, 1995). At low frequencies, time of arrival or phase difference is measured, the so-called *interaural time difference* (ITD). At high frequencies, the measurement switches to the *interaural intensity difference* (IID). The two mechanisms derive simply from the nature of sound itself.

To get significant absorption of sound, or waves in general, we need structures to be able to vibrate at the frequency of the wave and thus structures of the same order of size as the wavelength. Hence, just as it is much harder to absorb bass energy in rooms, not much bass energy gets absorbed by the skull, making the intensity mechanism much less accurate at low frequencies.

At high frequencies, where the wavelength is very short, small movements of the head can change the relative phase of the signals impinging on the two ears. This in turn makes computation of arrival time difficult, making the intensity method more accurate.

Taking the distance between the ears as 23 cm (Moore, 1997) and the speed of sound as 335 ms, sound takes 690 µs to cross from one side of the head to another. Thus an interval of this magnitude has to come directly from one side of the head and thus be localised entirely to one ear. A wavelength of 23 cm corresponds to a frequency of 1500 Hz. Considering pure, single frequency, tones, the indication of direction comes from the phase difference between the two ears. As the frequency rises above 1500 Hz more than one wavelength will fit within the head. We get a stroboscopic-like effect where the ambiguity now arises as to which side is which. Hence we need the second mechanism.

Unfortunately, the head does not really start to cast a very big acoustical shadow at 1500 Hz. Thus up to about 1800 Hz, localisation is quite poor. The IIDs do not locate well until then and never quite reach the performance of ITD. The absorption by the head may be as much as 20 dB at frequencies above 1.8 kHz. With two such mechanisms, with a gentle transition between them, it should be possible to balance one against the other. We thus have a *trading value*, an intensity difference which will compensate for a time difference. The trade-off is not straightforward. Depending on frequencies, stimulus type and duration, the values range from 1 µs per dB to over 100 µs.

The localisation from intensity seems to require around 0.5–1.0 dB difference to detect a change in angle. By around 10–12 dB, the stimulus is located entirely at one side, consistent with the absorption of the head.

7.3.1   Accuracy of azimuthal location

The best localisation occurs straight ahead, where the *minimal audible angle* (MAA) is about 1 degree. Performance falls dramatically off axis. If the distance between the eyes is d and the minimum resolvable angle is θ, then the difference in distance sound travels in getting to one or the other ear is *d sinθ*, which gives an ITD of 12 μs. This is consistent with our findings that the minimal time interval discriminable in visual movement is about the same and consistent with the other time constants for hearing (§7.5).

In angular terms these differences translate into a MAA of around 1 degree, which from the previous argument reaches this optimum in the straight-ahead position.

Stimuli vary in their localisability. Instead of the MAA we can measure directly, say using headphones, the minimum time differences detectable. A great deal of this psychophysical work was done several decades ago and still holds good. One might think that sharp transients would be the easiest to localise, but this is not the case. Figures quoted by Moore (1997) give sinusoids and noise bursts around 9–10 μs whereas clicks were closer to 30 μs. The best performance came from noise bursts of 700 ms duration at 6 μs. For click stimuli the low frequency components seem to be most important. Removing high frequency has less impact than removing low and masking noise is more damaging at low frequencies. This is reminiscent of the situation in vision: stereopsis is one of the most precise measurements the visual system makes, but it uses the lower frequency M channel rather than the higher frequencies in the P channel.

Finally, these localisation results are for static sources. Our ability to track sounds in space is surprisingly limited. It is referred to as binaural sluggishness. When we come to look at creating 3D sound fields for movies and virtual reality, this is quite relevant. Hearing a train moving around us in 3D in a cinema is something of which we are not very discriminating. Vision tends to dominate our impression of where sounds are and where they are going. Just think how we always localise the sound of peoples' voices to their heads on the screen and are quite oblivious to where the cinema sound is coming from.

7.3.2   The pinna and monoaural localisation

Why are ears such a funny shape? How come they vary so much from the almost invisible tiny ears of a wombat to the huge floppy appendages of an African elephant? Many mammals have deformable steerable ears, to better amplify and capture the sounds to which they are attending. In human such control is at best minor if not totally vestigial. But the complex shape of the pinna still plays a big role.

Part of the behaviour of the pinna is simply its absorption properties. Sounds from behind have to pass through it, whereas those from the front do not. The absorption of the pinna dictates that this effect only operates above around 2 kHz. Below this frequency the errors in blind conditions, between judging whether something came from front or behind increases significantly (Grantham, 1995).

The ratio of the spectrum of the incident sound field and the captured sound is referred to as the *head related transfer function* (HRTF) (Moore, 1997). Although diffraction and absorption in the head influences the HRTF, its primary determinant is the pinna, particularly at higher frequencies above 6 kHz. Different elevations of a sound source cause different sharp peaks and troughs in the sound field. The use to which this is put might seem a little surprising. Testing a subject with tones or noise bursts reveals that the frequency content determines the perceived direction – regardless of where the sound actually came from. So, the hearing system has learned to expect certain filtering operations in the HRTF and infers direction accordingly. If we cheat on the signal, the result is a spurious directional percept.

These vertical directional sensitivities are demonstrated using narrow band noise. For noise centred at 300 Hz or 3 kHz, the sound appears to come from the front, whereas for 1 kHz and 10 kHz it comes from behind, and 8 kHz makes the sound appear to come from above the head (Zwicker & Fastl, 1990).

The vertical localisation which the pinna achieves is also frequency dependent, seeming to require frequencies above 7 kHz (Grantham, 1995). This threshold corresponds to a wavelength of around 4 cm which is roughly the size of the ear. So these higher frequencies will resolve the little folds of the pinna. In the end, though, the vertical localisation is not that good, around 4 degrees.

But the hearing system is not stupid. It adapts rapidly, within a few seconds, to the acoustics of the environment and the susceptibiity to deception falls accordingly. One might wonder, given the huge variety of head shapes and sizes, how much the HRTF is a very specific individual thing. It seems to be quite possible to fool people with the HRTF derived from somebody else. This holds out great promise for sound reproduction, particularly over headphones.

Much of the psychophysics of hearing, frequency response, masking and so on is quite well established, the crucial experiments being decades old. But some of the new work coming out on the effect of the pinna and other experiments on masking with narrow noise bands is a beneficiary of technology. Until digital signal processing arrived, making sharp peaks or notches in an audio spectrum was well nigh impossible. Now it's easy. It's also not far off being amateur technology. Digital signal processors already exist in the consumer market, albeit at a price. They will get cheaper and problems of room acoustics or the 'in-the-head' sensation of headphones will be a thing of the past. Beyer Dynamic offer such software with some of their headphones (Dynamic, accessed 2010).

Another, successful application of HRTF ideas is from the company Sensaura, which built artificial ears/ear canals to simulate the frequency notching that occurs from different directions. They then went further and added cancellation signals to correct for cross-talk between right and left channels when listening via loudspeakers. The technology is supposed to be incorporated into the Microsoft XBox games console (Harvey, 2002). One new application mooted is in the plain old telephone service. Many people prefer telephone, audio-only calls to defective video conferencing in which the audio is distorted or suffers time delays. Yet a multi-person conference call, particularly if not everybody on the other end of the line is known to a participant can be confusing. Sensaura aim to create 3D sound which will separate the people in space.

Lake Technology developed an implementation of these ideas and Qantas and Singapore airlines have adopted the technology for in-flight entertainment. Dolby licensed the Lake patent and have on-licensed Motorola, Texas Instruments, Sanyo and others to make processing chips to implement the filtering (Fox, 2000).

The pinna provides information about elevation, but also weak information about distance. But our perception of the distance of sound sources is, like our tracking of movement (§7.3.2.1), relatively weak.

7.3.2.1   Tracking moving stimuli

The visual system is very good at tracking moving objects, but the auditory system is much less effective. If a stimulus is moving backwards and forwards at just 2.4 Hz, the binaural threshold starts to rise. When the frequency gets into the 10–20 Hz range, the ability to track the stimulus has almost disappeared. This is referred to as *binaural sluggishness* (Grantham, 1995).

A related phenomenon is binaural adaptability, where the threshold for detecting angular difference rises as the rate of stimulus repetition increases.

7.3.3   Reverberation and the precedence effect

In the real world, particularly indoors, we rarely hear just the direct signal from a sound source. It is usually mixed in with reflections from surrounding objects.

With stereopsis we saw that there was a significant correspondence problem to be solved. In the real world, lots of different noises impact upon our ears all the time. Sorting out which goes with what is again a very non-trivial exercise, again solved by animals long before it was solved by man. But with hearing there is yet another problem. The same sound may arrive several times in quick succession coming via different routes. The ear has a mechanism, referred to as the Haas effect, in which any similar sounds arriving are merged together as coming from the same source. We may be aware that the sound is mixed in with many reflections – we just do not hear them as distinct, unless the time gap is sufficiently long to generate an echo.

The Haas effect is rather complicated. To a first approximation there is a magic time interval of about 5 ms. If sounds arrive within this interval, only one sound is heard. But this interval may vary, getting longer with complex stimuli such as music, and the percept, first sound, second sound or composite also depends on circumstances. Moore (1997, p. 232) untangles the variations. Normally we associate the location with the first sound, but the second sound does have some influence. If it is louder (which it would never be for a reflection) then it may even dominate the location.

The effect disappears if the time interval between the two sounds is really short, less than 1 ms. In these circumstances, a single average location is inferred.

7.3.4   Physiology of binaural hearing

One of the animal world's great exponents of precision hearing is the barn owl (*Tyto alba*). It can successfully swoop and catch mice in total darkness based on sound alone. The owl has multiple spatial maps, derived from ITD and IID computations (Carr & Konishi, 1988). In the *inferior colliculus* these maps are combined. But whereas the usual paradigm for combining neural signals is summation, here the operation is multiplication, or, an AND gate. These neurons fire if and only if there is both an ITD and IID signal (Peña & Konishi, 2001). Belt and braces[5].

Finding three separate maps for spatial localisation of sound should not surprise us by now. But the multiplication is interesting. Although other sensory systems have implicated multiplication, such as the H1 movement-detecting neuron in fly, this is one of the first clear and conclusive demonstrations of multiplication in wetware (Helmuth, 2001).

There is another important difference between owl and man. Section 7.3 described how in man intensity and time difference operate in different frequency bands. Elevation information is much lower and comes from frequency notches created by the irregular shape of the pinna. But the owl has a precise 3D map of space, which requires two directional coordinates at each point. It gets one elevation (vertical) from intensity and the other azimuthal (horizontal) from time differences. Because of the importance of hearing to the behaviour of the barn owl, we know rather more about its auditory physiology than we know about mammals. However, there, too, evidence is accumulating for separate maps of spatially tuned cells. In rhesus monkey part of the auditory cortex just adjoining the primary area, referred to as the lateral belt, is a spatial map. It also projects to a different part of the prefrontal cortex, situated dorsally (Tian *et al.*, 2001). This fits in with the what/where paradigm for cortical streaming, evidence for separate streams in auditory as well as visual processing.

In the cat, the *lateral superior olive* (LSO) has tonotopically arranged cells with a high percentage, around 40% being inhibitory. The LSO handles frequencies from 0.2 kHz to 44 kHz, with the majority above 1kHz. The primarily locational role is in IID. Animals which have significant ultrasound hearing such as cat, have a well-developed LSO.

On the other hand, the *medial superior olive* (MSO) processes interaural time delays and is operative at low to middle frequencies with primarily excitatory neurons. There is an organisation of cells corresponding to interaural time delays, providing physiological support for the Jeffress model (Helfert & Aschoff, 1997). However, this is somewhat controversial. Brand *et al.* (2002) find that inhibition is crucially important for computing auditory time delays within the MSO but do not find such a clear topographical layout. The inhibition result does have one interesting implication: since inhibitory processes seem to decay the most rapidly with age, spatial hearing is likely to fall off first, i.e. loss of discrimination in cocktail parties, an effect which many people describe. Thus we see that the auditory system *streams* data in separate channels for processing IIDs and ITDs.

---

[5] The owl has a smaller head than man, so the frequency at which the ITD clues become ambiguous is higher. But the absorption of the head, and hence the acoustic shadow, should have roughly the same variation with frequency.

7.4   Artificial systems

Artificial spatial systems for hearing and vision have had quite different histories. The search for stereo sound began in the early days of recording. Alan Blumlein developed the theory and patent for the first important system in the late 1930s and the first stereo recordings date from about 1958–59. Stereo and now surround sound have been features of audio and cinema ever since (§7.4.4.2).

With vision the story is different. Although artificial stereo systems was designed and built going back a long way, use in the popular domain has been sparse until 2010. Films were occasionally released with stereo using red–green glasses. Part of the reason would seem to be the cost over perceived gain. Coloured glasses are just not very good. The state of the art shutter glasses alone cost a few hundred dollars still and require special ancillary hardware.

The situation changed radically in 2010. James Cameron's blockbuster movie, *Avatar*, was distributed in a 3D version and was an enormous hit. At the same time NVidia and other graphics cards came out with automatic stereo frame generation and Samsung introduced the first LCD monitor with a 120 Hz frame rate. To get smooth stereo requires a frame rate of 60 Hz, meaning that the left–right sequence has to be at double this, i.e. 120 Hz. For LCD monitors, achieving this frame rate was extremely rare until the Samsung launch.

7.4.1   Artificial stereo vision

The first challenge of providing artificial stereo is to make the stereo images in the first place. Traditional photographic methods require two cameras, or very careful control of placement for two shots with the same camera. Various attempts to market stereo cameras to the amateur market have been rather unsuccessful.

With computer-generated imagery, creating the two views is straightforward. It doubles the rendering time, however, and increases the amount of storage space required. But from the perspective of widespread home use, there is another problem – the stereo display itself.

7.4.1.1   Stereo glasses

Three-dimensional cinema has been a niche market for several decades, but the difficulty of making the films, the projection equipment and need for auxillary glasses all mitigate against its widespread use. Computers, however, have started to change things.

Assuming that we have two different stereo views to project, the low cost solution is to use glasses with different colour, usually red and green lenses. We then build the stereo pairs by superimposing them with different colour casts. Apart from what this does to the accuracy of our colour perception (actually not very much), there has to be a lot of overlap between the two images. Thus the depth information is relatively weak. However, only one projector is required once the film has been made. Cardboard red–green glasses are commonplace, given away with books and toys.

A more effective but more complicated approach is to use Polaroid glasses. Somewhat more expensive to produce, in theory they can produce much better depth information. The two lenses have the axis of polarisation at right angles. Thus with light from the source polarised in one direction, only one of the lenses will transmit.

Getting two sources of different polarisation is not so easy. For cinema-type projection, we need a screen which will preserve the polarisation in the reflected light. Such screens can be built but are more expensive and not normally of the same material as cinema screens. Then the projector has somehow to display the frames for each eye with opposite polarisations. Two cameras, two projectors are workable but expensive. One projector with interleaved frames and a filter switching system is also workable, but making the film is considerably more complicated. There is also at least a 50% light loss from use of the filters – and projectors rarely have too much light!

Computer-generated Polaroid displays are also possible, with filters across the monitor. Generating interleaved frames is simple in software. But the stereo glasses of choice for computers and virtual reality now use LCDs. A radio or infrared signal is received by the glasses which have two LCD filters. These can be transparent or black and completely opaque. The LCDs are switched between the two states by the airborne synchronising signal. Such glasses are currently fairly expensive, hundreds of dollars, and so are not yet available to the mainstream gamer community. At the time of writing in 2010, the prices are just starting to fall.

Multiple projector systems tend to be expensive and wearing special glasses is inconvenient. But new technology is just coming to market, with Sharp offering laptops and computer displays based upon it. The idea is based on parallax. If we put a vertical bar pattern in front of a screen, then each eye will see a slightly different part of the screen. Thus the two images required for stereopsis could be put onto a single screen behind such a pattern. By continually moving the pattern, the viewer does not perceive the stripes. This will only work for one head position, however. To get around this, infrared lights and sensors track the position of the viewers' eyes to compensate the image for head movement. Unfortunately this still can only work for one person, so is most appropriate to computer users rather than cinema displays (Alpert, 2002).

7.4.2 Three-dimensional stereo games

At the time this book goes to press, 3D stereo is just taking off in the games world, with the Sony Playstation 3 seeing its first year of successful game releases. Generating, or capturing, stereoscopic data requires the setting of *two* parameters: the *interaxial distance*, E, the distance between the eyes; and the *horizontal image translation*, also referred to sometimes as the convergence, C, see Figure 7.5.

A stereo camera has two individual cameras, one for each eye. If the cameras point straight ahead there is no convergence, and all the imagery will appear in front of the projection screen. If the cameras are toed in towards each other, the convergence increases and some of the objects in the scene may now appear behind the screen. In general this is less fatiguing. With imagery captured with parallel cameras, *horizontal image translation* may be used in post-processing to adjust the relative depth.

Together the convergence and interaxial distance determine the perceived scale, S, as in the empirical formula (Bickerstaff *et al.*, 2011), eqn 7.1:

$$S = \frac{C}{E} \tag{7.1}$$

Designing stereo games brings a whole range of new challenges, since poor construction can lead to player confusion and fatigue. Sony advocate a number of design principles for high-quality 3D stereo games (courtesy of Sony Computer Entertainment Europe (Bickerstaff *et al.*, 2011 and pers. comm.)):

- *The left and right images are being sent to the correct eyes*, a seemingly simple precept, but the effects are disconcerting if there is an accidental switch.
- *Each eye contains the same elements*. Generation of 3D graphics content can involve complex multi-layered images, which could lead to mismatch in details in some layers in the two eyes.
- *The images represent the same moment in time*. With fast moving content, such as car racing games, it is possible for the generated frames to be slightly mismatched in time, causing stereo errors.
- *The parallax is within a comfortable range* is a more general issue which applies also to television broadcasts (such as sport). Industry standards set the maximum amount of allowable parallax for comfortable viewing.
- *There is sufficient parallax to show as 3D. There is no vertical parallax*, which is something that could arise from the synthetic content in games or through camera misalignment.
- *There are no sustained window violations*. When an object leaps out of the screen, if it is too close to the window edge, it will be clipped as it comes out. Thus objects which project outwards from the screen towards the viewer need to be fairly central.
- *There are no sustained depth conflicts*. Text logos, gun cross-hairs can get separated from the objects to which they should be attached, or can end up, say, in the plane of the screen rather than on some surface in the image.
- *Rapid changes in parallax are minimised*. In a racing game, it might be tempting to cut between different viewpoints, different virtual cameras. In televising sport, shifts from aerial views to player-centred views are common, but this technique does not work so well in 3D, as it can induce fatigue in rapid switching of stereo matching. In real life, movement of the eyes (and corresponding depth and focus) is under conscious control.
- *Some image cross-talk may be acceptable*. Using shutter glasses, to get absolutely clean left–right images would mean that the display would have to go completely blank after a frame with no lingering after-image. New monitors have appeared to better meet this requirement (§7.4). Cross-talk introduces artefacts but some slight mixing may be tolerable.
- Use of wide angle lenses and minimal use of high contrast also help to produce good stereo quality.

Figure 7.5 Two images of a shark with different interaxial distances (represented in a single image in each case). Courtesy of Sony Entertainment.

7.4.3   Reproducing 3D sound fields

Creating stereo vision is straightforward. For natural scenes we just need two cameras (Figure 7.5). For synthetic graphics we need to calculate the disparities for each point in a scene directly – hard work but doable. In stereo vision systems, we feed images to the right and left eye separately, using stereo glasses or other devices. But in sound not only do we normally allow each ear to hear sounds from both the left and right channel (e.g. each ear hears both loudspeakers), but the head itself modifies the sound field.

In principle what we want is to capture the sound field at the position of a hypothetical listener. If we assume that we want this information at a single point, then we need amplitude and direction in three dimensions, six pieces of information. With the exception of ambiosonics (Gerzon, 1973) until recently two-dimensional fields have been the only interest. After all if you want to capture a band or an orchestra, the height dimension is not normally important. Mind you there are examples in grand opera, where height might be useful, such as the final scene of Verdi's Aida with Amneris above ground and Aida and Ramades in the cave below.

It would seem, then, that we just place microphones at the listener position, and many stereo early recordings of the 1960s followed Alan Blumstein's original patent (§7.4.4.2). In fact one company, Calrec, created the *Soundfield* microphone, to accurately capture the entire 3D sound field at a point. Many people, the author included, regard these recordings as amongst the most realistic spatial representations. Yet, the techniques drifted to other, superficially more attractive, but actually less accurate methods, known as multi-miking.

The difficulties with the single capture approach are practical. For a start, venues sound different whether or not an audience is present. People absorb a lot of sound. But if the recording is done live, then there are extraneous noises such as coughs, and the performance becomes a single take, without the flexibility of studio sessions.

As auxiliary equipment got better, able to record many tracks and mix them together, using numerous microphones, attached to instruments, in front of performers or at large in the studio or hall itself became the norm. Thus the recording at the end of the day becomes a work of art in its own right. It is the mix the recording engineer judges most desirable. The recording and its reproduction coevolve into something the listener enjoys. It may not be that accurate.

7.4.4   Regenerating the sound field

At the time of writing recreation of the 3D sound field is in a state of flux. In the home theatre world, the main standard is 5.1, but already 7.1 and even more loudspeakers are being trialled. The symbol 5.1 indicates 5 channels plus an additional subwoofer channel. Remember that low frequencies do not provide very good directional information, hence most systems use just a single subwoofer, hidden away somewhere. The five channels then constitute a left, right and centre front channels and two rear channels.

The way information is encoded for these channels is to some extent proprietary (Harris, 2003; Hawksford, 1998). The codecs which define how the audio signal is compressed and decompressed are not fully specified by the standard. Fairly widespread is Dolby *ac3* which has a bandwidth of up to 384 kbs and follows previous cinematic practice of encoding the signal at the edge of the film around the sprocket holes.

DTS began as a small independent company aiming to produce a higher level of sound quality for movies and music. Their big break came with the sound track for *Jurassic Park,* using what they refer to as the *coherent acoustics* codec, with a compression ratio of about 4:1. Unlike other systems, DTS does not encode the sound onto the film, but has merely a synchronisation track. The sound is stored on CD-ROMs operated independently. With the imminent adoption of digital cinema, these distinctions will become unimportant. DTS uses adaptive differential pulse code modulation with 32 discrete sub-bands and very little pyschoacoustic modelling. The 5.1 encoding is around 1.5 Mbps, but films such as *Dances with Wolves* were encoded at around 7 Mbps.

Finally after many false starts we may eventually see surround sound for standalone music reproduction. Both the DVD audio standard and SACD (super audio compact disc) have surround-sound modes, although the take-up of 5.1 SACD by manufacturers of players has been minimal. DTS is available on a limited number of CDs, but will soon be phased out in favour of DVD (Harris, 2003). Effective though these systems are, we are yet again in a format war, between Sony/Philips (SACD) and the rest (DVD). In the 1970s this was a disaster, essentially destroying quadraphonic sound. SACD hangs on, but DVD became the dominant medium, now being slowly replaced by higher resolution and storage capacity, *blu-ray*.

7.4.4.1   Headphone solutions

The obvious way to isolate the ears is to use headphones. As we saw in §7.3.2, headphones have their own problems, because they do not allow sound to flow properly around the pinna. We can ignore this and assume, as we discussed (§7.3.2), that it can be fixed by digital signal processing using a HRTF (§7.3.2).

Headphone listening took off with the Sony Walkman and the advent of personal stereos and is now ubiquitous, but fed by iPods and MP3 players. But the pressure for accuracy from this (huge) market was minimal, and the recent surge of popularity in low-grade MP3 does not suggest that people are becoming more critical. However, good 3D headphone sound dates back over 20 years, when the BBC, amongst other organisations, ran experiments in *binaural hearing.*

Since the acoustic shadow of the head and the pinna strongly influence the perception of direction, the way to make a binaural recording is to use a *dummy head.* The life-size model of the head is constructed to have the same acoustical properties as a human skull, and the microphones are embedded within the ear canals. This way each microphone records the actual signal the ear would receive.

Dummy head recordings can be very impressive. Yet the location problems are the same as for the Soundfield microphone (§7.4.3) and as we have seen the demand is not there within the music domain.

However, with computer games and virtual reality headsets, there is a demand for spatially accurate audio and it is once again an active research area.

7.4.4.2   Historical notes

Blumlein worked for EMI before World War II and in British patent No. 394325 laid the foundations for stereo sound still used today. Unfortunately he died in a plane crash in 1942 and his patent languished until the 1950s, the first stereo recording being released in 1958. He used two microphones, with a cosine intensity sensitivity, i.e. the intensity falls as a function $cos\theta$ where θ is the angle made with the axis of the microphone. The microphones were lcoated in the same place, and thus referred to as a *coincident, crossed-pair.* Blumlein demonstrated in his patent that two loudspeakers playing back this signal would recreate a convincing representation of the original sound field.

Many enthusiasts regard the Blumlein approach as still superior for stereo to many more recent innovations. However, recording studios, recognising the need to cater for a very wide range of playback scenarios, from radios to cars, frequently add other microphones, close up to singers or instruments, or even directly off the instruments themselves. Although the resulting mixes are dramatic and effective, the *natural ambience* captured by a cross-paired microphone is often missing.

The 1960s saw a number of attempts to go beyond two speakers, most of which were commercial failures. The costs of extra amplifiers and speakers were not offset by a dramatic improvement in subjective experience. One such simple system was the Hafler approach in which two rear speakers, usually smaller and cheaper, were fed the difference between right and left speakers. The idea was to subtract out the direct signal and leave just the ambient reflections. It produces a subjective enhancement, but not in any way an accurate one.

Alongside some quadraphonic systems which were briefly marketed by the major record companies, such as SQ from RCA and QS from Decca, a mathematician at the University of Oxford, Michael Gerzon, set about developing an accurate model of recreating a sound field based on human perceptual theories. Regrettably Gerzon died before the advent of home cinema and DVD where the system could finally be implemented and marketed.

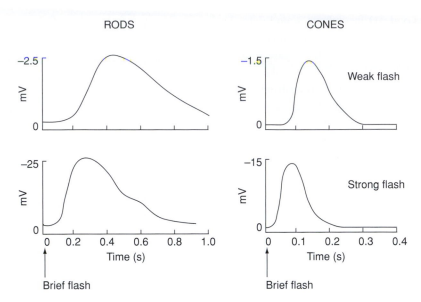

Figure 7.6 Temporal responses of turtle rods and cones at 20 °C. Human cones (at 37 °C) are around three times faster. (From Barlow & Mollon, 1982, p. 153.)

7.5   Movement

Movement information travels largely via the primate M pathway, to motion processing areas such as V3 and V5. It *is* concerned with spatial location of objects, extracting them from the background and of course tracking them as they move around. Some of the requirements of the movement tracking are similar to stereopsis – the need to solve the correspondence problem – and they share the same pathway to the brain.

The processing of movement information is every bit as complex as that of colour, in many ways more so. What is more, it is an absolutely core strategy to many simple animals such as insects. But our basic themes of information content and processing do not need the details of how we accomplish all this processing. We just need to analyse the limitations and perfomance ceilings.

A major limitation to motion is set right at the beginning in the photoreceptors. It takes time to integrate the visual signal and produce an output from the rods and cones. Just as in a digital camera the shutter speed may be hundredths of a second, the time to download the image off the CCD to the memory card may be much longer and severely limits the average rate at which photographs may be taken.

Figure 7.6 shows the time course of the photoreceptor responses to a flash of light (Baylor & Hodgkin, 1973; Woodhouse, 1982). In both cases the time to peak is longer for the weak flash than the strong one. Also the cones have a generally faster response. This makes sense. They operate in brighter conditions, and have less need for a long integration time (equivalent to a slow shutter speed) than the rods.

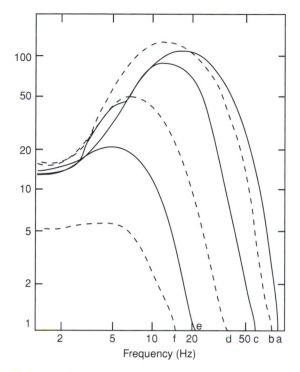

Frequency (Hz)

Figure 7.7 Flicker sensitivity as a function of frequency. (From Barlow & Mollon, 1982, p. 155.)

This temporal filtering leads to a loss of contrast as temporal frequency increases. Figure 7.7 shows how the contrast sensitivity to a flickering source varies with flicker frequency. The limit, the *flicker fusion frequency*, varies between 30 and 60 Hz. We are actually more sensitive to stimuli with some flicker or movement associated with them, with a peak around 5-10 Hz. As the light level decreases from (a) to (f), going by a factor of ten each time starting with bright daylight, the maximum frequency we can see goes down. In other words we have less movement sensitivity in the dark.

Figure 7.8 shows the end result for the *critical flicker fusion frequency*, again as a function of light level.

In bright light, the maximum frequency we can resolve is 50–60 Hz, which is the refresh rate we set for television etc. As the level falls, so does the fusion frequency. But something interesting happens as the rods start to come into play. At the blue end of the spectrum the flicker fusion frequency extends out to low light levels, but disappears at the red end of the spectrum as the cones cease to operate. This shift in spectral balance with light level is a general property, known as the Purkinje shift.

7.5.1   Structure from movement

Movement is a very strong clue to object identification and spatial properties. We can easily identify moving figures from just patterns of moving lights or dots, so-called

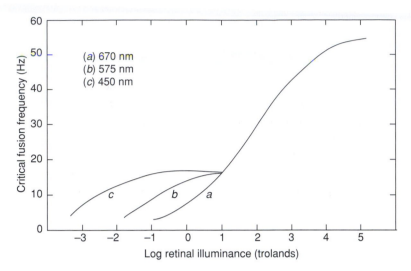

Figure 7.8 Critical flicker fusion frequency. (From Baylor & Mallon, 1982, p. 156.)

structure from motion. Yet motion is also a strong clue to distance through parallax and optic flow.

Vanduffel *et al.* (2002) used fMRI to compare monkey and human to ascertain which areas respond to patterns of moving lines creating a 3D percept. There are close analogues but not exact one-to-one comparison between the species. They find activity in the MT/V5 (dorsal) pathway similar to human but with discrepancies in the parietal area. Connor (2002) notes, however, that fMRI sometimes does not show up areas which are known from single cell studies to have 3D properties, perhaps because to get a significant fMRI signal requires a large population to be simultaneously active.

Long-range apparent motion occurs when two stimuli appear in sequence but are spatially separated. It is often thought to be closely related to real motion and activates corresponding visual areas, notably in the where pathway. However, Zhuo *et al.* (2003) show that it is also associated with global form perception, that the tokens may be more higher level structural forms, and activates the ventral what pathway, and main activation is the anterior temporal lobe, a long way from MT and the where pathway. Short- and long-range AM are fundamentally different.

7.5.2   Optic flow

As we move along a path, the world flows past us. The visual system uses the flow of visual information in the retina to determine rate of movement and other properties of the direction and path.

This processing occurs in the temporal cortex, notably in the medial superior temporal (MST) area (Bradley, 2002).

7.6   Compression

The MST encodes both the instantaneous heading and the path. It may also integrate current positional information too (Froehler & Duffy, 2002). Having taken a good look at the various pathways of the visual system, we can now see how this information is put to good use. There are three common standards for compression of image and video data. The GIF (Graphics Interchange Format) standard, subsequently PNG (Portable Network Graphics), is one of two major standards on the web for static or simple animated images. The other is JPEG (Joint Photographic Experts Group), just entering a major revision at the time of writing. Video is much more complicated, with a wide range of compression expansion algorithms. For our purposes it will be enough to look at the principles of how we make use of correlation between frames in a movie sequence to get extra compression.

7.6.1   GIF and PNG

GIF image compression relies on run-length encoding, discussed in §4.2.5, and although widespread will ultimately give way to the newer non-proprietary standard PNG. The GIF encoding method does not exploit features of the human visual system, but it does use similar principles to the predictive coding we discussed for the retina.

The story behind GIF is rather an interesting one and something of a warning about proprietary formats. The GIF compression technique was patented, but the patent essentially lay dormant, passing from one company to another in a series of takeovers. Its importance was minimal until the advent of the web. In the early days a large proportion of images on web pages were GIFs and a large number of tools grew up to create and manipulate the format.

At this point, the then owner of the format, Unisys, realised it could make serious money out of royalties on the patent and proceeded to do so. The need for a non-proprietary standard brought about the PNG format from the world wide web consortium. PNG is gradually gaining ground over GIF, having the advantage that it supports full 24 bit colour.

7.6.2   JPEG

JPEG[6] does exploit properties of the human visual system, both in colour and spatial resolution. JPEG tends to perform better than GIF on images with a full tonal range and is now the preferred choice over GIF on the web for such images.

In JPEG the image is first transformed into luminance and colour components, *each treated separately.* Each image is now split into blocks. A discrete cosine transform (DCT) is applied to each block and the coefficients are quantised according to their frequency, as discussed in §3.9.3.

---

[6] JPEG is often used too in the sense of a file format, but it is actually a data compression scheme. The file format is strictly called JIFF, the JPEG Image File Format.

JPEG can be used without any data loss, but is more normally used with varying levels of compression and quality loss. One of the disadvantages of the block-wise transform is that the errors tend to show up first along the block boundaries, giving an unpleasant tiling appearance. Thus JPEG 2, coming into operation at the time of writing, dispenses with the DCT and uses wavelets instead (§3.8.2). The wavelet transform offers the same level of compression, but, since the wavelets themselves decay rapidly, but smoothly, rather than being sharply truncated, they produce smoother boundaries.

7.6.3   MPEG video

Compressing moving images is a complex and commercially lucrative affair. For our purposes there are two interesting things:

- The frame rate itself, and temporal frequency filtering, are a direct property of the human visual system.
- The image data itself, regardless of the human observer, has a great deal of frame-to-frame redundancy. Apart from when the camera jumps, each frame will usually contain the same background and often the same objects as the previous one. Thus all this information does not need to be stored in every frame. Differential storage offers huge economies beyond the compression we get for each individual frame.

7.7   Envoi

The perception of objects or sounds in space requires some very clever computation. In each case it requires the solution of a correspondence problem, which seems to operate with very low level primitive elements. In both binaural hearing and stereopsis the spatial and temporal resolution are the highest possible in either sense. Although it might seem easier to match objects, features, particular sounds between left and right, matching takes place with very primitive tokens before any pattern recognition takes place. The matching of tokens in real moving images also uses this strategy of matching primitive tokens rather than objects or higher level entities. Apparent motion, however, does seem to make some use of object properties.

8 The properties of surfaces: colour and texture

Colour is a power which directly influences the soul.
                                                    Wassily Kandinsky

8.1   Overview

Two aspects of the *what*, or *ventral* pathway, are colour and texture. In the biological world colour and texture are properties of surfaces which help in identifying objects, often rapidly. This realisation that colour did little in defining the boundaries in an image (§8.5.2) and the shape of things is relatively recent, the last few decades, whereas interest in colour itself goes back to the time of Newton and before.

Colour is essentially concerned with *what* things are, the nature of surfaces and the objects they represent. In mammals it plays a minimal role in form analysis, separating objects from one another or in analysing their shape. It travels exclusively along the P/K-cell pathways from the retina (§6.6.2) to the area V4 where colour, as opposed to wavelength of incoming light, is extracted (§8.4.2).

Colour does not carry anywhere near as much information as the monochromatic channels and subserves nowhere near as many functions. Yet it fascinates us, with writings from the ancient Greeks onwards, which Wade discusses in his book on the history of vision (Wade, 1998). It seems to generate emotional connotations. We talk of red with anger or green with envy. Kandinsky based his abstract art on elaborate theories of colour, while some of the great abstract expressionists, such as Mark Rothko, relied almost entirely on subtle shades of colour to generate a deep impact out of pictures with little figural content.

Colour vision has its beginnings in the photoreceptor pigments. In humans there are three distinct pigments, the so-called short, medium and long wavelength pigments which we often refer to as blue, green and red. This makes man a *trichromat*. As we shall see later this common description is not strictly correct. Colour is a percept and the anatomical correlates of colour vision, cells tuned to colour rather than wavelength of incoming light do not appear until deep inside the brain in area V4.

It is often more difficult to demonstrate that an animal has a particular sort of behaviour than to show that the anatomy or physiology implies that it should have. So it is with colour vision. Animals which seem to have no detectable colour vision in practice may have more than one photopigment and the potential for colour vision. Cat, for example, seems to be completely colour blind, but does in fact have a few long wavelength cones (and has some very low resolution colour vision) (§8.3.2).

Along with most mammals it is at least theoretically a *dichromat*. Recently, a genetic experiment in mouse retina has revealed that *limitations in the diversity of photopigments* might drive the lack of trichromatic vision. Jacobs *et al.* (2007) show that if mice are given a long wavelength pigment by genetic modification, they exhibit additional colour discriminations, even though they do not have the retinal cells which mediate such discriminations in primates such as humans.

The information streaming perspective reveals several interesting features:

- Colour involves a trade-off with achromatic data (§8.3.2).
- Spatial/chromatic interactions, which arise because the red and green receptors are assumed to produce identical achromatic signals, general noise. This is noise not in the colour, but the achromatic channels (§8.5.1.1).
- It is essentially a *relative* measure of the intensity in one wavelength band relative to another. So it's a differential quality with a correspondingly higher error (noise) level.
- Competitive evolution has led to extreme colours, from the plumage of birds to the petals of flowers (§8.6).
- Illumination noise, arising from uncertainties in the illumination.
- Full trichromatic colour vision occurs relatively rarely in the mammalian kingdom and people can sometimes be red–green colour blind without realising it. Nevertheless man-made applications are growing and machine vision may make much more use of colour than animals do.
- Colour is one stream of visual information, the separation of which begins in the retina and becomes more precise at higher levels of processing. It is subdivided into at least two distinct streams between the LGN and V1: blue–yellow and red–green terminate in different cortical layers (Chatternee & Callaway, 2003).

Texture is another property of surfaces, characterising the pattern and the surface smoothness or roughness rather than the details of the individual elements which make up the surface. Fast processing of surface information also helps in rapid object identification, and texture processing often has *preattentive* characteristics (§8.8). Texture is useful in both the where and what pathways. Because, in addition to telling us about what the surface is, grass, fur, skin etc., it can also tell us about surface orientation and distance from the viewer (§7.2.5) and thus feeds into spatial analysis as well as object recognition. Recent work shows that surface information gleaned from texture converges on the same pathways used by stereopsis.

A new and interesting finding is the integration of cross-sensory information. Surface roughness is not only something seen but also something felt as well. There are a number of interesting information theoretic questions surrounding texture, but they are not totally resolved:

1. How accurate a clue to surface orientation and shape is texture? (Some of these issues are discussed in Chapter 7 on stereo and from a cross-modality perspective in Chapter 12 on integration.)

2. How accurate a clue to surface identity is texture? This is really difficult to formulate let alone answer. It is easy enough to say distinguish sand from grass. But what about different types of sand or grass? Part of the difficulty lies in the merging of recognition of the texture as a statistical ensemble of many elements versus categorising individual elements (such as identifying the species of individual leaves in a grassy field).

Section 8.2 deals with the physics of colour, §8.3 deals with the anatomical and neural aspects and §8.6 with evolutionary aspects. Section 8.7 takes a brief look at some of the numerous applications of colour. Section 8.4 discusses the pathways spectral (colour) information follows on the way to the brain and Section 8.5 adds some additional details on colour psychophysics. Section 8.8 concludes the chapter with an information processing perspective on texture.

8.2   Colour physics

An animal uses vision for many things, one of which is the identification or characterisation of objects. The intrinsic properties of the surface of an object include the dyes, pigments and other light absorbers, on or near its surface, its translucency and so on. They modify the surface reflectance, the amount of light reflected at every wavelength. Thus the variation of absorption and reflection with wavelength of light, $S(\lambda)$, is a powerful indicator of the surface nature and properties. For object recognition, the extraction of this reflectance information is crucial, and it drives the need to abstract the properties of the surface from the light coming from it.

But when we look at an object, the light entering our eyes is determined by the spectral reflectance of the object and the spectral distribution of the incoming light. So if we shine a red laser onto a blue object, basically no light comes back, since the blue object reflects only blue light and thus appears black.

Now in terms of visual information processing, the spectral distribution of the reflected light is not that interesting. It is the reflective properties of the surface which matter. Thus the colour we *perceive* should be independent of the incident light. Is this a red or a green apple? So, visual systems have evolved a strategy referred to as *colour constancy*, to make objects appear the same colour regardless of illumination (§8.4.2).

We use language mostly to describe the surface properties rather than the properties of particular light reflected[1]: we refer to grass as green, the sky as blue. In fact, we don't have *conscious access* (Snyder *et al.*, 2004) to knowledge about the illumination, except in very general terms such as the light is redder at sunset. But there is philosophical debate over the meaning of subjective experiences such as the colour red. It would take us too far off course to enter this territory, so this book adopts some practical simplifications. Surfaces of real-world objects we shall be happy to refer to by their colours. When we come to pigments and photoreceptors, we should strictly refer to their wavelengths of

---

[1] Of course sometimes the characteristic of a semi-transparent object is the light it transmits. But for simplicity we shall work just with reflectance. Transmitted colours of dissolved substances can have much sharper spectra than is usually found in solids, but the difference is not usually important for the discussion. It is relevant, though, to oil droplets (§8.3.1.2), which achieve their sharp spectral tuning through using pigments in solution within the droplet.

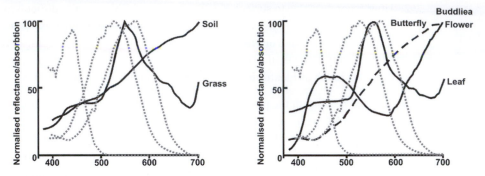

Figure 8.1 Some examples of natural reflectances. Dotted lines show cone spectral sensitivities. (After Osorio & Bossomaier, 1992.)

absorption, from the long wavelength red end of the spectrum to the short wavelength blue. For ease of reading we shall be pretty casual in our mixing of wavelength and colour terms, only drawing the distinction when there might be ambiguity. Hopefully, by the end of the chapter the nature of and need for these distinctions will be clear.

8.2.1   Reflected light and photoreceptor spectral types

The light reflected from a surface, $R(\lambda)$, at wavelength λ is not directly proportional to the surface reflection, $S(\lambda)$, but depends also on the spectral distribution of the illuminant, $I(\lambda)$ and is equal to:

$$R(\lambda) = S(\lambda)I(\lambda) \tag{8.1}$$

The signal from the photoreceptor s_c with sensitivity $P_c(\lambda)$ is then:

$$s_c = \int P_c(\lambda)S(\lambda)I(\lambda)d\lambda \tag{8.2}$$

where c enumerates the spectral types (long, medium, short etc.). The photoreceptor response depends on the light entering the eye and is not solely a function of surface reflectance. Many animals get away with two or fewer spectral types and humans have three. It might seem puzzling that there are so few. Osorio and Bossomaier (1992) recorded the spectral sensitivity of a range of natural objects, from sand to leaves and found a convincing explanation. The spectral reflectances of natural objects are very smooth functions. Figure 8.1 shows some examples. They do not vary dramatically and we would gain little by going to a larger number of spectral types, if we adopt the perspective of general information gathering from the environment.

In a subsequent study, Chiao *et al.* (2000) studied not only the reflected light but also considered how the various forms of illumination (see Figure 6.2) reflected the principal components found. Again three components were sufficient to capture 98% of the colour variation. (The situation for fish in coral reefs they found is dramatically different.)

On the other hand there is a downside to having too many spectral types, discussed in §8.3.2. Animals may be essentially monochromatic or may go as high as tetrachromatic

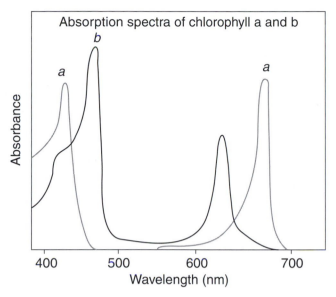

Figure 8.2 Absorption spectrum of chlorophyll. It occurs in two forms, a and b. (Courtesy of Paul May, University of Bristol.)

(§8.3.1.3). But there are examples of distinctive biological colours, from flowers to fish, where colour has a specific signalling role (§8.6.2) where such arguments would not apply.

Across a range of vegetation, the leaf (chlorophyll) reflectance lies between about 550 nm and 590 nm, complementary to the absorption (Figure 8.2). Thus the primate/human L-cone is particularly good for capturing the maximum amount of light from leaves.

When a photopigment molecule absorbs a photon, it enters an excited state, which decays releasing energy and generates a voltage, the process of visual transduction. After absorption, *no information remains about the wavelength of the photon*. The cone itself is colour blind. This is the *Principle of Univariance*, coined by William Rushton (1972). The wavelength of the photon will determine the probability of its absorption. But the cone outputs just a single voltage. It cannot distinguish between many photons at a wavelength where the absorption probability is low, or a smaller number at the peak sensitivity.

The first limitation on colour vision occurs right here at the absorption stage: the filtering properties of the pigments in the photoreceptors and, we shall see in §7.5 the temporal properties also are constrained. Two distinct filtering mechanisms have evolved. The first of these which dominates mammalian vision is variation in the pigments themselves. But birds and reptiles have an additional mechanism: oil droplets inside the photoreceptor also restrict the wavelengths incident on the photoreceptor pigment (§8.3.1.2). The photoreceptor pigments may also be themselves different, but oil droplets allow tuning of the low wavelength cut-off.

8.2.1.1   Super sharp colours in nature

The statistics of natural scenes, which have proved an important building block to understanding visual information processes, do not usually include some rare colours which have very sharp spectra, some geological, some biological.

- Most solids have diffuse spectra, even though the energy levels of individual component molecules are precisely defined and well separated, essentially because of the multitude of interactions arising from close proximity of atoms or molecules. However, some gems and crystals have sharp colours since they are essentially one gigantic molecule.
- Interference[2] can produce extremely sharp spectral bands, sometimes only a few nanometres wide. It requires very thin materials, with thicknesses a fraction of the wavelength of light. Some insects have wings of multiple ultra thin layers, which produce sharp irridescent colours. Dragonflies are one of the most beautiful examples.

There are sharp spectral transitions in some pure minerals, but the brightest, most saturated, colours of the natural world are frequently biological, such as the bright colours of flowers, or the plumage of birds.

8.3   Anatomy and physiology

Spectral information follows similar pathways to other visual information, along the ganglicon cells to the thalamus and on to V1 and V2 before ending up in V4. But, unlike other sensory data it remains multiplexed with spatial and temporal information out from the retina and through the thalamus. It does not actually become *colour* until V4.

Spectral discrimination starts at the photoreceptors containing different pigments. Preprocessing in the retina gives us a variety of difference signals, yellow from blue, red from green and so on, referred to as colour opponency (Plate 8.2).

8.3.1   Spectral sensitivity of the cones

Human visual sensitivity vision extends from below 400 nm to above 700 nm. The rods, active in low light conditions have a different sensitivity to the cones, the former referred to as the scotopic, the latter as the photopic sensitivity (Figure 6.3).

The human retina has three different cone photopigments. Each cone has just one, giving it a sensitivity to a particular range of wavelengths of visible light. In insects sensitivity sometimes stretches into the ultra-violet while some snakes and other reptiles

---

[2] Where an absorber has thickness comparable with the wavelength of light, interference effects can dramatically enhance or suppress different wavelengths. Such properties lead to the use of thin-film reflectance coatings in optical instruments to reduce flare. Some animals also use interference effects for making reflecting coatings.

are sensitive to the near infrared, although the latter is less of a spatially precise visual sense. Plate 8.2 shows the spectral sensitivities.

One thing that is immediately apparent is the huge amount of overlap of the red and green pigments, with the blue quite disjointed. Recall that the blue cones are also different, with lower spatial resolution (§6.5.1.2). But why should the red and green be so close together? There are a variety of possible reasons. For example, the two pigments may have evolved from a single pigment in which mutation has gradually pushed them apart. But §8.3.2 considers constraints on the information in the achromatic channel which force the pigments close together.

Precise psychophysical work inferred that the genes for photoreceptor pigments lie on the X chromosome, from colour differences between men and women. Neitz and Jacobs (1986) showed that the long wavelength cone is dimorphic. Heterozygous females exhibit different spectral tuning arising from having both genes present.

8.3.1.1 The blue cones

The blue cones do not fit smoothly into the model of a trichromatic system with a mixture of three different cone types. There are far fewer blues cones, probably fewer than 5% of the total and they are absent from the central 20 min of arc (100 μ) of the fovea (Bumsted & Hendrickson, 1999; Rodieck, 1973). They are also smaller and thus are likely to saturate more rapidly.

Using the formula for spatial frequency cut-off (eqn 6.5), the sampling frequency of the blue would have to be considerably higher based on comparing the peak sensitivity of the long/medium wavelength system at 550 nm with the blue at 420 nm. In fact it is the opposite way round, with the sampling frequency for blue *much* lower. This is to some extent alleviated by chromatic aberration. The spatial frequency bandwidth of the short wavelength image on the retina would be lower as a result of chromatic aberration defocus, since the focal length is set for red/green. But this is a somewhat weak argument, since evolution has achieved some amazing optical designs. Achromatic doublets are easy to make and it would seem that if the evolutionary pressure was there, chromatic aberration would not be a problem.

The situation seems to be rather that the blue cones play a much weaker role in many of the spatial vision processes. They do not contribute to stereopsis, for example, and probably make a minimal contribution to the achromatic channel (Lennie, 2000) (and of course are absent from the area of greatest spatial acuity).

We shall see that this impacts on the transmission of blue cone signal in the ganglion cells (§8.3.1.1), the contribution of the blue signal to luminance and the psychophysical mechanisms which may be isolated by adaptation (§8.5).

8.3.1.2 Oil droplets

There is another spectral sensitivity mechanism present in vertebrates, especially in reptiles and birds. The cones of these animals often have coloured oil droplets at the base of the inner segments.

There were several theories as to what oil droplets might do, at first some disputing their role in colour vision (Rodieck, 1973). Vorobyev (2003) reviews several such possibilities:

they may protect sensitive biological tissue from the damaging effects of ultra-violet light; and they may improve spatial vision through removing out-of-focus blue light. But he shows that there is an information-theoretic trade-off.

Oil droplets do have the advantage in principle in that they can provide sharper spectral cut-offs[3]. In a primitive species with hard-wired behaviour patterns, where the important environmental features were unchanging during the lifetime of an individual, this could be an advantage. The sharper spectral cut-off may imply a loss of sensitivity, since photons absorbed by the droplet are thrown away and do not contribute to the visual signal. But many animals which have oil droplets such as birds and reptiles operate in bright sunlight conditions. Vorobyev (2003) found that in bright light oil droplets are beneficial as they allow the discrimination of more colours. But in dim light the loss of photons becomes important and reduces discriminability. For three species of birds, he finds the number of discriminable colours can be more than a factor of two in bright light, with even a small gain one log unit down. Note that this is a pure colour argument and does not consider the impact on spatial vision of reduced quantal catch.

The use of oil droplets is analogous to the construction of a photographic film. A quality colour film has a dozen or more layers. Some of them contain silver halide, sensitised by different dyes, so contributing different colour layers to the image. Others serve only to remove light of a particular spectral band to avoid light from that band exposing layers further down.

Oil droplets also exist in marsupials, but they are colourless. Their role seems to be acting as a lens, concentrating light into the pigment containing the outer segment of the cone (Ahnelt & Kolb, 2000).

8.3.1.3　　Beyond trichromacy

There are huge numbers of animals which go beyond trichromacy, having four distinct pigments. Many birds are tetrachromatic, so are many fish, reptiles and amphibians. In many cases this extra pigment subserves an ultra-violet channel and this has been demonstrated behaviourally in birds such as chickens.

Goldfish (*Carassius auratus*) is a much studied experimental animal (quite good behaviourally) as is turtle. Both are tetrachromats (Neumayer, 1998). The turtle (*Pseudemys scripta elegans*) has an ultra-violet pigment but behaviourally (hard, somewhat leisurely, animal) shows trichromacy. Similarly, the tiger salamander (*Ambystoma tigrinum*) has four cone pigments (one ultra-violet) but behavioural studies have not conclusively shown that it fully exploits all four.

8.3.1.4　　Colour blindness

There are many possible ways that colour blindness could arise, from problems at the receptor level through to deficits in the brain itself. However, by far the most common type is loss of a spectral type, usually the long wavelength (*protanope*) or medium wavelength (*deutanope*) wherein there is only one pigment instead of two. The eighteenth-century chemist, John Dalton, observed strange variations in his perception of colour and deduced

---

[3] This results from the absorption properties of dyes in liquids rather than solid matrices as noted in §8.2.1.1.

he was colour blind; however, it was much later that Hunt *et al.* (1995) demonstrated he was a deutenope. However, colour blindness can be quite hard to diagnose and some people are colour blind without knowing it.

Since the coding for these pigments is on the X chromosome, colour blindness involving these cone pigments (red–green colour blindness) is most common amongst men, with up to 10% of the population being colour blind. It is also possibe that some women could be tetrachromatic with two green pigments.

Using the deformable mirror approach (§6.5.3) to photoreceptor imaging (Roorda & Williams, 1999) some rare causes of colour blindness have been discovered. People may be missing an entire spectral type. The retina just has holes where there should be cones. This form of colour blindness would have significant implications for spatial vision, since the sampling array has been reduced in density.

To determine if somebody is colour blind the principle we need to exploit is isoluminance, but this is hard to bring about (§8.5.2). So it is necessary to make up signals where there are many conflicting spatial cues. Thus colour blindness tests are made up of grids of dots of varying colours with patterns embedded within.

Even though colour blindness may be relatively unimportant in many tasks in the natural world, it can prove a problem with coloured photographs and diagrams in journals. Recently some journals have started to use software to map red to magenta to improve contrast for people with red–green colour blindness (Holden, 2002b).

8.3.2 Information in the colour signal

From the Principle of Univariance (§8.2.1), it follows that all the information about the spectral component of the incoming light comes from just a few averages. These are the averages of the incoming spectrally diverse signal over the spectral sensitivity of the cone, according to eqn 8.2. Thus the *colour* signal is a comparative one, between different photopigments. But as discussed shortly, the spectral sensitivities of the pigments overlap a great deal and the colour information is in the difference between two signals which are very similar in magnitude. Thus having the pigments this close together does not seem to be good for colour vision. The precision in the colour signal, like the difference between two large numbers will be poor. The colour signal is noisy and will be most useful in bright light, and it is not surprising to find that colour vision disappears at low light levels. Less obvious, but consistent with the Fourier Analysis arguments of earlier chapters, is that the spatial resolution of colour vision is much lower, with a practical cut-off around 10 cpd[4] (but see §8.5.1.1 for a more precise upper boundary). Similarly we might also expect that animals, such as cat, specialised for nocturnal hunting, would lack significant colour vision[5].

---

[4] One environmental possibility for this advocated by Tom Troscianko *et al.* (2009) is that the tuning is set by the need to pick (ripe) fruit at arm's length.

[5] Whether pets have colour vision is a frequently asked question. Cat has far more rods than cones, to support its nocturnal activity, thus colour vision would be at best poor. In fact it can discriminate blue from green and grey, but the smallest object for which it can discriminate colour would be about the size of a credit card (Loop & Bruce, 1978).

But this begs the question of why the photopigments have not evolved to overlap less. The sampling arguments of §6.5.2 and §3.6 provide at least part of the explanation. The same cones which provide spectral sensitivity also provide the achromatic signal. Since the maximum discriminable spatial frequency close to the diffraction limit (§6.2.3) is at the sampling frequency of the cones treated as if they were one homogeneous array, each must contribute a signal to the achromatic channel.

Using the signals from *both* L- and M-cones for the purpose of the achromatic channel gives an immediate increase in resolution, almost a factor of square root of two[6]. So, if for a monochrome signal, there is a gain in resolution, but since a red or green cone would give exactly the same signal, there is no error.

But the light coming in to the eye does vary in wavelength. Thus a variation in signal between an L-cone and an M-cone may be a variation in brightness *or* it may be a variation in spectral content. This uncertainty is thus a source of noise. Obviously, the further apart the pigments, the greater will be this source of noise. In an extreme situation, we could create a pattern incident on the cones which has the same brightness everywhere, but has red over the red cones and green over the green cones. The cones will thus signal vigorously a fine-grained pattern, which could be interpreted as an erroneous monochrome pattern[7].

In a trade-off situation of this kind, there will be optimal values, although these would be criterion dependent. But the biological evidence suggests it is not revelant: animals within the same environment can show quite different spectral tuning (Snyder *et al.*, 1988).

So, for humans, how big is this chromatic noise? Its value will obviously depend upon the sort of environment we inhabit and other factors. One approach is to average over a range of natural scenes. Osorio *et al.* (1998) did this and obtained an aggregate value about 40 times less than the signal power. So, taking this figure at face value, in bright sunlight, Osorio *et al.* conclude that the chromatic noise becomes important. There is a possible ramification to this. Some evidence suggests that there might be clumping of L- and M-cones (Hofer *et al.*, 2005). This would create areas of low chromatic noise for the achromatic channel, but it would also impinge on the maximal colour resolution. Given that this is not that high anyway, and probably lower than the maximum one might expect from the density of receptors, the psychophysics are reasonably consistent (§8.5.1.1).

Now we can return to the optimal sampling for the human retina. Photon and intrinsic receptor noise are independent of the optics. Thus by increasing the contrast at high frequencies there is an increased signal to noise ratio, even though some additional aliasing noise appears. But the chromatic noise *is* affected by the optics – it is actually derived from the image, the signal itself. Thus *no matter what the optics, the signal to noise ratio at each and every spatial frequency remains the same.* So, if the chromatic noise dominates, there is no value to undersampling. In fact here the aliasing noise is

---

[6] We have twice as many cones per unit area. But because the mosaic is hexagonal and the spectral types are more or less distributed spatially *at random*, an exact figure is hard to determine.

[7] In fact such a pattern would be rather difficult to create without laser interferometry, for the same reason that we cannot see the human cone mosaic through an ophthalmosocope (§6.4.2.4).

a disadvantage. Thus primate-like retinas, where animals operate in bright light, should have approximately matched sampling as do humans.

8.4   Colour pathways

The P-cell pathway carries the spectral information to the brain as discussed in §6.6.2. Through the LGN, V1 and V2 it remains *spectral*, defined by the wavelengths entering the eye. Finally in V4 it becomes *colour*, more or less independent of the illumination.

8.4.1   Ganglion cells

The receptive fields of the P cells are cylindrical with a different spectral sensitivity of centre and surround. They are thus referred to as *spectral-opponent*, or, less accurately, *colour-opponent.* Recall that at this level, the signals depend on the wavelength of light incident on the photoreceptors and *not* on the colour of objects.

The centre usually receives input from just a single cone and thus has a precisely defined spectral sensitivity. The surround covers a larger area and thus receives input from multiple cones. As such there are two possibilities for its spectral sensitivity. If it connects to all cones in the area, then it will average to white with some variability arising from the randomness of the spectral types in the cone matrix. Alternatively, it may connect specifically to one particular cone type, giving it a distinctly opponent character, say 'red' centre and 'green' surround.

In fact most combinations occur except that blue appears only as an excitatory centre. So blue always appears in opponent cells with a blue centre and red/green or other surround. This atypical encoding accounts for the anomalous π mechanisms alluded to in §8.5[8].

It also appears that the blue cones do not contribute to the M-cell pathway. Changes in the blue component of an object dramatically affect its hue but not its perceived brightness.

Why do we have so many cells seemingly devoted to the colour channel, a channel with low spatial resolution and low information content? For a while it was perhaps thought that the P-cell pathway handled colour and colour alone. But this is clearly not the case. The fine structure of texture patterns (§8.8) on surfaces, essentially the high spatial frequencies, are all handled by the P-cell pathway and only the P-cell pathway has one cell per cone, and thus can transmit the maximum sampling frequency of the cone mosaic.

The idea of colour opponency could be a little confusing. It does *not* mean that the cell does not respond unless there is colour contrast in the signal. If the cell receives a white signal in the centre, which stimulates all cones equally, a red-green opponent cell will behave more or less the same way as a cell which had no differential spectral sensitivity. Plate 8.1 summarises this multiplexing of information.

---

[8]   Because the density of blue cones is much lower, it would need a large surround to get enough blue cones and achieve any sort of circular symmetry. Whereas the centres of the P cells may be fed by just a single cone, making a blue centre quite feasible.

The lower contrast sensitivity and slower dynamics of the P cells might be misleading. Remember that they are more plentiful than the M cells. In fact, Chaparro *et al.* (1993) show that *colour is what the eye sees best.* By looking for the lowest contrast flash against a background, they found that colour stimuli required the lowest threshold for detection, at least a factor of five better than the optimal pure luminance stimuli. Colour is also perceived before motion by around 80 ms (Zeki & Moutoussis, 1997).

8.4.2   Colour constancy

This process of extracting *colour* from different wavelengths of light is a process which happens fairly deep inside the brain. Semir Zeki at University College London was one of the pioneers of brain physiology of colour vision and further details may be found in his his seminal book, *A Vision of the Brain* (Zeki, 1993). Following Zeki, recordings from cells in V1 respond to actual wavelengths of light, in other words corresponding to the colours as we *perceive* them. Regardless of the colours of the patches to which it is directed, if we adjust the illumination so that the *reflected light* is long wavelength the cell fires.

Thus, the radiation coming off the surface depends on the illumination. If we take two patches of different colour, but adjust the spectral balance of the incident light, we can arrange for the light reflected from both patches to be the same. We find that cells in the first stage of visual processing, V1, are sensitive to the wavelength of the light, not the colour of the surface.

Cells in the optic nerve and first stages of vision have various sorts of *spectral opponency* where they are activated, say, by long wavelength light in the centre and suppressed by medium wavelength in the surround. Plate 8.3 shows a double opponent cell.

When the signal reaches cortical area V4, cells are at long last sensitive to colour, as shown in Plate 8.3 although it has not been completely established that this is the only role for V4 cells (Conway, 2009). Now a cell responds to a yellow patch, regardless of the wavelength incident upon it[9]. It does not respond to the other colours.

Colour cells occur in regions dubbed *globs* where cells of similar colour sensitivity are grouped together in cortical columns (Conway, 2009), affirming Barlow's conjecture that cells of similar properties will be grouped together (§2.3.2).

Edwin Land was one of the great figures of industrial science. He made his first fortune out of Polaroid for filtering polarised light, now ubiquitous in optical systems, including commercial sunglasses. He made his second from instant photography, but still managed fundamental contributions to colour vision in what he called the *retinex* theory. Land developed the use of rectangular colour mosaics, called Mondrians[10] to study colour constancy.

---

[9] With the slight proviso that there is a reasonable mixture of wavelengths in the incident light. A single wavelength blue laser would just generate black.

[10] Piet Mondrian was an abstract expressionist of the early twentieth century who painted many pictures consisting of rectangular tiles of different colours. See http://www.pietmondrian.org.

In one experiment with Mondrians, Land used projectors with adjustable brightness and interference filters to generate different illumination. Such experiments could be carried out rather more easily today, with digital displays. These filters enabled him to produce three beams of light, each with a narrow bandwidth of 80–100 nm. The centres were approximately 440 nm (S), 540 nm (M) and 620 nm (L) (short, medium and long wavelengths respectively). If we project these individually onto a white square in a darkened room, we see blue from the S projector (think **S**apphire), green (think **M**aple leaf) and red (think **L**ondon bus). However, if we embed this square in a mosaic of squares of many different colours it looks white. This is the essence of Land's experiment.

Pointing the projectors at any given coloured patch and measuring the reflected light from the patch, it is possible to adjust the projector brightness of each projector individually to give the same reflected light signal. For a blue patch, say, most of the reflected light will be blue, so the L and M projectors will need to be much brighter to give the same reflected light. In this case, since the incident light is in effect yellow. Yet when all the projectors are turned on at the same time a white patch does in fact look white, *not* yellow.

The brain uses an average of the wavelengths coming in to correct for the wavelength of the illuminant information. In fact this averaging process may occur in one hemisphere and provide information to the other hemisphere. Thus if the Mondrian is presented to the right eye it enables correct inference of, say, a purple colour in the left eye (the display to each eye is kept separate from the other by optical means, so one eye sees only one display (Plate 8.5)). Thus the effect occurs inside the brain after fusing of information from the two eyes.

However, in patients with a damaged or destroyed *corpus callosum* which connects the two hemispheres of the brain together, the right eye would not get the reference information and the colour judgement is erratic (Land *et al.*, 1983).

In fact the retinex theory although powerful conceptually is not a terribly good algorithm and many people have improved upon it. Maloney and Wandell (1986) formulated a method of deriving the incident light distribution and then, by inversion, surface reflectances. Their model also fits in with the lower spatial sampling density of the blue cones: if the ambient light varies slowly across the scene, then a smaller sampling set at one wavelength suffices to construct the ambient illumination. Human colour constancy is not perfect, since cross-talk between the cones of different spectral types itself creates errors (McCann, 2005).

8.4.2.1 Information loss from colour

So, recapping the discussion so far, there are three main sources of spectral/colour noise:

1. The loss of spectral resolution from having just three spectral types. Although most of the spectral variance may be captured in just three degrees of freedom, there is evidence that humans can make discriminations for numbers as high as eight (§8.5).
2. The loss of spatio-chromatic information from pooling multiple spectral types to create the illuminance channel, forcing significant overlap of the receptor absorption curves (§8.3.1).

3. The uncertainties in the illumination. If the illumination is not known exactly then it is effectively a source of noise, but for common terrestrial scenes, three broadband components capture a lot of the variance.

8.5   Colour psychophysics and behaviour

Understanding how colour works at a neural level requires sophisticated experimentation, the use of fine spectral probes for absorption, tools for mapping genes on chromosomes and recording of the signals from individual cells or populations of cells. Yet a surprising level of understanding of the end result of colour processing has been obtained at the psychophysical level. Trichromatic theory began in the nineteenth century. Thomas Young (1802) proposed that there were three colour types and Hermann Helmholtz (1896) suggested that they responded to short, medium and long wavelengths of light (see Figure 8.1).

Walter S. Stiles was one of the great thinkers and pioneers in colour vision. Before micro-spectrophotometry he elucidated different wavelength mechanisms of colour vision with surprising accuracy.

Stiles' procedure was to use adaptation, using the Principle of Adaptive Independence (§8.2.1). The idea was to wash out all but one mechanism by adapting it to a background signal and then observe the remainder by looking at the threshold for detection of small patches of monochromatic light in the centre. So if we adapt at wavelength μ with a test flash at wavelength λ, then the threshold will depend on both these wavelengths. We can plot the threshold intensity of the test flash against the background field intensity.

It's easy to see what happens when we vary the wavelength of the test flash. If all the other colour mechanisms have adapted, then we see only the sensitivity to the test flash. As we vary its wavelength, the variation in detection threshold will give us the spectral sensitivity of this mechanism. As we change the wavelength the plot of sensitivity against background just shifts up the y-axis without changing shape, the *second Displacement Rule*. If the shape were to change, it would imply that the test flash and background field were not independent. If we vary the wavelength of the adapting field, this time we see a shift along the x-axis, again without a change of shape, *the first Displacement Rule*.

Using these adaptation procedures, Stiles was able to identify psychophysically, with reasonable accuracy, the three primary sensitivity peaks (he estimated 440 versus 419, 540 versus 531 and 570 versus 559 nm). He referred to these as π_3, π_4, π_5. However he also found two other mechanisms in the blue, π_1, π_2, which are now known to be spurious. But we now understand how they arise: the blue mechanism has different ganglion cell encoding to the other wavelengths (§8.3.1.1).

Although a lot of evidence (§8.2.1) points to three components being enough to capture all the variations in colour signals we might need, in fact our discrimination can be somewhat better. One approach has been to use the Munsell colour chips used in the paint and colour industries. These cover a wide range of reflectances, but they are engineered to vary smoothly to ensure stable colour reproduction. For example,

Buchsbaum and Gottschalk (1983) showed that just three frequency components suffice to cover many Munsell chips on the chromaticity diagram.

Nascimento *et al.* (2005) took an empirical approach by getting observers to make forced choiced decisions on whether a forest or urban scene was the original or one represented with a reduced set of principal components. It transpired that up to *eight* such components were required. This might seem confusing at first. Just because there are just three cone spectral types, it does not mean that these spectral types of necessity match the first three principal components of natural scenes. It would be interesting if they did, but bear in mind that lots of animals inhabit the same environment with diverse colour vision. So, the principal components of natural scenes need to be weighted in some way according to their biological importance (Snyder *et al.*, 1988).

The challenge of finding a way to represent natural scenes with the minimum number of degrees of freedom is a challenge for artificial systems, at first photographic film and the printing industry and subsequently television and computer displays. Section 8.7.1 takes up this topic in more detail.

8.5.1 Colour discrimination

Gouras (1991) gives the following approximate figures for colour discrimination. The number of distinct hues is around 200. The number of levels of saturation is around 20, in other words about 12 bits of information. Digital cameras are now approaching this level (in 2010) at the professional end of the market. The best film scanners already operate at this level.

A crude estimate of the colour bandwidth, from a psychophysical perspective, requires a spatial element on top for this hue/saturation discrimination. The spread function for a bandwidth of 20 cpd will be around 5–10 minutes, so let's assume around 10 patches per degree or 100 patches per square degree. Since colour vision falls off towards the periphery as more and more rods take up space, we take a 30 degree field and 50 ms integration time and we end up with around 2 Mbs.

8.5.1.1 Colour resolution

Since any one spectral type has only about 70% of the resolution of the whole cone matrix, we should expect that the resolution of coloured (e.g. red–green) gratings, which are isoluminant, would be lower than monochrome. Measured performance is worse than what you would expect based just on receptor density. Maximum spatial frequency made with natural images is around 10 cpd. But this value is influenced by numerous factors such as chromatic aberration. Sekiguchi *et al.* (1993) used laser-generated interference fringes on the retina and found much higher values, at least 20 cpd and possibly closer to 30 cpd.

Efforts to demonstrate colour aliasing have not been very successful. However, Williams *et al.* (1991) proposed that a well-known psychophysical phenomenon is a demonstration. Brewster's colours are the transient appearance of colour in fine black and white gratings. They argue that existing explanations based on eye movements are inadequate and propose that the effects arise from chromatic aliasing.

8.5.2   Isoluminance and spatial vision

Implicit in the discussion so far is that we have three spectral types and trichromatic vision. But if the luminance (brightness) component is separated out, we have just two pure colour channels. One might then ask how much spatial and shape information we can get from them. Intuitively it is tempting to think that there would be a lot, such as in the strong contrast of a red flower against a green leafy background. But our intuition leads us astray.

Imagine bright red and green bars adjoining each other. We see a strong edge between them. Now gradually adjust the light level from each until the intensity, the number of photons coming from each is the same[11]. At this point we have *isoluminance*. The two sides of the edge differ only in the wavelength of light and not in intensity. As we approach this point, the edge surprisingly becomes harder and harder to define! This is really surprising the first time one sees it. One thinks of colour as defining edges and shapes very strongly.

In principle[12] we can take out all the luminance variation in an image and leave just the variations in colour, the isoluminant condition. As we have noted, stereopsis, movement, form analysis and contour detection are virtually independent of colour. So, what we find is very fuzzy outlines and considerable difficulty in selecting out objects from a scene. But the extent of combined processing is more complex than first thought, and, for example, geometric illusions have now been demonstrated in isoluminant conditions (Hamberger *et al.*, 2007).

Fortunately isoluminant stimuli are rare in nature (or rather our visual mechanisms have evolved to attach a low priority and low performance to things which are rare). A corollary is that the visual system tends to assume major colour variations occur at luminance boundaries. Thus, subjectively colour appears to 'bleed' to boundaries. Livingstone (1988) points out that these strategies of human colour vision have been exploited by artists, such as the Impressionist schools.

8.6   Evolutionary forces in colour

Matching of sensory information processing to lifestyle and environment is widespread across all sensory modalities. But with colour a lot of research has focused on very specific tasks for which it might be useful, and these have been seen as drivers for spectral sensitivity. General purpose vision may not derive that much from colour as evidenced by how common red–green colour blindness is amongst human males (§6.5.3, §8.3.1.4). Nagle and Osorio (1993) showed that the red–green difference signal is minimal in many natural scenes. One such task which has attracted a lot of interest is the discrimination

---

[11] The number of photons is an approximation to the achromatic channel, but is adequate for the discussion here.

[12] It is difficult to construct perfectly isoluminant stimuli because the spectral distribution of cones varies across the retina and thus the effective spectral sensitivity varies too.

of fruit. When fruit ripens it usually changes colour, but the shape and size, and possibly the luminance, change little if at all.

Given that the long and medium wavelength cone spectral sensitivities overlap considerably, where should the peaks in the spectrum lie? One big factor is the peak reflectance of chlorophyll (see Figure 8.2) at 550 nm. The L–M cones are not very good at distinguishing leaves from the background. The contrast with the S-cone is much better for this. But the colour contrast against chlorophyll would be advantageous.

To demonstrate this, Osorio and Vorobyev (1996) examined the advantages of trichromacy over dichromacy for the identification individually of fruit and leaves and of the detection of fruit against leaves. The latter showed the largest advantage with a peak detection in their experiments of 85% compared to a best performance of 46% for the dichromat. By modelling the detection as a function of wavelength of the M-cone, they were able to show that the value of trichromacy diminishes dramatically if the wavelength of the M-cone rises above 540 nm, with an optimum around 520 nm.

But fruit can often be distinguished in the yellow–blue channel, and it has other indicators, such as its odour, when it is ripe. Dominy and Lucas (2001) propose an alternative which seems plausible for chimpanzees (*Pan troglodytes*) and colobus monkeys (*Colobus guereza*) in parts of Africa. The red–green channel is especially good for distinguishing young leaves from older less desirable ones. Thus it may be leaf foraging which is the driving factor.

But the spectral peaks of New World primates (Catarrhini) are at 533 and 565 nm. Molecular evidence implies that trichromacy has *evolved several times*, arriving at these wavelengths. Thus there is some doubt about the completeness of the model. Is it that biochemical constraints limit the wavelength of the M pigment, or that if we were to average overall stimuli with useful colour contrast we would end up with these different figures?

Against the possible multiple evolution of trichromacy stands the question of why it has not evolved in other mammals outside of primates. Wässle (1999) argues that the limitations were not in the evolution of an additional cone pigment, but in the neural architecture, the P-cell (midget) system. He argues that this neural pathway evolved to provide acute spatial vision and stereopsis, such as for the challenges of living in trees. With this system *already* in place it became possible to wire up individual cones. Before the midget system the receptive field centres would have covered several cones, thus averaging out the spectral information.

8.6.1 Colour implications for evolution

The presence of oil droplets in mammalian cones has been very useful in establishing evolutionary pathways and taxonomy. Placental mammals do not have oil droplets. But marsupials and monotremes do. Since the oil droplets are present in the *out-group* reptiles from which mammals developed, the oil droplets are considered a more primitive condition (Johnson, 1986).

These animals have double as well as single cones. Some Australian marsupials have double cones with oil droplets in each, similar to birds. In monotremes such as the platypus, *Ornithorhynchus anatinus*, there are single cones containing oil droplets or double cones with oil droplets in just one. By tracking these developments it has been possible to build up models of the evolutionary patterns of mammals and in some cases reclassify species (Johnson, 1986).

Trichromacy, as in humans and many primates, is almost uniquely characteristic of these species. The more common pattern in mammals is dichromacy, although far from all species have been studied. Such dichromats tend to have a short and long wavelength cone. For example, Neitz *et al.* (1989) found two pigments in dog with peaks at 429 nm and 555 nm and cite other evidence for dichromacy in pig, shrews etc. This yellow–blue discrimination is suitable for distinguishing leaves from background in natural scenes as shown by Osorio and Bossomaier (1992).

For a long time rodents, such as rats, were thought to be at best dichromats, i.e. having just a single-cone pigment. However, Jacobs *et al.* (1991) discovered that house mice (*Mus musculus*) do in fact have a second cone, but its peak sensitivity is in the ultra-violet at 370μ or less. Subsequent experiments showed some ultra-violet behavioural sensitivity to ultra-violet rays in mice, while other rodents such as rats and gerbils also exhibit some ultra-violet sensitivity. One possibility suggested by Chaévez *et al.* (2003) for the rodent *Octodon degus* is that the urine signalling marks are visible as well as detectable by odours. Kestrels may also use their ultra-violet sensitivity for detecting fresh urine trails (and hence proximity of their prey) (Viitala *et al.*, 1995)! From high up in the air, visual would presumably be more effective than olfactory detection.

8.6.2   Special signalling mechanisms

Specialised signals occur across sensory modalities, sonic echo location in bats and dolphins (§5.4, §11.4.3, §11.41), chemical pheromones in many species (§9.4.1) and infrasound for territory marking in tigers and elephants (§5.4.2). Colour is no exception. Not unsurprisingly we find some examples in colour vision. The bright plumage of birds and the striking markings of tropical fish exemplify elaborate animal signalling behaviour. Many species have different colours for male and female, often with the male the more colourful. Sometimes, as in the Australian King Parrot, the full bright red plumage of the male takes years to develop.

Apart from the bright colours of males as indicators of sexual prowess, there are also some interesting examples of colour being a direct indicator of health. Carotenoids, the chemicals which give carrots their bright orange colour, are important to immune system functioning. Healthy birds can spare some carotenoids, just for colouring in crests and beaks. If the bird is fighting off an infection, however, these chemicals are needed for immune system use and thus the beaks and plumage may lose colour. Thus beak colour becomes sexually selected as in blackbirds (*Turdus merula*) (bright yellow) (Faivre *et al.*, 2003) and zebra finch (*Poephila guttata castanotis*) (Blount *et al.*, 2003).

8.6.3   Insects

Many insects have a symbiotic relationship particularly with plants. They get nectar in exchange for moving the pollen from plant to plant. The plant may go to some trouble to make itself attractive to just some variety of insects. The route to the nectar may be easy for insects of a given shape and size. But this effort would go to waste if the insect were not able to find the flowers in the first place.

8.6.4   Making your own light: phosphorescence and fluorescence

Deep in the ocean, there is little light available from the sun, so some fish make their own: they phosphoresce. With such a precise signal, other fish may tune in to it, but predators probably find it useful too.

In very deep water it used to be thought that fish were not sensitive to long wavelengths. But really deep down live some amazing creatures. At depths of 2000 metres strange jellyfish-like predators, longer than the longest terrestrial snake (siphonophores, *Cnidaria*), are blind, yet they exhibit red phosphorescence, thought to attract fish. This area is still very much under investigation (Haddock *et al.*, 2005).

Why are fish so brightly coloured? Again, we have similar behavioural mechanisms to birds. Tropical fish on reefs, where the water is shallow and there is plenty of light, display amazing colours and patterns. They intermingle and swim at different speeds, often turning and darting in and out of cover very rapidly. In such an environment a rapid, definite cue, such as a distinctive coloured pattern may be highly advantageous. However, this is not the only driving force. Sludge dwellers, such as carp, are also tetrachromatic (Osorio, pers. comm.).

However, because water filters light to some extent, colour constancy is not necessarily easy to achieve. Flourescence, which produces bright colours at particular wavelengths, independent of the wavelength of incident light, is thus a useful strategy. The mantis shrimp, *Lysioqullina glabriuscula*, a large (22 cm) shrimp, takes in light around 440 nm (excitation spectrum) and fluoresceses at a maximum of around 524 nm, accounting for up to 10% of the light emitted and reflected in this yellow band (Mazel *et al.*, 2004).

There is an interesting corollary to the use of fluorescence – absolute wavelength rather than light corrected for incident wavelength is important. Thus *stomapods*, which include mantis shrimp, have at least eight spectral classes in the range 400–700 nm (Mazel *et al.*, 2004) and presumably do *not* correct for the spectral content of background illumination.

Budgerigars (*Melopsittacus undulatus*) use fluorescence as a sexual display. Their feathers absorb ultra-violet light and emit bright yellow, with wavelength optimally matched to their cone sensitivities (Arnold *et al.*, 2002).

8.7   Artificial colour

The capture and reproduction of colour has a long history and much of the theory and technology is quite mature. However, the computing power and appropriate communication protocols have only recently started to approach the limits of human discrimination. The 2010 standard HDMI 1.3 now supports *deep colour* which is essentially up to 16 bit colour, up from 8 bit. It also suppors xv.colour ('IEC', 2006) and improved colour space (§8.7.6). However, we have to go back to the first part of the twentieth century for the core standards in colour encoding.

8.7.1   Trichromacy and representation of colour spaces

The mathematical implication of trichromacy is that there are only three independent degrees of freedom in colour space, corresponding to the three colour pigments. Thus it should be possible to represent any colour by mixing together just three lights of different spectral content. But, mixing lights is *additive*, which introduces various complications.

The first extensive colour matching experiments were done independently at almost exactly the same time by Wright and Guild (see Broadbent, 2009 pers. comm., for a description of their work). Since there are only three receptor types (ignoring the possibility of tetrachromatic females), then any colour should be representable by just three values.

Suppose the spectral distribution functions are P_i with $i = 1, 2, 3$ for L-, M- and S-cones respectively, R_i, the spectral sensitivites of the cones themselves. Then the signal produced in cone j by primary i is a matrix A with elements a_{ij} given in eqn 8.3:

$$a_{ij} = \int P_i(\lambda)R_j(\lambda)d\lambda \qquad (8.3)$$

Suppose we want to match a light with spectral distribution $I(\lambda)$. Then the cones' signals, as a vector c_i are:

$$c_i = \int I(\lambda)R_i(\lambda)d\lambda \qquad (8.4)$$

So, we need coefficients q_i for each of our primaries to produce the same cone signals, i.e.

$$c = Aq \qquad (8.5)$$

which allows us to find q from eqn 8.6:

$$q = A^{-1}c \qquad (8.6)$$

But there is no guarantee that all the values of q_{ij} will be positive. To achieve a match if one of these turns out to be negative will require *adding this light to the sample light*.

In the 1920s there were not very many options for getting a bright single wavelength light in the laboratory. The only practical way to do this was to use individual spectral lines. The standard primaries at the National Physics Laboratory (NPL) used 700 nm (hydrogen, red), 546.1 nm (mercury, green) and 435.8 nm (mercury, blue). Wright however used primaries of 650 nm, 530 nm and 460 nm, obtained fron a Tungsten light source dispersed prismatically to produce narrow bands. With narrow primaries such as these, which we can approximate as Dirac delta functions (§3.3.2) at wavelengths λ_i, the cone signals c_i are simply given by eqn 8.7:

$$c_i = \int \delta(\lambda - \lambda_i)R(\lambda)d\lambda = R_i(\lambda_i) \qquad (8.7)$$

However, the international community, through the *Commission Internationale de l'Éclairage* (CIE), agreed on a standard set of lights, or primaries, early this century

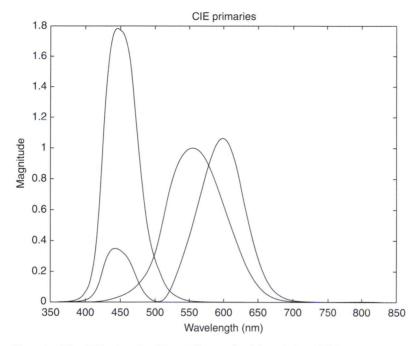

Figure 8.3 The CIE primaries (blue, yellow and red from left to right).

in 1931. Each is a wavelength distribution, $X(\lambda)$, $Y(\lambda)$, $Z(\lambda)$. If we are given any light mixture, $I(\lambda)$, then we can compute the amount of each of these primaries we need to faithfully represent the colour and intensity (Foley *et al.*, 1996). Figure 8.3 shows the CIE primaries. Note the small blip of the red primary in the blue region, which is necessary to match as many colours as possible using addition only.

To find how much of each we need, we use a set of bi-orthonormal functions, denoted $\bar{x}$, $\bar{y}$ and $\bar{z}$. Bi-orthonormal functions are a variation on the orthogonal sets described in §3.2.3, defined according for functions X_i, $\bar{x}_j$ to eqn 8.8:

$$\int X_i \lambda) \bar{x}_j d\lambda) = \delta_{ij} \tag{8.8}$$

So if

$$I(\lambda) = \sum_i b_i X_i(\lambda)$$

Then, multiplying by $\vec{x}_j$ and integrating we get:

$$\int I(\lambda) \bar{x}_j d\lambda \sum_i b_i \int X_i \bar{x}_j = b_j$$

In other words we get the weighting factor for primary X_i by integrating the source $I(\lambda)$ with $\bar{x}_i$, writing $\vec{x}$, $\vec{y}$, $\vec{z}$ for i = 1 . . . 3.

Thus the amounts, L, M, S are given by:

$$L = k \int I(\lambda)\vec{x}(\lambda)\,d\lambda \qquad (8.9)$$

$$M = k \int I(\lambda)\vec{y}(\lambda)\,d\lambda \qquad (8.10)$$

$$S = k \int I(\lambda)\vec{z}(\lambda)\,d\lambda \qquad (8.11)$$

where k is a constant.

Now these three values implicitly contain the luminance as well as the colour. So we have different ways of describing a colour signal. We can normalise out the intensity (luminance) and end up with three *tristimulus values*, as in eqn 8.12. We can also represent the signal as a triple of *hue, saturation and luminance*. We shall now consider each in turn. One of the useful things we shall get out of this analysis is a way of quantifying the capacity of devices to adequately represent colour signals, be they colour prints, computer monitors or photographic film:

$$x = \frac{L}{(L + M + S)}; y = \frac{M}{(L + M + S)}; z = \frac{S}{(L + M + S)} \qquad (8.12)$$

As we saw in §8.3.1.1, the blue system contributes very little to our estimate of brightness and the L and M are effectively pooled. In insects, for example, the M channel is the channel used for all luminance-based tasks. Thus the tristimulus values are mapped to one luminance component ($Y = M$) and two chromatic components (x, y).

At maximum brightness $L + M + S = 1$ which we now project on to the L, M plane to give the map of visible colours shown in Plate 8.4.

To find the x tristimulus value, for a wavelength $\lambda = 500$ nm, requires the Dirac delta function (§3.3):

$$l = \int \delta(\lambda - 500)\vec{x}(\lambda)\,d\lambda = \vec{x}(500) \qquad (8.13)$$

If we calculate the (x, y) values for each wavelength of the visible spectrum, we get the curve shape in the figure going from 700 to around 400 nm. These are the maximally saturated colours. For the CIE primaries no colours lie outside the horseshoe-shaped region. To see this, imagine mixing two wavelengths together. The tristimulus value is going to lie somewhere along the line joining them. So *any* mixture will give tristimulus values inside the horseshoe (Plate 8.4), the chromaticity diagram.

We can also take a value for white light, C, where, approximately, $x = y = z = \frac{1}{3}$; which will lie roughly in the middle. The reference point used for white is called **D65**, which has a correlated colour temperature of 6902 °K (Figure 8.4) (Davis, 1931). Colour temperature is the black body temperature with closest colour match in human perception, used to specify the colour of broad spectrum light sources. It is the temperature of a black body (§6.2.1) which has the same subjective colour as the light source. D65 approximates midday sunlight.

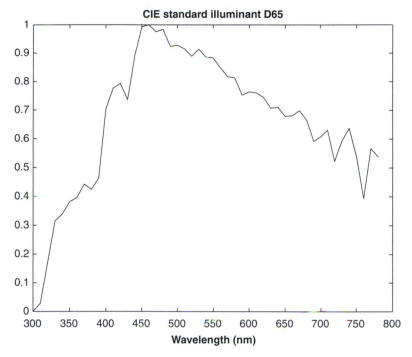

Figure 8.4 CIE D65 (reference white for midday sunlight).

So if we start at yellow on the chromaticity diagram and move in towards white, the colour gets less saturated. As we go through white, we get the unsaturated complementary colour which reaches full saturation (blue) when we reach the horseshoe curve. There is a similar line from red to its complement cyan.

What happens with green is interesting though. Now the line through white does not actually hit the horseshoe at all. The complement of green, majenta, is a *non-spectral colour*. This has an easy intuitive explanation. If we subtract green from the middle of the spectrum it leaves a hole with red and blue either side. So it is not altogether surprising that majenta should need at least two wavelengths to represent it rather than just one.

As we mix together the extreme wavelengths of the horseshoe we get all the hues along the straight line joining them together. If we add any additional wavelength, then it will pull the hue into the horseshoe. So this line represents the limit of the non-spectral hues and all visible hues fit within the closed figure (horseshoe plus line).

If we have a colour, C, somewhere in the middle of the chromaticity diagram, we can plot a line (the dashed line in the figure) from C to the edge. The wavelength, Q, where this line hits is the *dominant wavelength* of the colour. Its distance from C indicates its saturation. The same distance in the opposite direction gives the *complementary colour*, Z.

Computer display devices usually have just three different light sources, usually referred to as red, green and blue. Each can be represented as a point on the chromaticity diagram, such as the points U, V, W. If we turn off, and just mix, U and V we get the range of hues along the line UV. Now turning on W brings the hue on the screen inside the triangle UVW. Hence all the colours which can be represented by this device lie inside this triangle, known as the *gamut*.

Most display devices do not come close to covering the whole of the chromaticity horseshoe. They are noticeably deficient in the deep reds and the blue–green area. To some extent, this is not as bad as it sounds, since our wavelength discrimination is weakest in the blue–green area at the top of the diagram. However, we are quite sensitive to deep reds and they usually make a big impact in reproduced pictures. Hence some companies are still striving to create displays with better gamuts.

For example, Genoa Color (Genoa, accessed 2010) has developed ColorPeak©tech-nology for display devices which adds up to three additional primaries (usually adding cyan, magenta and yellow to red, green and blue). They claim 35% better coverage of the gamut and an increase in brightness of 40%.

8.7.2   Capture

In the first decade of the new millennium, silver halide photographic film finally gave way to digital cameras. Many years of research and development have produced very high perceptual quality in colour reproduction through film, particularly colour transparency. Digital cameras have some inherent limitations for which computer post-processing partially compensates. As with vinyl records, many years of tuning performance to human perception and exploiting the soft limits of analogue technologies presented considerable difficulties to the early days of digital photography.

The best colour films may have 20 or so layers of varying spectral sensitivity and function. Why have so many one might wonder? It is a somewhat complicated story. Silver halide's greatest sensitivity is to blue light and it needs to be sensitised by adsorbing appropriate dyes to the surface. Unfortunately, although organic molecules may achieve strongly saturated colours, they frequently absorb elsewhere too.

Achieving the same spectral sensitivity as the human observer, particularly in the red, is extremely hard, necessitating a carefully tailored series of layers, each taking a nibble at part of the spectrum.

With digital cameras the situation is a little different, but is much closer to the human eye. Unlike a film, the digital camera has an array of detector elements, which is usually rectangular. On top of this we superimpose a coloured mask, where each receptor ends up with a particular colour filter sitting over the top of it. A typical mask is the Bayer mask, which uses a 2×2 grid with one red, one blue and two green sensors in a mosaic of regular square tiles.

The Foveon X3 sensor is a new design of chip, which has a receptor mosaic more like an insect than a mammal. It is trichromatic, but instead of a mosaic of different colours superimposed on top of broadband receptor elements, each receptor has itself three colours built in. Thus it suffers none of the colour aliasing/chromatic noise effects discussed above. The first camera to use this chip is now on the market, made by Sigma, but it does not yet outperform existing systems acccording to current reviews.

We could capture the same information as the human eye if we used the same spatial distribution of colours and the same spectral sensitivity for each of the filters as the corresponding human cones. But we don't know the exact spatial distribution. It seems to be random, with perhaps some degree of clumping of individual spectral types (Roorda & Williams, 1999). Copying the spectral sensitivity is also not easy, for the same reasons that we have with sensitising dyes and filters in colour film.

8.7.3 Hue and saturation

An alternative representation of colour space is in terms of hue (H), saturation (S) and brightness (V). This is an intuitive representation which accords well with how we think about the quality of colours. Saturation is the vividness of the colour. Technically it is the amount of white light mixed in – a lot of white gives a washed-out colour, no white, the maximum possible colour vividness. Saturated colours occupy the perimeter of the gamut diagram. As we move in towards the white point, D65, they become less saturated.

Hue is the mixture of different components, the degree of blueness, redness of the colour. Brightness and saturation vary from 0 to 1. Hue does not have a simple linear relationship and thus is represented on a circle. Thus the coordinate representation is cylindrical and overall representation is a cone. Brightness is the cone axis, saturation the radius and hue the 'latitude' around the cone surface.

This aligns rather better with our intuition. We judge colours as saturated or dilute and hues as mixtures and we think of them somewhat independently.

The cone does not have a circular base, but a hexagonal one, constructed as follows. We take the RGB cube and look along the diagonal axis, going from black to white. There are six corners, three at distance 1 (red, green and blue) and three at distance $\sqrt{2}$, red + green (yellow), green + blue (cyan) and blue + red (magenta). Thus as we go round the six corners of the hue hexagon we go through the sequence red (0), yellow (60), green (120), cyan (180), blue (240), magenta (300), values in degrees. Complementary colours (e.g. yellow, blue) are 180 degrees apart.

8.7.3.1 Names of colours

We have many names for colours, although not nearly so many as the colours we can discriminate of course. However, some of these names do not actually refer to hues, but rather level of saturation of a hue. So, for example, as we decrease the saturation and brightness of orange we end up with brown.

8.7.4 Colour by reflectance (CMYK)

For representation of colour on a reflecting surface, such as a colour printer or a photographic print, we have to go through an inversion process. Instead of adding lights to make a given colour as we do on a colour monitor, we subtract components from incident white light. Thus we use the complementary colours, cyan, magenta and yellow, to red, green and blue.

Dyes are not perfect, however. Getting three dyes to add up to a pure black with no trace of colour cast is extremely difficult. Hence the reflective colour process frequently adds a fourth component, K, or black, denoting specifically a (pure) black ink.

8.7.5   Digital cameras

The advent of CCD detectors in video recorders and digital cameras brought with it similar principles to the primate retina. The array of photodetecting elements has, over the top of it, a coloured mask, passing each detector's input through a red, green or blue filter.

Thus we might ask how the optimal mosaics compare with biological vision. Companies such as *FillFactory* specialise in the construction of such masks and outline a range of structures. The different colours may be arranged in stripes or repeating clusters. Pseudo-random, which is what we *think* is the biological organisation, is disadvantaged by artefacts arising from bleeding from one receptor to another.

Another distinguishing feature is the more equal treatment given to blue, because the applications of these cameras may go beyond the provision of images for human viewing. The issues of chromatic resolution remain moot.

8.7.6   Robotics and games

Although most animals rely on passive colour vision, using light (ultimately) from the sun, in robotics it's quite a different story. Man-made narrowly tuned signals are easy to make with lasers or light emitting diodes (LEDs). Thus a whole new range of possible uses for colour arises, going beyond what may happen in general perception to the specific signalling found in phosphorescence and animal plumage.

Some things are found more easily by colour. A Coca Cola tin, especially after it has been drunk, has a complex, irregular shape but a distinctive colour. Imagine the problem of finding Coke cans containing beer and other soft drinks. Insects presumably do this with flowers, which is why colour vision is well developed.

In virtual environments, the first primitive senses used by animats are likely to use colour as a primary mechanism. Its simplicity and directness, without the need for complex form processing, will make it computationally feasible much sooner than any sort of full visual system.

High definition television and vastly increased computing resources, such as video memory, have meant the quality of colour has been rapidly increasing. For several decades premium colour meant 8 bits for red, green and blue, 24 bits in all. In fact the widely used GIF image still uses only 8 bit colour with a mapping of RGB values to 256 colour map levels.

Display technology has also been improving, with greater saturation now available from some of the best LCD and plasma screens than used to exist with CRT. There are now a lot of evolving standards which go beyond 8 bit RGB. Some, such as xv colour ('IEC', 2006) claim a gamut 1.8 times that of earlier monitors.

8.8   Texture

Another important property of surfaces, useful in identifying objects, is texture. Section 7.2.5 discusses how texture gives important spatial clues for distance and surface shape. But patterns are important in themselves. It is useful to be able to identify a patch of tiger in the grass, even if the whole animal is not visible! Texture is frequently three-dimensional, it may feel rough or smooth, isotropic or directional and the somatosensory and visual systems work together in surface identification.

Many animals do not have mirror smooth surfaces: fur, hair, skin patterns, for example, all create a textured surface, about which the animal can do little, although there would be evolutionary pressure to minimise visiblity to predators. Texture often helps an animal to blend into the background and camouflage is an important element of body markings and colour.

Thus it is not surprising that the visual system has rapid, *preattentive* texture discrimination. Some characteristics of the visual world activate this preattentive processing: they pop out after very short exposures and show a parallel rather than serial search mechanism. In other words the processing time grows slowly with the number of elements, whereas it would grow roughly linearly if each element had to be checked sequentially.

Bela Julesz (1975) pioneered the characterising of texture patterns which show this immediate jump out character. Some textures do, some don't, but the details are complex. Treisman has studied many other kinds of preattentive stimuli, of which *colour is one* (Kandel & Wurtz, 2000; Treisman, 1986, p. 502).

The precise characteristics of what makes one texture preattentively discriminable from another are still not fully understood, despite over three decades of work. There have been two common approaches:

1. **Using spatial frequency** or the power spectrum to characterise the patterns. This is immediately amenable to information theoretic analysis, since the noise in each frequency band can be estimated from photoreceptor and other contributions. It has the disadvantage that texture discriminators may not always have a signature in the power spectrum (second order statistics).
2. **Using spatial statistics**, such as the density of particular shapes, distributions of orientations of tokens of some kind. Although this can be analysed for discrimination thresholds, it gets rather texture specific and is not easy to generalise.

8.8.1   Camouflage

Is colour good at breaking camouflage? Perhaps it's easier to match spatial patterns than colour – this would be a relatively low resolution task. Camouflage abounds in the animal kingdom. For mammals and birds it is usually a static pattern or mixture of colours. Troscianko *et al.* (2009) discuss a range of such mechanisms. A common strategy is to make the pattern so distinctive that it distracts attention from, or makes it difficult to see, the outline of the animal.

An individual zebra, *Equus burchellii*, looks very distinctive. But in a herd of zebras the high contrast stripes make it hard to tell one from another, in turn making it hard for a predator to target a single animal. Other antelopes when running in herds make random, wild jumps and direction changes, again making it difficult for a lion to single out her prey.

Some animals, however, have gone much further. Cuttlefish can create *dynamic* patterns on their surface to help them blend in with the background as they swim across the water. How they do this is computationally very interesting, since they are effectively making a dynamic model of the statistics of the background and generating a pattern of similar statistics.

8.8.2   Filling in holes

In almost every second of our waking lives, the visual system is actively *filling in* the image content of the blind spot (§6.3.2.1). It's useful to know how it does this, because this has a definite artificial application in patching images. How many brilliant photographs have been ruined by an inconveniently situated road sign?

8.8.3   Textures in computer games

Whereas colour representation is a fairly mature art, the computer games industry has been very active of late in producing richer textures for greater realism. Of the two problems faced, one is the generic one of getting good models of textures of all kinds, from grass to garments, and the work on scene statistics and capturing of natural scenes falls into this category. Camouflage (§8.8.1) and filling in (§8.8.2) both rely on this understanding. The other is rendering these textures in a computationally efficient manner.

Edwin Catmull (1974) came up with the first approach to texture mapping in his PhD thesis at the University of Utah. Given that computers were orders of magnitude less powerful at that time, the crucial issue was to produce a convincing texture without adding to the complexity of rendering. In the virtual worlds built for computer games, the usual approach is to represent the world as a three-dimensional (3D) mesh of planar polygons like a Buckminster Fuller dome. Each polygon has a colour and a battery of techniques allow this mesh to be *rendered* to a two-dimensional (2D) image from a particular viewpoint with a particular illumination distribution.

Catmull's solution was to add a texture skin to each polygon. So the texture has no 3D structure of its own. But the excellent camoulage achieved by animals (§8.8.1) testifies to just how effective this can be. A technique called *normal* or *dot3 mapping* was to stay with the goal of not increasing the number of polygons in the mesh, but, now to add a 3D normal (micro-orientation of the surface) at each pixel (Blinn, 1978).

Moving on two decades, *parallax mapping* (Kaneko, 2001) uses a trick of displacing the points on a texture slightly to create an effect of depth parallax, a trick exploiting our perceptual assumptions (§7.2). Further advances in adding occlusion effects add yet further realism (Tatarchuk, 2005).

One of the most recent developments occurs in the game MotorStorm, with so-called four-dimensional (4D) textures, which can change over time, a technique supported by the architecture of the Sony Playstation 3. The importance of textures is exemplified by the effort Sony and Microsoft have gone to in their respective games consoles to provide hardware support specifically for texture.

9 The chemical senses

Nothing revives the past so completely as a smell that was once associated with it.

Vladimir Nabokov

9.1 Introduction

The importance of the chemical senses varies throughout the animal kingdom. Humans are dominated by audiovisual stimuli. Not only is this reflected in the amount of brain capacity devoted to chemical senses, but it is also reflected in language. In a wide-ranging study of languages across the globe, two-thirds to three-quarters of words denote sensory experience or function referring to hearing or vision (Wilson, 1998).

We think of our senses as quite distinct, although there are cross-over effects, referred to as synaesthesia. People with synaesthesia get strong percepts of another sense from the stimulus of one. Thus a particular smell or musical tone may invoke a distinct colour (§12.3.2).

Chemical senses are at their most developed in olfaction, yet there are chemical sensors throughout the skin and internal organs. In fact chemical sensing might be considered the first to evolve of all the senses, being present in simple unicellular organisms such as bacteria and protozoans. Some are able to sense and move along chemical gradients. Taste and smell have obvious similarities and synergies, but the detector systems are not confined to the tongue and nose. Even bacteria emit a range of chemicals which impact across eukayrotes, plants, fungi and animals (Dunkel *et al.*, 2009).

Our subjective experience of smell tends to be one of gathering environmental information. Think how different it is for many other mammals. Think about how dogs sniff each other when they meet. Think about the way cats are dominated by the smells in their territory, particularly the highly specific urine markers of other cats. There is another lifestyle out there, a lifestyle in which molecules, *semiomolecules*, are synthesised to be passed by one animal or plant to another as a signal. It might be a signal to mate, a signal that fruit is ripe and ready to eat, a signal of where other members of the group have found food.

Humans are not completely out of it, but their awareness of these signalling mechanisms is low. Controversy surrounds whether humans have the special organ in the nasal area, the *vomeronasal organ*, specialised for pheromones, found in many other mammals.

The yawning/snarling type of behaviour seen in lions, called the *flehmen* response, is a mechanism to draw air over the vomeronasal organ. Recently, though, some experiments have demonstrated equivalent functionality. It's a bit like the appendix, a relic of bygone times. But there or not, it certainly is not accessible to consciousness. We take this up further in §9.4.1.

Our knowledge of taste has made rapid strides quite recently. Usually regarded as a primitive sense, greatly dependent upon its coupling with olfaction, its complexity may have been underrated.

Many of the principles of olfaction also underlie taste. Taste and smell have a lot in common, but taste has been very much the junior partner in assumed importance and knowledge. We might think of the difference between them as smell being airborne but taste solution or solid delivery, but this is simplisitic: aquatic animals use their sense of smell underwater. The differences lie in the transduction mechanisms (§9.3, §9.5.1) and the associated neural pathways. Both are increasingly important as artificial sensors from pollution to drugs and explosives (Corcelli *et al.*, 2010).

Thus, from the perspective of humans interacting with a virtual world, audiovisual dominates. But from the perspective of an animat, a robot or an entity in a virtual world or computer game, we should not neglect the chemical senses. After all it's not only bloodhounds who are interested in scents. Many insects are dominated by the chemical environment and numerous computer algorithms exploit the smell-driven social behaviour of ants.

9.2   Physics of olfaction and taste

The first challenge for taste and smell is to transport volatile molecules, or just volatiles[1] to receptor sites. Diffusion through air is slow, and concentrations are frequently very low. So, olfaction is intricately coupled with breathing, with the nose specialised for forcing the air across the detector surfaces. Once inside solution, specialised molecular transport mechanisms come into play (§9.3.2). Nevertheless, the diffusion process is a slow one relative to the capture of light or sound.

With hearing and vision we had stimuli to which we could attach quantitative measures. We also found linear behaviour which enabled us to work in the Fourier domain achieving a simpler perspective. The chemical senses do not permit such luxuries.

The transduction of the chemical senses relies on molecule matching by receptor proteins, not dissimilar to the immune system. At worst we might have no way of reducing the dimensionality from just the number of different receptor proteins – rather a lot to consider. Fortunately the specificity does not seem to be this precise.

Chemists have many scales for ordering and classifying molecules. We can measure a molecule's polarity, its weight, its electronegativity, its aromaticity. But neither olfaction nor taste fall on any of these established scales. There is some evidence that odour

---

[1] Usually molecules of high vapour pressure and low molecular weight (Dunkel *et al.*, 2009).

varies with molecular weight and chain length, but this cannot explain all variability. Unfortunately we do not know how to classify arbitrary molecules according to the receptors they activate.

The intuition is that any molecule has shape components which will fit into a receptor, the *stereochemical theory* (§9.4). Peterlin *et al.* (2008) demonstrate this with a range of aldehyde molecules in rat olfaction and a detailed model of receptor-odorant interaction is now developing (Reisert & Restrep, 2009).

The three-dimensional (3D) structure of the odorants is a significant factor in the recognition process since stereoisomers may smell quite different. D-carvone, for example, smells of spearmint, while the L isomer, L-carvone smells like carraway (Kandel *et al.*, 1991, p. 512). Odour similarities like this are quite common. The distinctive smell of almonds comes from small traces of cyanide, although you would have to be seriously addicted to almonds to get poisoned. Benzaldehyde has exactly this same smell yet it is a completely different molecule and far less toxic.

Although empirical work with animal sensors will yield increasingly precise information about the relationship between G proteins and the molecules which bind to them, data mining is also proving a useful tool. Dunkel *et al.* (2009) mined publication databases looking for chemical structures and odour attributes to assemble the *SuperScent* database. This can be searched by chemical structure.

Thus measuring the stimulus diversity is in the hard/unknown basket at present. It is not surprising that smell and taste are largely associative – we link smells with memories and emotions. It is also not easy to assess the physical limits. Moths are sensitive to pheromones, for example, at around 10^3 to -10^4 molecules, or just five molecules in 10 cm^3 of air (Angioy *et al.*, 2003; Anon, 2003). In vision we know that single photons will activate a rod and the indications are that just one or two pheromone molecules can activate a moth olfactory neuron. Vertebrate olfaction is somewhat less sensitive. Thus the limitations are set by transduction noise rather than quantal limits, which is thought to be a general property amongst vertebrates (Lowe & Gold, 1995). The human nose has a threshold of around eight molecules for detection and less than a hundred for a conscious recognition of a smell (Burton, 1976, p. 108).

9.3 Olfactory anatomy and physiology

Olfaction begins with the capture of molecules from the air by specialised olfactory neurons in the nose, in the *olfactory epithelial layer*. A huge number of such neurons enter the *olfactory bulb* where extensive preprocessing occurs. The *lateral ofactory tract* carries this processed information into the brain, both directly to memory and emotional areas, and to the thalamus and hence to the cortex. In fact, it was only quite recently, in the 1960s, that Powell (Keverne, 1982) discovered this additional, non-thalamic pathway.

Figure 9.1 outlines the pathways.

Despite the relative simplicity of olfaction, we shall see that many of the information processing issues are unresolved. The development of artificial noses rests more on empirical data than theory. Besides issues of just how data is encoded at the periphery,

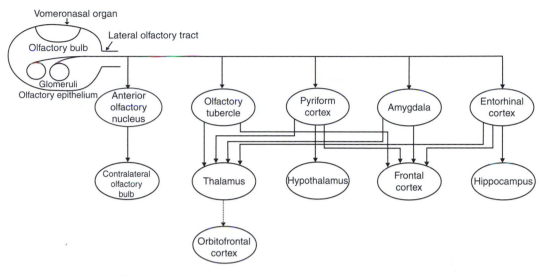

Figure 9.1 Olfactory pathways.

olfaction exhibits complex neural dynamics for pattern storage and recognition, which are again not fully understood.

Even though there are still issues to resolve in theoretically mapping odours to molecules, there are now databases of human odorants, such as *Flavornet* (Flavornet, accessed 2010), *ScentBase* (ScentBase, n.d.) (floral) and a general database of pheromones and other semiochemicals, *Pherobase* (Pherobase, n.d.).

9.3.1   The front end

The nostrils provide a degree of amplification of the odorant signal before it reaches the olfactory epithelium. Air sucked in through the nose is usually channelled through a restriction in the pathway to concentrate the airflow over the transducer area. Large noses, like large ears, usually indicate a well-developed sense of smell, although sensitivity is independent of airflow! There are other olfactory specialisations in the animal kingdom. The forked tongue of the snake is designed to pick up odours.

9.3.1.1   Stereo smell?

We have two eyes and ears for directional information processing and it seems that the same is true to a much weaker extent for smell. Early experiments done by von Békésy, quoted in Keverne (1982, p. 423), find a human differential sensitivity of about 10% in concentration between the nostrils or about a 1 ms difference in arrival time. Much later results from Rajan *et al.* (2006) do give rats stereo information from smell (i.e. directional information is available from a single sniff), but the sensitivity is far lower with a neural timing selectivity of around 50–100 ms (around one sniff of 125 ms).

Sharks also obtain directional information from olfaction, by inferring a source from properties of swirls of odour plumes in the water, so-called *eddy chemotaxis*. However,

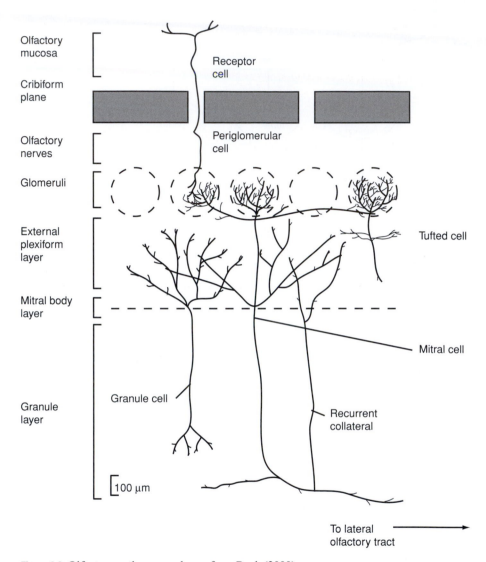

Olfactory
mucosa

Receptor
cell

Cribiform
plane

Olfactory
nerves

Periglomerular
cell

Glomeruli

External
plexiform
layer

Tufted cell

Mitral body
layer

Mitral cell

Granule
layer

Granule cell

Recurrent
collateral

100 μm

To lateral
olfactory tract

Figure 9.2 Olfactory pathways, redrawn from Buck (2000).

this mechanism requires integration from the lateral line, which picks up vibrations (and serves as the hearing mechanism for a lot of fish), but also serves in the detection of turbulence (*rheotaxis* (Gardiner & Atema, 2007)).

9.3.2   Olfactory epithelium

Figure 9.2 shows the components of the front end of the olfactory system. At the back of the nose is a thick, mucous covered layer of cells, the olfactory epithelium (OE). It varies in area from species to species. Man has about 2–4 mm^2. Dog as you might expect as about $18\,\mathrm{mm}^2$, but cat is even better at $21\,\mathrm{mm}^2$, obviously with some variations between species.

Bowman's glands in the nose produce the mucous covering the surface. The mucous plays an important role in absorbing the odours and clearing them away. Diffusion of molecules through this mucous will takes time, but the process may not be so simple. *Olfactory binding proteins* are thought to be involved in the transport of hydrophobic molecules and also in protecting the cells from excessive quantities of odorants – a front-end adaptation mechanism[2]. This is analogous to the impedance matching we saw in the ear. The mucous and binding proteins serve to capture and concentrate odours onto the receptors. The binding of an odorant to a receptor causes the release of a G protein and a cascade of events leading ultimately to spike activity. The transduction system uses different cascade components compared to vision and unlike vision, the mammalian olfactory epithelium is fully developed at birth.

The OE has four distinct zones, with different receptor types scattered randomly through the zone. However, each receptor type is confined to a single zone, at least in mouse (Buck, 2000). But this same structure appears in the olfactory bulb. Its function, if any, is as yet unknown (Buck, 2000). As Freeman (1999) points out, the turbulence of the incoming air will make the pattern of stimulation of the receptor array different for every sniff. Thus we should expect a random receptor distribution.

In itself this would not be a problem. We would just learn to associate whatever pattern of stimulation that occurred in the epithelium with the corresponding odorant. But, unlike all other neurons in the brain or sensory periphery, olfactory neurons have a short life: they change over every 60 days or so (Buck, 2000). Worse still, their replacements are more or less randomly chosen from the entire repertoire of receptors. So the spatial distribution of odorant cells in the epithelium seems to be changing all the time! There has to be some mechanism by which a new cell in the olfactory epithelium knows which cells to connect to higher up the chain. We do not yet know what it is, but §9.3.6 reveals that the mechanism is innate.

Embedded within the epithelium are about 50 000 olfactory neurons, the *transducing* cells. The olfactory neurons are bipolar, with dendrites ending in the olfactory epithelium and axons projecting to the olfactory bulb. Inside the epithelium, the dendrite grows into a blob from which a dozen or so cilia project into the mucous. The cell projects through a bony structure, the *cribiform plate*, to the next processsing stage, the olfactory bulb, of which there are two, one for each nostril. As shown, there are several distinct layers within the bulb, each of which has an information processing role, before the *lateral olfactory tract* (LOF) carries information away to the brain.

9.3.3   Receptor proteins and their genetics

There are about 1000 genes, 1% of the human DNA, making olfactory genes one of the largest single gene families known (Axel, 1995). The olfactory receptor cells contain proteins sensitive to classes of molecules, similar to the way antibodies are specialised

---

[2] Similar *retinol binding proteins* exist in photoreceptors (§6.5.1).

to recognize particular viruses or other invaders, but they are not so specific. As we saw in §9.2, the shapes of molecules or pieces of molecules determine receptor sensitivity, so that two molecules such as benzaldehyde (single ring aromatic compound) and cyanide (tiny two atom molecule) smell the same, they are effectively *olfactory aliases*.

Pig has an unfortunate olfactory alias itself, forcing it into a frustrating and never fulfilled search for partners underground. Truffles, the ultra-expensive and highly prized gourmet delight, have a smell which, to pig, is extremely similar to pig sex pheromones. Thus pig is a willing ally in the French search for the delicacy, although the frantic digging is not just loyalty to, or fear of retribution from, his owner (Ackerman, 1990).

Sometimes defective genes may make an individual insensitive to some odorants, a pathology known as *anosmia*. Strokes and other injuries may also cause anosmia, distinct from *hyposmia*, the diminished sensitivity we get from a variety of conditions, including genetic factors and blockages in the nasal passages, say, from a cold. Surprisingly about 50% of the population are hyposmic to urine, while a few are insensitive to much more dangerous chemicals, such as cyanide.

This is *metaphorically akin* to the aliasing phenomena discussed in Chapter 3. But it is fundamentally different, since there is no 'odour space' where we can represent the vast array of stimuli as points on a series of coordinates[3]. We can see this more clearly by referring back to the way birds and reptiles transduce colour. Their photoreceptors contain a myriad of colour droplets, each with its own photopigment. But there are still only three or four pigments and the visual system *does* decode colours as points in a metric colour space. If we had thousands of different colour droplets, each representing a colour, red, green, purple, fuschia, whatever, and if we looked for activity in the fuschia droplet for presence of fuschia, we would have something closer to smell. So the spatial patterns on the bulb are points in the space of different components, unlike the topographic map of colour in the retina.

We do not yet know how many different types of olfactory receptor (OR) are present in each olfactory neuron. Earlier work had suggested as many as a couple of dozen, but current opinion favours just one – the *one OR-one neuron* rule. Clever genetics has shown that the receptor protein itself causes negative feedback to the genome, preventing the occurrence of multiple OR in a given cell (Serizawa *et al.*, 2003). What is particularly cute about this mechanism is that if a defective gene gets initially selected, it does not generate a viable protein for feedback and another gene thus gets selected and the neuron gets a receptor protein after all (Lewcock & Reed, 2003).

But what is interesting, and still not well understood, is how the identity of the odour is transmitted. In hearing and vision we have spatial information, on the retina or the basilar membrane, which defines important characteristics of the signal of any given cell. In the olfactory epithelium, there are distinct hot spots, with high sensitivity to particular odorants. But in no way is the olfactory epithelium a regular odorant map,

---

[3] There is a slight qualification here: an odorant might stimulate several receptor sites and be somehow represented as the sum of the activity at each site. But the number of dimensions of this space (the number of receptor proteins) would be so huge compared to the number of odours we recognise that this is not a useful avenue to explore.

at least according to present understanding. A further complication is that odorants can inhibit one another, leading to complex perceptual properties of mixtures (Oka *et al.*, 2004).

A regular organisation of the olfactory epithelium into patches corresponding to particular odours (receptor proteins) perhaps would not work very well anyway. The air currents in the nose generated by sniffing are turbulent with erratic variations of concentration. Thus the overall sensitivity is likely to be higher by distributing receptors across the whole area. Add to this the regular growth of new cells with different receptor sensitivity, and we can see that the signal coming out of the olfactory epithelium is a somewhat variable feast. The bulb is even worse (§9.3.5)!

We do not know the number of receptor proteins precisely. In mouse and rat it is around 1000, in humans around 750. In zebrafish and catfish, which are popular experimental animals for olfaction, it is much smaller, probably around 100.

In humans there is an additional consideration. Many of the genes, between 38 and 76%, are *pseudo-genes*. Such genes do not code for full peptide sequences and thus might be dysfunctional. Thus there has been considerable degeneration of the human sense of smell over time (Keverne, 1999; Mombaerts, 1999). The perception of human pheromones is also made more dubious in that most of the vomeronasal genes so far identified are also pseudo-genes. This issue is still somewhat controversial (Mast & Samuelsen, 2009). An interesting perspective on this is the space pressures in the brain which may have caused this loss of olfaction, an issue discussed in §12.4.10.

9.3.4 The olfactory bulb

The olfactory neurons project through the cribiform plate to the next processsing stage, the *olfactory bulb*. As we noted earlier, transport of odorants through the mucous is an inherently slow process, so fast nerve transmission at this stage is not vital. Thus we might expect the olfactory axons to be unmyelinated and this is indeed the case. Lack of a myelin sheath allows a greater packing density and saves on the cost of additional support cells. The neurons are bundled together in groups of 10–100, the bundles surrounded by Schwann cells.

Around 50 million unmyelinated nerve fibres arrive in the olfactory bulb and enter 2000 or so glomeruli in mouse. Inside each of these subunits there are two categories of cells, the first of which are responsible for information processing within the bulb itself, while the second carry the bulbs output:

1. The *granule cells*, short-axon cells and *periglomerular cells* connect within the bulb and the *external plexiform layer* (EPL) and have a local inhibitory function.
2. The mitral and tufted cells project out from the bulb in the lateral olfactory tract. The tufted cells are found in the EPL, while the bodies of the mitral cells are located in their own layer.

Within each glomerulus we have convergence of about 1000:1; 25 000 fibres coming in with about 25 mitral cells going out. As shown in Figure 9.2, the first stage of processing is the bulb in which the olfactory neurons make contact with the *apical dendrite* of the mitral and tufted cells. This contact is excitatory. Each mitral cell must therefore make contact with 1000 olfactory neurons. Each type of olfactory neuron synapses onto a single glomerulus, while each mitral cell has a primary dendrite into just one glomerulus too (Buck, 2000).

Since a mouse glomerulus receives input from just one or two receptor types, the 25 output cells represent considerable redundancy. Replicating cells is not a very efficient way of providing a higher signal to noise ratio (which goes up as just the square root of the number of cells). There is some subtlety awaiting elucidation.

The situation in humans is still unravelling. Maresh *et al.* (2009) show that compared to mice with over 1000 odour receptors, in humans there are only around 350. Yet these converge to a larger number of glomeruli, approximately 5500, with considerable variability. Thus the convergence ratio goes up to 16:1.

The periglomerular cells are located within the glomeruli and their axons project out to other glomeruli. They receive excitatory input from the olfactory neurons, but their principle action on the mitral and tufted cells is inhibitory.

The granule and short-axon cells occupy a thick layer and project into the EPL where they make many contacts between the mitral cells, again, inhibitory in nature. The mixture of recurrent excitatory and inhibitory connections is essential to create the oscillatory carrier waves which characterise the bulb dynamics (§9.3.5).

This whole network of cells is rather analogous to the retina, where a range of cells preprocess the photoreceptor output before the ganglion cells of the optic nerve. In the retina, however, the relationship of the ganglion cell output to the photoreceptors is fairly straightforward – the spatial topography remains the same and the intensity/frequency tuning is simple. In the olfactory bulb it is a quite different story.

The projections out from the bulb to the olfactory cortex now seem to be genetically determined. Work in Buck's lab has shown recently that the patterns of connectivity from glomeruli to cortex are identical across mice of a particular species. Work by Luo and colleagues shows a similar result in fruit flies. The pathways out from the glomeruli develop in sequence, one glomerulus at a time and thus ensure a reproducible organisation of the pathway.

9.3.5   Coding in the olfactory bulb glomerulus

Although the chemical senses are in many ways the simplest, the coding mechanisms relate less readily to engineering concepts and are still some way from being understood fully. Hence, in this section we capture some of the important ideas and trends in theory, but, alas, do not have any final answers. A great deal of research here is carried out on rabbit, but the principles seem to hold for other common experimental animals such as rat and cat and are likely to hold for primates and humans too.

In the absence of stimuli, the neurons of the olfactory bulb gently tick over in a chaotic pattern of activity, a primed, waiting state. Sniffing an odour synchronises this chaotic activity into a global oscillation of activity across the bulb, in the so-called *gamma* range of about 20–80 Hz. This *carrier wave* is common across the whole bulb and arises from negative feedback. But the amplitude varies from place to place, known as *amplitude modulation.*

Now, we might assume that these amplitude patterns would be the memory for particular odours, but this is not the case. The patterns drift with time and other olfactory experience over a period of days. There is no correlation between areas of bulb and individual odorants. In fact, a lot of the bulb may be lost through disease or surgery but odour discrimination remains.

Friedrich and Laurent (2001) propose an alternative redundancy reduction mechanism. Olfaction is slow, so, time is put to good use. The mitral cell outputs steadily change with time and exposure over a period of one to two seconds decorrelates the odour signal, similar to the predictive coding ideas we saw for the retina (§4.2.5). The odours move gradually apart to make the signals from each different odour more distinctive and therefore more reliably detected.

The bulb is a convenient point at which to assess the bandwidth of olfaction. With 2000 glomeruli, each representing just one or two receptor proteins, we can calculate the number of states using information theory, using identical analysis to the one we pursued for vision. If we let the signal to noise ratio of a glomerulus be ζ, then if there were no correlation between the glomerula signals, the information capacity would be approximately $H_{max} = \frac{N}{2} log_2(1 + \zeta^2)$ for N glomeruli.

But the signals of the glomeruli are strongly correlated because many odours stimulate more than one receptor. But unlike photoreceptor arrays in vision, where the correlation is local, here the correlation is more random across the bulb. Nevertheless some evidence is emerging that closely related molecules may correspond to neighbouring glomeruli (Tsuboi *et al.*, 1999). However, we still see the same effect of correlation. H_{max} is reduced to roughly $\frac{1}{2} log_2 R$ where R is the correlation matrix.

9.3.6   More on genetics

A major breakthrough in olfaction occurred just over a decade ago. Buck and Axel (Mombaerts, 1999) reasoned that the G-protein mechanism should require a particular class, referred to as 7TM, of receptor proteins. It thus became possible to search for these protein sequences in the genetic code.

Not all G proteins are actually expressed in the nose, however. An olfactory receptor protein, designated hOR17-4, appears in the testes and is used by sperm in *chemotaxis*, the chemically mediated direction of swimming, naturally enough, towards the egg (Spehr *et al.*, 2003). The nature of the attractant, and whether it comes from the egg itself, remains to be determined (Babcock, 2003).

This genetic approach enabled much progress to be made in understanding olfactory coding. Modern genetic techniques have given us a good understanding of the wiring between the epithelium and olfactory bulb. One approach is to attach marker genes to genes under investigation. When the gene which expresses a receptor protein is active, so the marker gene also becomes active. In elegant experiments Axel (1995) and coworkers used a marker gene which would cause the neuron in which it was expressed to turn blue. Thus blue pathways are clearly visible. This and other evidence demonstrate that the olfactory neurons sort themselves out so that each of the 1000 or so receptor molecules connect to just two specific glomeruli in the bulb, even if the spatial path to be followed is a tortuous one (Treloar *et al.*, 2002). Unlike vision and auditory map organisation, which is driven by early experience, the glomerular map forms without sensory activation (Dulac, 2006). The mechanism by which they follow this path is becoming clear too. Barnea *et al.* (2004) show that the receptor proteins (in mice) are themselves expressed on the dendrites and the ends of the axons. But the details are still not entirely clear, since it may not be actually the protein itself but other messages from a cascade triggered by the protein (Imai *et al.*, 2006).

9.3.7   Complexity of odours

Bringing these elements together is tricky, and our speculation here could turn out to be wrong. Stripping away the complex dynamics of the implementation, what we have is an *associative memory* (§2.4.5). But the strategy seems to be something like this. Each odorant forms a dynamic pattern of activity in the olfactory bulb. These patterns are not constant. They change with time, and are influenced by other odorants sniffed in-between.

The dynamic pattern is *an attractor* in dynamical terms. Associated with the attractor is the *basin of attraction*. If a dynamical system enters the attractor basin, it will ultimately end up on the attractor after some time. The attractor basins undergo regular distortion and the attractors themselves change. This drift and flexibility is probably advantageous in the processing of mixtures of chemicals which may not always be quite the same.

But, by some coupling mechanism we do not fully understand, a given attractor will generate the same dynamics within the brain. Thus recognition of an odour is stable over the lifetime of the animal (Freeman, 1999).

This attractor model gets very complicated when we have mixtures of chemicals. Psychophysical and imaging studies indicate that the signal and processing of mixtures is similar to that of single odorants, i.e. it is the learned pattern not the chemical complexity which is important.

We would get a combinatorial explosion if the bulb and cortex were to have different attractor states for each possible mixture. This suggests that odour processing is sequential. A consequence of this is that adaptation has to occur very rapidly. But it may be the case that some odours, perhaps through genetic predisposition, form very strong attractors, from which it is hard to escape. Some smells linger!

The number of neurons and the neural capacity seems to be way above what we might expect for the number of smells we can discriminate. Yet, the simple Hopfield model would predict a number of attractors roughly equal to the number of neurons and this is what we deduced above. In effect, the time dimension is being utilised in the definition of the attractor, soaking up a lot of potential bandwidth. We are getting much more of the pattern recognition processing at the front end, rather than deep inside the brain.

Unlike vision and hearing, smell (and taste) are highly non-linear. As we saw in Chapter 3 the characteristic of non-linear systems is that a stimulus will change the system itself. Our receptivity to odours is very dependent on history, what we have sniffed in the immediate past (§9.4). Odours may also change their character with concentration and we do not yet have any precise mathematical models for making a smell percept from a precisely defined mixture.

9.3.8 Sparse coding models

Olfaction is similar across a big part of the animal kingdom, and insect smell has much in common with mammals. Its strong similarity in structure in insects and mammals belies the likely separate evolution. The olfactory system is *much* simpler than other senses such as vision, with just two as opposed to up to 100 major brain areas (Singh, 2003). Insects can thus be very good models for studying coding as we found in the discussion of spike entropies and information (§4.6.2). The locust has been quite useful in understanding the transition from population to *sparse* coding.

The analogue to the mammalian nose in locust is the antennal lobe which contains around 90 000 olfactory receptor neurons feeding around 1000 *projection neurons* (PNs). PNs project to the mushroom body which contains around 50 000 neurons known as Kendal Cells (KCs) (Perez-Orive *et al.*, 2002). The neural dynamics in the two are quite different.

Perez-Orive *et al.* (2002) show that the PNs have high resting rates of discharge. When an odour is detected, large groups of them fire and there is oscillatory locking of the spike trains. Thus each individual PN is itself very undiscriminating to odours. Not so with the KCs, which are much quieter at rest. They fire only a few spikes in response and are much more discriminating to odours. Thus the actual memory uses a sparse code. Each odour is represented by a small population of cells out of a big population and each KC acts like a coincidence detector of incoming spikes.

At present the views of olfactory coding are thus quite divergent and much remains to be unravelled.

9.3.9 Olfactory pathways into the brain

As we hinted earlier, olfaction has two distinct pathways into the brain, and, interestingly, these differ between primates and other mammals. The differences concern the vomeronasal organ, the specialised part of the sensory apparatus which handles pheromones. At the time of writing it is debated as to the existence/status of this organ in man. In mouse there is not only a distinct vomeronasal organ in the nose, but also an accessory olfactory bulb to which pheromonal information is fed. Activity of cells in the olfactory bulb is linked to direct face to face or face to genital touching, but whether this is mediated by the somatosensory system is not clear as yet (Luo *et al.*, 2003).

The olfactory cortex is distributed, consisting of the five areas shown (Figure 9.1). The first pathway is the usual sensory pathway we have seen before – to the cortex via our old friend the thalamus, the sensory relay station. The lateral olfactory tract travels to a relay station, the *olfactory tubercle*, from whence it proceeds to the *medial dorsal nucleus* of the *thalamus*. From the thalamus it then proceeds to the *orbitofrontal cortex*.

The indications are that it is this pathway that is responsible for our conscious sensation of smell. Stroke damage to this pathway will create anosmias.

Attempts to distinguish conscious awareness pathways substantiate this thalamic pathway but not unambiguously. Sobel *et al.* (1999) spefically examined the issue of blind smell with a putative human pheromone. Using fMRI they found activation in several brain areas at both a low and high concentration. There was weak awareness of the high concentration, no awareness of the low. The thalamic areas showed the greatest differential activation.

The second pathway diverges before the olfactory tubercle and heads to a region known as the *pyriform cortex*. Recordings of cellular activity here imply that this area is part of the associative memory which begins with the olfactory bulb. It is usually referred to as *allocortex*, appearing somewhat different to most of the other cortices. From there it heads to the *amygdala*[4] and the *entorhinal cortex*[5]. The amygdaloid complex is associated with emotional responses, particularly fear. From the entorhinal cortex, the pathway continues to the *hippocampus*, the seat of long-term memory creation[6].

The amygdala project to many areas of the cortex and receive input back. They also receive projections from the thalamus. Thus for the other senses, the fear response goes via a sensory signal up to the thalamus and then back to the amygdala. In olfaction, the route is more direct. This additional pathway is undoubtedly an evolutionary left-over. But, with the thalamic pathways having developed, it could have just atrophied as no longer needed. But, smell is inherently a slow sense. Maybe, just as using a direct thalamic pathway for a fear response will be faster than using the cortex–amygdala pathways, this direct route will give a faster response to the smell of an encroaching tiger or other threat to life.

As with all the senses, there are fibres which project back to the periphery. In the case of olfaction, they seem to be able to enhance the sensitivity to particular odours, e.g. to food when the animal is hungry.

We can now offer a suggestion as to why we can detect but are not conscious of human pheromones. Unlike the many odours we learn through our lifetime, a pheromone is a specialised molecule, both generated and detected by the human being itself. It is thus quite plausible that this signalling system pre-dates man and has survived intact from our ancestors of long ago.

It may go back so far, that the information is processed only with the memory and emotional centres of the brain and not passed onto the cortex via the thalamus. Thus we experience these atavistic attractive impulses without the intervention of our conscious brain.

---

[4] Another quaint name, this time coming from the Latin word for almond, the anatomical appearance of this brain area.

[5] This general area was at one time called the limbic area but is now referred to as the rhinencephalon, meaning, literally 'smell brain' (LeDoux, 1998). However, in higher mammals, smell is only one of many functions.

[6] The hippocampus is involved in long-term memory for many brain functions. It would be misleading, though, to consider it the location of long-term memory – those locations are scattered throughout the brain. It is more of a control centre controlling the laying down of long-term memory and establishing context.

This view implies that perfumers adding boar pheromone to human scent are not far from the mark: our pheromones are likely to be quite close to those of other mammals!

9.3.10 Sensing dietary deficiency

Many animals are good at sensing dietary deficiencies. Cats, for example, have to get the amino acid taurine regularly, otherwise they suffer serious harm. But this acid is not that common, driving cats to continually search for a variety of food, as any cat owner will testify!

But this sensing of missing amino acids does not go via the conventional chemical senses, taste and smell. It involves a direct pathway, where absence of the amino acid causes a chemical cascade inside the neurons of the anterior pyriform cortex (APC). Output from the APC drives the animal to seek suitable food (Hao *et al.*, 2005).

9.4 Psychophysics of olfaction

Our understanding of perception in the sense of smell is somewhat less developed than for the other senses. Smell is clearly very non-linear, adaptive and has a strong hysteresis. Thus we become accustomed to smells very quickly. The first glass of wine always has the best bouquet. Our perception of an odorant depends on what we have recently sniffed. In fact the whole gourmet experience rests on supplying foods in the right combination and in the right order!

Experiments with mixtures of odours show some linear behaviours, some non-linear. Binary mixtures tend to create the same impression at different concentrations providing the ratio is kept constant. But the percept is somewhat non-linear, switching abruptly from dominance by one to dominance of the other (Brodin *et al.*, 2009). But there is an even less linear behaviour. One chemical such as undecal may act as an *antagonist* in that it blocks a receptor for another molecule but does not itself generate a strong response from that receptor, creating an asymmetry in the response (Brodin *et al.*, 2009).

From our study of hearing and vision, we might look for a set of primary smells, like the primary colours. So far we have found no such set, and, when we look at the anatomy, it will become clear that this is a risky search. On the other hand, for artificial systems, such a system would be extremely useful. Imagine how useful it would be to be able to send the bouquet of a great wine over the internet for an online wine course. But realise at the same time how improbable this is, given the huge resources, experience and expense tied up in the greatest wines. We shall look at progress on artificial olfaction later (§9.6).

J.E. Amoore (1971) proposed a stereochemical model of smell: odours with a particular shape have a corresponding smell, e.g. musks are disc-shaped molecules while peppermint are wedge-shaped. The stereo-specificity of odorants is clear (§9.2) but the idea of classes and shapes has not held up in more recent work. Amoore's classification had seven basic odours: minty (peppermint); ethereal (delicate fruit-like scents); floral; putrid (rotting food); resinous (camphor); musk resinous; and acrid (smoke, vinegar).

Studying human anosmias reveals 8–30 independent or primary odours. But whereas in vision we can relate primary colours directly back to photoreceptor pigments, we have no such information for olfaction. These primary odours may have no direct neurological correlate, but they are essential tools of the perfume industry (Burton, 1976).

Perfumers, like wine connoisseurs, have a professional need for well-developed smell. They are thought to be able to recognise some 10 000 different odours.

SuperScent (Dunkel *et al.*, 2009) uses an odour tree with 29 base odours with subcategories (such as types of fruit) with a total of 121 groups to map 2147 compounds at the time of publication.

Surprisingly though, the scales used by professionals seem to match rather poorly those of amateurs (Jellinek, 1992). For unidentified perfumes, amateurs produce different groupings of the perfumes to experts. Expert categories include floral, chypre (from Cyprus) and oriental common to most of the perfume houses. Aldehydic and green are also commonly used. However, in the experiments reported by Jellinek, the groupings of a set of ten perfumes by amateurs were quite different to professionals. Furthermore, only the category floral seemed to be universally agreed upon. Chypre was a complete mystery to many, as it would have been to the author!

It seems fair to say that our understanding of the representation of smell is about where our understanding of colour was before trichromacy became established. We just don't have a good model for the real dimensionality of smell.

In vision, a person's face may take on many different views. The light reflected from it will change in wavelength dependent upon the illumination. Little variations in colour, little blemishes, facial expressions all add together to make a very complex pattern recognition task. The complex geometrical pattern analysis does not take place in olfaction. But an odour is just an odour, even though it may be made up of many components: fruity smells come from clearly identified esters. *Penguins smell starkly penguin, a smell so specific and unique that one succinct adjective should capture it* (Ackerman, 1990).

Conversely, memories for smells last a lifetime, and in some animals early imprinting is a key to future life-long behaviour[7] and associated genes were identified recently and operate generically *after* the olfactory neurons (Remy & Hobert, 2005) (§9.5.2). As we know (§9.3.9) olfaction projects directly to the emotional and memory senses of the brain. Smells are powerful triggers, sometimes memories of very long ago. As Diane Ackerman (1990) puts it:

Smells detonate softly in our memory like poignant landmines, hidden under the weedy mass of many years and experiences. Hit a tripwire of smell and memories explode all at once.

Ongoing research seems to suggest that some bad (dangerous) odours and some, but by no means all, pleasant relaxing smells are innate in humans (Merali, 2006).

The pioneer of olfactory neuroscience, Walter J. Freeman, had a particular philosophical take on this in a recent book (Freeman, 1999, pp. 120–121):

These findings have a profound implication. The only patterns that are integrated into the activities of the brain to which the sensory cortices transmit their outputs are those patterns they have constructed within themselves . . . They are not direct transcriptions or impressions from the

---

[7] Imprinting occurs across the animal kingdom, including worms such as *C. elegans* on which much basic neurological experimentation is carried out.

environment inside or outside the body. All that brains can know has been synthesised within themselves, in the form of hypotheses about the world and the outcomes of their own tests of the hypotheses' success or failure and the manner of failure.

This pattern recognition starts to make a little more sense when we realise that most of the time the important stimuli are blends. Strawberries do not smell of one particular chemical. The bouquet of a great wine is a mixture of which we understand very little.

9.4.1   Signalling

Pheromones[8] abound in the animal kingdom. They can be sexual attractants, found in species as diverse as moths and higher primates. They may mark territory in the way that cats mark territory with urine or by rubbing their faces and releasing chemicals from the glands in their cheeks. Elephants secrete sex pheromones in their skin. The Indian practice of *abhynanga* is a rubbing down of female elephants to extract the aphrodysiac to attract the male. However, only in the last few years has evidence for pheromones in birds been found. Hagelin *et al.* (2003) found social odours in a seabird, the crested auklet (*Aethia cristatella*).

Chemicals may also communicate between species. Racoons are exceptionally sensitive to the point at which fruit, such as grapes, are ripe to eat as a result of the chemicals released by the fruit. A particular cactus in the Bahamas flowers once, overnight, and vast numbers of sphinx moths swarm in attracted by the scents which are the key to pollination (Ackerman, 1990).

Blends of chemicals are often used as attractants, since they convey some evolutionary flexibility. A fruit, such as a strawberry, has a distinctive smell, made up of a range of different odours (although in some cases a particular compound may be dominant). In a fine wine, balance is often the key to excellence, where no one characteristic dominates the others. These different components act a bit like basis vectors, allowing for the generation by genetic mutation of a large range of possible odour complexes. A change to one or more component can give a different response, allowing for diversification of intra-species signals or symbiotic relationships. It was therefore interesting to find that some orchids rely on just a single molecule, in fact the actual wasp sexual pheromone, to attract specific pollinating wasps. Since there are some 300 such orchids, the explanation for evolutionary diversity is still awaited (Barnea *et al.*, 2004).

Artificial pheromones are being exploited for taming premenstrual syndrome and premenstrual dysphoric disorder in women (Zandonella, 2001). These conditions are thought to arise from hormone fluctuations, which are stabilised by pheromones. A California company, Pherin Pharmaceuticals, has developed a pheromone-like chemical, PH80, which seems to have the right effects. fMRI scans also show that PH80 causes activation in the same parts of the hypothalamus as premenstrual syndrome.

Olfaction plays an important role in navigation in many species. Two mechanisms have been advanced (Burton, 1976). One is a precise memory of the natural odour of the environment, the quality of the water, and a sort of odour map which will guide them up

---

[8] From Greek *pheren*, carry and *horman*, excite.

the right streams. The guidance may be pheromonenal – relying on very tiny quantities of pheromone which linger in the stream beds and banks and the water itself from season to season. Pheromones also seem to operate in lamprey eels, where navigation to spawning sites is instinctive.

In salmon, though, navigation cues are learned. Salmon return to streams to spawn and have an exceptionally accurate memory for where they came from. Shoals return there year after year to spawn. Hasler and graduate student Scholz demonstrated first the imprinting of odours and the use of odours as a navigational mechanism (Barinaga, 1999)[9]. By exposing salmon with two distinctive chemicals, not part of the natural environment, morpholine and phenethyl alcohol before migration, the salmon would preferentially (as high as 90%) pick the stream marked by this chemical when they returned to spawn several years later.

The *imprinting* of the odour to home in salmon occurs at the time of surges in the release of thyroid hormone. The first surge occurs immediately after birth, imprinting the area where they hatched. The second surge occurs during the smolting phase, about a year after birth, just before the salmon begin the migration.

The imprinting is a peripheral, rather than a high level phenomenon. Recent evidence (Barinaga, 1999) points to increased activity in the olfactory epithelium in response to the imprinting chemicals. Similar results are known for mice, while rabbits also display imprinting for the food their mothers eat while pregnant and nursing.

Fish also use odours for navigation to locate food sources. Dimethylsulfoniopropionate (DMSP) is released by phytoplankton and algae around coral reefs and serves as a lure for various sorts of fish (DeBose *et al.*, 2008).

Birds were long thought to have very little sense of smell, based partly on the seeming lack of olfactory neurons in the brain[10]. However, Bang and others in the 1960s discovered that although forest-dwelling birds might have as little as 3% of the brain devoted to olfaction, seabirds often had a great deal more, as high as 37% (Malakoff, 1999). Antarctic prions, *P. Lachipltila desolata*, are able to recognise both their own odour and that of others such as their (life-long) partner (Bonadonna & Nevitt, 2004).

Since then many other new results have come in. Pigeons, well known for their homing instinct, seem to be dependent to some extent on olfaction. Inhibiting their sense of smell greatly reduces their navigational accuracy and it now appears that it may be odour topographic maps rather than magnetic information which guides long-range navigation (Gagliardo & Wild, 2006). But perhaps most interesting is the long-range tracking behaviour of albatross and other ocean wanderers. They feed from patches of shrimp-like prey, which in turn feed on phytoplankton, and track the chemicals they release. They are now thought to navigate via an olfactory map (Malakoff, 1999).

---

[9] Fish also navigate by sound. Many of the fish which make up reef populations locate reefs to inhabit by the sound they make. Cardinal fish (*apogonids*) track the sounds made both by fish (low frequency) and shrimps (high frequency) whereas damsel fish (*pomacentrids*) prefer high frequency (Simpson *et al.*, 2005).

[10] Since chemical senses go way back in evolution, one might ask if any animals lack a sense of smell. In fact whales and dolphins gave up olfaction in the evolution of their blowholes. They do, however, retain cranial nerve zero (most likely for pheromones) (Fields, 2007a).

Corsican tits surround their nests with various aromatic herbs to ward off mites and other invaders. Recent experiments demonstrate that it is the smell that guides the selection of herbs rather than colour or mechanical properties (Holden, 2002a).

Paper wasps, *Polistes fuscatus*, recognise kin from the smell of the nest. The variety of plant material picked up from the local environment makes the odour of the nest quite specific. This odour gets absorbed into the skin of the wasp at birth and provides it with a nest specific odour. Similarly the carpenter ant, *Camponotus japonicus*, has a sensilium within the antenna specialised to recognise blends of cuticular hydrocarbon, which are nest specific. Non-nest mates promote aggression (Ozaki *et al.*, 2005).

Sea squirts (salps, tunicates) are brainless (Parker, 2006). They live in colonies on rocks in which all members have identical genetic makeup. It turns out that a gene in the histocompatibility complex (HC) area is responsible for whether a new tunicate is allowed to settle close to an existing one or is repelled by toxic release. House mice, *Mus musculus*, recognise smells which are again coded for in the HC genes. They choose mates which are different to avoid inbreeding.

Pheromones are not usually single compounds but mixtures. There is a clear advantage to using a mixture. Varying the mix creates a degree of flexibility in signalling and adaptability without the need for new pheromone coding genes. There are strong indications that the vomeronasal organ has evolved (and atrophied) independently of the main olfactory system. The genes, of which there are only about 100, code for quite different proteins, although the G-protein transduction mechanism still operates (§9.5.2).

9.4.1.1 Human pheromones?

So what about humans? We shall see later that the physiological and genetic evidence is somewhat negative. But the behaviour seems more positive (Small, 1999). At the time of writing, evidence is accumulating for non-conscious judgements about other people's odour, but whether it is a distinct pheromone mechanism or a general olfactory mechanism is unclear.

A group of male students were asked to sleep for several nights in the same T-shirt. Presumably these were among the more attractive males or the girls in the other half or the experiment paid well. They were asked to sniff the T-shirts and rank the men in terms of attractiveness! This bizarre experiment produced interesting results. The girls were clearly sensitive to the histocompatibility of the men. They were able, without realising it, to judge the genetic makeup and mating suitability from the odours.

A follow-up study showed that the sensitivity varies with hormonal state. Women who were on the pill or lactating showed a different preference. They were more interested in the same genetic makeup, perhaps because now their need is for the protection and support of kin. Again they were unaware of what was going on. In §9.6.2 we see an example of an artificial nose sufficiently sensitive to detect MHC variations in mice and may be used for human MHCs. This leads to interesting speculation as additional premarital tests before the contract is signed!

Very recently Ivanka Savic-Berglund at the Karolinska Institute in Sweden has taken a brain imaging approach, showing that men and women have different, non-conscious sensitivity to sexual odours. She observed the activity of the hypothalamus as testosterone and oestrogen wafted across subjects of both genders. Brain activity was much greater for the hormone of the opposite sex, oestrogen for men and testosterone for women (Savic-Berglund, 2010).

Customs still abound throughout human races which involve an exchange of body smells. Rubbing noses, hands in the armpit, deep kissing may all serve to transmit chemical information about one another. But if real information is transmitted, it is non-conscious. Unlike Al Pacino in *Scent of a Woman*, most of us can't even recognise specific perfumes let alone unique body odours.

One possible evolutionary advantage to sexual sensitivity would be for a male to be able to detect when his mate was ovulating, a time to be extra vigilant for possible interlopers. T-shirt experiments, again, reveal that men may be able to tell. Although experiments to determine if one can tell ovulating from non-ovulating women have proved negative so far, experiments where men have to make a judgement based on T-shirts worn by the same female before and after ovulation yield positive results. Men prefer the odour of the shirt worn during ovulation! Again, this is a non-conscious discrimination, without being able to specifically say, one smells like vanilla, the other like ripe strawberries (Holden, 2001).

Keeping in mind that the genetic/physiological basis for these findings is still hotly disputed (whether in fact there is any vomeronasal-like sensitivity left in humans at all), some recent findings in mice make sense of this data. Leinders-Zufall *et al.* (2004), in an impressive demonstration of lateral thinking, guessed the molecule class which activate part of the vomeronasal organ (VMO) in mice. The VMO has two distinct regions, apical and basal. The apical, with about 150 associated genes, expresses a particular type of receptor known as V1R with a particular G protein, $G\alpha i2$ which is sensitive to mouse pheromones, urine and similar things. But the basal region has more genes, expresses receptor V2R with G protein $G\alpha o$. Identifying what region might be sensitive out of all the possible molecules a mouse might come across is pretty challenging. It turned out that the MHCs were the triggers needing around 50 receptor types to represent all of them for mice (so there might be other molecule classes too). Thus we get a low level physiological mechanism for how we *might* be able to make non-conscious reproductive and immunological characteristics.

The last part of the story is cranial nerve zero. This nerve is smaller than the other cranial nerves, and was sometimes thought to be a branch of the olfactory nerve one. It now seems that this is indeed a seperate nerve and has a strong role in sexual behaviour (Fields, 2007a). It is still operational in humans. Liberles and Buck (2006) found evidence for a new class of receptors, so-called TAAR (trace amine-associated receptors), in mice. These receptors are activated only after puberty. They found that six of these receptors are live genes in humans!

9.4.2   Information content from psychophysics

If we can smell about 10 000 odours then we might say that the information content is around $log_2 10^4 \approx 13$ bits per sniff. This is way short to be compatible with the amount of neural wetware associated with olfaction.

At the other extreme we could imagine each of these 10 000 odours is a basis vector. So our estimate per sniff would now be 50 000. But this assumes that we can accurately discriminate the components in mixtures and their relative proportions. We certainly have great difficulty in naming components and judging level and there is a great deal of hysteresis. In fact we can only discriminate and identify 3–4 components of relatively simple mixtures even when they are well learned (Livermore & Laing, 1998).

The intensity variations to which humans seem to be sensitive are about 2×10^5 across multiple receptors, giving about 36 bits (Dusenbury, 1992). The number of smell components we can manipulate is very hard to quantify. At the simplest level we would say that it was somewhat equivalent to the approximate primary odours, say 20. This would give us 100 bits of information per sniff. This is far too low.

A better estimate for a basis might be to use the number of receptor proteins coded for in the genes. An upper estimate would be around 1000, giving us about 5000 bits per sniff. This corresponds to roughly 1 bit per neuron which is much lower than we get in other modalities. However, such a space would be extremely sparsely occupied, thus this does not seem a useful viewpoint. Whatever way we look at it, *olfaction is a low bandwidth sense* (Laurent, 1999).

9.5   Taste

The tongue in addition to holding the taste buds has many cells which are touch sensitive. The tongue is like a super-sensitive somatosensory organ. It has highly developed tactile and chemical sensors. The texture, the surface properties, of things we put into our mouths is also important to monitoring what we eat.

There are four or five categories of taste: salty; sweet; bitter; sour; and unami. The latter, a sensitivity to glutamate (occurring, for example, as the flavour enhancer monosodium glutamate or MSG), is still somewhat controversial.

Taste has an immediate survival value. Both bitter and sweet, at least, appear innate. Bitter is indicative of alkaloids, chemicals plants specifically manufacture to deter being eaten. Some of the most poisonous naturally occurring chemicals are alkaloids, such as the berries of yew trees and the deadly aconitine from monkshood, Britain's most toxic plant.

We seem to find fat an essential component of desirable food, from salad dressings, butter in sandwiches to the rich realm of French sauces. Although after the movie *Super Size Me* we might be a little wary of overconsuming fried food, fat is both an essential requirement to bodily tissues and a highly efficient fuel in terms of energy per gram. A new finding (Anon, 2005) of a receptor protein, CD36, in mice, sensitive to lipids suggests that we may sense them directly, rather than through their effect on the texture of food.

Plants have evolved all sorts of mechanisms for pollination. In some cases they solicit animal vectors, such as insects. In other cases they specifically need to exclude being eaten by animals. Thus the extreme toxicity of plants serves this purpose. The innocent looking fruit of the East India tree contains the alkaloid strychnine, well known as a viscous poison. Alkaloids usually have a strong bitter taste, hence our dislike and wariness of bitter tastes. As Dr Strangelove said[11], '*there is no point in having a doomsday weapon, if you don't tell anybody about it*'.

Thus, from an evolutionary point of view, one might expect to develop quite a strong discrimination for bitter. After all, strychnine is a deadly poison, whereas the quinine in tonic water is relatively innocuous. Yet until a couple of years ago, bitter was assumed to be just a single taste signal, an issue we take up in §9.5.2.

Sour is likewise a danger signal: it is often indicative of food which has gone off. A sip of sour milk, taken by mistake, is an experience which lasts an unpleasantly long time. This is an important characteristic of taste. It has an exceptionally strong associative memory for bad effects, a memory which will correlate events over days. Rats are hard to poison because they take small samples of prospective food and watch patiently for ill effects over a long period. Conversely, one of the unfortunate side-effects of chemotherapy is loss of appetite. The drugs often create feelings of nausea, which in turn creates negative association with a lot of different foods, exacerbating undesirable weight loss in patients.

9.5.1   Capture and transduction

In taste the tongue provides additional surface area for the taste buds. In both gustation and olfaction, a liquid medium, mucous in the nose, saliva in the mouth intervenes between external stimulus and the receptors. Tastants must dissolve in the saliva before they can be detected.

There are approximately 100–150 taste cells inside a single taste bud with a variety of functions. Some of the cells inside the bud serve support functions such as mucous generation. The taste cells recycle every ten days in young people, although there are indications that the rate of regeneration falls with increasing age.

Taste buds are located in papillae, little nipple-like protuberances, of which there are several types:

- *Fungiform* papillae have a mushroom-like appearance and contain rather small numbers of buds, usually not more than five.
- *Circumvallate* papillae form a sort of circular raised mound with a moat around it; the buds are in the walls and are washed by mucous within the moat; they contain thousands of buds; they are large, situated towards the back of the tongue and quite sparse, around 12 in total (Smith & Margolskee, 2001).
- *Foliate* papillae also contain thousands of buds.

---

[11] Stanley Kubrick's film, *Dr. Strangelove*.

- The *filiform* papillae, in fact the most numerous type, contain no taste buds and serve a mainly tactile role. The texture of food is also important, which is why soggy potato chips are unappetising (Weiss, 2001). Some cells code for the sensation of fat on the tongue (perhaps with a particular preference for cream . . .).

As we shall see (§9.5.3), the taste buds at the front of the tongue supply a different nerve to the brain to those at the back. Unlike the other senses, we do not have a clear idea why these distinct pathways exist. Each taste bud may contain different receptors, so it is not a segregation based on sensitivity. In mouse bitter receptors appear in cicumvallate and foliate papillae, while they are rare in the fungiform papillae (Sugita & Shiba, 2005). The pathways to the brain hold no clue and as yet sufficient information is not yet known about the dynamics of the cells in the relevant cranial nerves to infer any functional distinction.

One possibility might be that the taste buds at the front of the tongue serve as an early warning system, for tasting unknown foods. Another is that they might subserve a kin recognition function, licking as opposed to eating. Bitter tastes evolved to indicate alkaloids in plants which are often extremely toxic (§9.5). Thus immediate ejection is needed requiring the fastest possible response. In time we might expect a more detailed understanding of the pathways into the brain to shed light on these enigmas.

9.5.1.1 Perpetuating a myth

Discussions of taste exhibit the making and perpetuating of a myth. Sometimes in science an idea becomes commonplace and is often repeated, even though it is completely wrong and was never thought to be correct. In the nineteenth century, the data pertaining to where the different taste sensations were located on the tongue was badly misinterpreted. This happens. Usually the critical process of peer refereeing of scientific and other academic work ensures that such errors are weeded out. Perhaps because it was an attractive, tidy idea, the theory that the tongue could be divided into regions for sweet, sour, acid etc. became commonplace. It is in fact completely wrong and was never justified by the data (Smith & Margolskee, 2001). However, textbooks continue to reproduce it to this day!

However, there is some degree of specialisation in the tongue surface. Experiments to transplant the nerves innervating the tongue show that it is not possible to generate taste cells in ordinary skin epithelium. Contrast this with the flexibility of the cortex where, as we saw earlier, we can grow visual cortex where auditory would normally be. Thus crossing the nerves over causes the nerves to carry different taste sensitivity, e.g. bitter instead of salty.

9.5.2 Taste receptors

Just as in olfaction, the absence or malfunction of some genes may lead to very specific taste loss. An example used in many studies is the threshold for detecting phenylthiourea, a bitter substance almost undetectable by some people. There is no clear relationship between the chemical class or structure of the tastant and the subjective taste.

The first stage of transduction is similar to olfaction: the *tastant* has to dissolve in the saliva, which transports it through the taste pore to the taste buds themselves. At the transduction stage, taste seems to be quite different to olfaction. The four/five taste categories create neural signals in different ways (Smith & Margolskee, 2001).

Each mechanism has the ultimate effect of depolarising the cell by closing potassium ion channels. Depolarisation causes the release of neurotransmitters and the generation of a signal in the attached taste neuron. All mechanisms seem to operate in every cell, although cells tend to have a bias towards one of the stimuli classes.

Salts and acids act directly on the ionic mechanisms of the cell, whereas sweet, bitter and unami go via specialised receptor proteins and the G proteins[12], such as *gustducin*, referred to as GPCRs (G-protein coupled receptors), labelled as TnRm, where n is 1 for sweet/unami and 2 for bitter. For sweet there are just three values for m (T1R1 and T1R3 give unami and T1R2 and T1R3 give sweet) (Sugita & Shiba, 2005). The four/five taste categories create neural signals in different ways:

Salts Salts have the simplest mechanism. Sodium ions enter the cell and depolarise the cell directly.

Acids Acids generate hydogen ions in solution. These ions act in several ways, all of which increase the total number of positive ions in the cells. H^{++} bonds to potassium channels on the cell, stopping the exit of potassium and also opening channels which admit other positive ions such as calcium. They also enter the cell directly.

Sweet In the case of sweet stimuli, an intermediate agent, the *secondary messenger*, closes potassium channels. Sweet stimuli stimulate receptors on the surface of the cell, which in turn split G proteins into two subunits. These fragments, in turn, stimulate an enzyme to generate the secondary messenger.

Bitter Bitter stimuli also operate via G proteins, but the secondary messenger in this case stimulates the release of calcium ions, so leading to depolarisation and release of neurotransmitter.

Umami: sensitivity to amino acids The umami sensitivity is still debated. It also acts via G proteins, but the secondary messenger mechanisms are not understood. However the genetics, at least in mice, are now becoming clearer. The *sac* gene is expressed as a particular receptor, known as T1r3, and controls most of the sensitivity to artificial sweeteners (mice like artificial sweeteners as well as sugar). In combination with another receptor T1r1 it conveys sensitivity to umami and with T1r2 for sweetness (Damak *et al.*, 2003).

For the taste components, the neurotransmitter was recently shown to be ATP (Finger *et al.*, 2005), although sensitivity to touch, temperature and menthol within the mouth use different transmitters. But at the time of writing the story is still complicated. The mice used in these experiments, which have had the ATP receptors disabled, still have sensitivity to sour and bitter tastes, such as citric acid and caffeine. It seems to be the case that suitable receptors exist also in the larynx outside of taste buds and the transmitters of these cells are different (Finger *et al.*, 2005).

---

[12] So-called because they are regulated by guanonsine triphosphate or guanine-nucleotide binding proteins.

The receptor/G-protein relationship is a quite recent discovery, around 1992. Determining the number of receptor proteins is still an ongoing task. For bitter a class of receptor protein genes has been indentified within just the few years (Barinaga, 2000a), along with evidence that they are made in the same cells as gustducin. This family has around 40–80 members with around 30 GPCRs in mouse. Thus present indications are that the dimensionality of taste is considerably lower than olfaction, in accordance with our intuition.

Sweet is still something of an enigma. In 2001 four groups independently reported locating a gene in mice which they assert codes for sweetness using a similar G-protein coupled receptor (Davenport, 2001). Variations in this gene distinguish mice which like sweet things from those which are indifferent. Experimentally isolating taste receptor proteins is difficult and the result is not completely certain, nor whether the same holds in people. So far, too, it is only one gene. Further details of perception of sweetness are eagerly awaited.

One additional piece of the taste jigsaw has fallen into place with the discovery that taste cells express most of these proteins[13]. Thus we have one generic response to bitter tastes, since each taste cell sends one signal to the brain, indicating that one or more of the many different receptors has caught a bitter tastant (Brown, 2001b). At the time of writing this is still contentious, however. Caicedo and Roper (2001) showed that of five common, strongly bitter compounds, two-thirds of taste cells responsive to bitter (about 18% of the total population of taste cells) responded to just one and a further quarter to just two.

Markolskec and colleagues who discovered the details of many of these taste reactions have now created a company, Linguagen, to exploit their understanding. Many naturally occurring food stuffs contain 'bitter blockers' which block the release of gustducin, so masking the bitter tase. They see two important areas of application. Many foods are loaded up with sugar or salt to mask bitterness. Obesity is now at almost epidemic proportions in the developed world and any neat trick to reduce the sugar levels in processed foods would almost certainly help. The other application is to mask really bitter, unpleasant drugs which have to be taken regularly for some conditions.

9.5.3 Coding

Coding follows a path similar to olfaction. We don't have one receptor per scent, not even one receptor per molecule type. It is spatial patterns of activity which characterise odours. Any given taste bud may be sensitive to several tastants.

Taste is carried to the brain via cranial nerves VII, IX and X (Keverne, 1982). Cranial VII nerve innervates the anterior (front) part of the tongue as the *chorda tympani*. Cranial IX, the *glossopharyngeal nerve*, innervates the posterior (rear) part of the tongue. It may come as a surprise, but not all the taste buds are actually on the tongue. They occur also on the palate and the top of the oesophagus. They are innervated by the cranial X.

---

[13] To be really precise there is a very large overlap, like spectral pigments, rather than complete identity.

Taste is still not the most researched sense. For example, it is still not completely clear if taste is synthetic or analytic (Smith & Margolskee, 2001). Sensitivity is also dependent upon levels of the brain transmitters serotonin and noradrenaline as discoved as recently as 2006 by Heath *et al.* (2006). Drugs which enhance serotonin increase sensitivity to sweet and bitter. Depressed patients often have reduced taste sensitivity and this selective sensitivity is now being explored in clinical trials as a marker for the most effective drugs for treatment (Canoon, 2008).

9.5.3.1   Receptor or cell for taste quality

Zhao and colleagues have added a further refinement of how we appreciate taste. Some cells code for sweet or bitter, seemingly independent of the receptor protein they express. By genetically modifying mice taste receptor proteins, labelled T1R1 to T1R3, they were able to make mice respond to the artificial sweetener aspartame, to which they are not normally sensitive, and to even like normally tasteless substances (Pagán, 2003; Zhao *et al.*, 2003). This is redolent of the introduction of a third spectral type (§8.1) and indicates just how flexible the pathways to the brain and subsequent processing really are.

9.5.4   Pathways

Like other senses, taste projects first to the thalamus and then on to the cortex. Again we see parallel streams. One stream goes to the parietal cortex and seems to respond to a mixture of taste and tactile sensations. The other pathway to the *insular* cortex seems to respond to taste alone. Sugita and Shiba (2005) show that segregation of bitter and sweet occurs with separate pathways leading to separate clusters in the gustatory cortex.

Other projections go to subcortical areas. A projection goes to the amygdalia, important for emotional reactions to taste. We get strong likes and dislikes, pleasure and disgust from what we put in our mouths. The last pathway projects to the lateral hypothalamus which is implicated in feeding behaviour, in other words, the controller of appetite.

Finally we note that taste and smell are in many ways more primitive than vision and hearing. We do not get any significant breakdown into components in the way that we extract edges and features from images. We also do not get the complex normalisation processes which go on in, say, colour constancy. In vision we have complex processing to get at the properties of objects independent of the variability of the light flux coming into the eye. The chemical senses are very fickle. Fine wines do not go with hot curries. We eat from savoury to sweet because of the non-linear adaptation which goes on, conditioning our current taste sensation on what we have experienced in the not too distant past. This is of course quite different to vision, where we try to process out factors such as the illumination to get at the surface properties of an object.

9.6   Artificial systems

As we have seen already, humans are predominantly audiovisual animals. Thus there is not the same driving force for representation of taste in virtual reality. But it still exists.

Imagine the advantages of being able to browse a wine web site and get a taste of the 1966 Grange Hermitage or a good vintage Chateau Latour.

On the other hand, taste sensors for robots are also potentially useful. In a way anything which analyses the contents of a solution could be considered taste. The distinction lies in the way complex mixtures are mapped into simple taste sensations. Thus we can imagine testing a batch of wine for signs of manufacturing defects.

So there is some activity. Neikurk at the University of Texas at Austin and the company Labnetics are developing small beads, about 150 μ in diameter, which change colour in response to simple tastes. Further development of this work may include rapid identification of antibodies. Hewlett Packard and Cyrano Sciences are developing e-noses for monitoring the quality of food and other consumer items (Brown, 1999).

Robot fish are under development at the University of Essex. They have small chemical sensors aimed at hunting down pollution. At Australian $29 000 each, sturgeon would have trouble competing on a price per kilo basis.

9.6.1   Chemical senses in computer games

We can see two ways in which smell might become a factor in virtual reality and games. In the search for greater realism we can add synthetic odours which the player can smell. Just think how the sensation of a live firework display is enhanced by the smell of gunpower. The smell of the sea, the mustiness of caves, the sense of place and involvement would grow as we add these primitive but highly emotional and evocative components.

In the second the artificial odours remain within the game and are sensed by the animats and non-playing characters in the game. At the game play level, being able to track or being tracked by robot bloodhounds is a dimension yet to be explored. At a more abstract level, good path finding and other algorithms have been developed from pheromone generation and tracking by ants.

Taste has at the time of writing moved from being a future possibility to a practicality for virtual reality systems.

9.6.2   More electronic noses

An early precursor to artificial smell is the scent-strip developed by the 3M Corporation. Tiny capsules of odorant are embedded in a covered strip. As the cover is peeled away the capsules are fractured, releasing the odour. Such devices appear in many places from advertisements to some of the films of John Sayles filmed in '*Odorama*'.

But real electronic noses tend to work in a rather similar way to biological olfaction, using polymers which change in some way when they adsorb or absorb a vapour, just as a receptor protein modifies membrane behaviour. The first is really very close – the polymer changes its conductivity, marketed in 2001 by company Aromascan. An alternative to changes in the conductivity of the polymer itself is to use a generic polymer containing a conductive material. As the polymer swells on absorbing a vapour so the conductivity changes. This system was developed at Caltech (Schmiedeskamp, 2001).

The other principal approach is to get the detecting polymer to change colour. Organometallic compounds frequently have strong light absorption and take up compounds to which polymers are often insensitive. Another approach is to dope polymers with fluorescent dyes, embedding polymer beads at the end of fibre optics (Schmiedeskamp, 2001).

An additional electronic nose is also being developed by David Watts, and is specialised for detecting landmines. The usual landmine explosive, TNT, gives off a vapour, dinitrotoluene, which the electronic nose detects. It consists of seven polymer-based sensors which fluoresce when they adsorb particular chemicals. The combination of intensities and colours uniquely characterises a wide range of possible molecules. But the calibration is not worked out for each possible smell, as would be theoretically possible. Instead a machine-learning algorithm is used.

Another highly sensitive electronic nose is under development at the University of Tübingen by Hans-Georg Rammensee. Again the strategy is similar to a biological nose. We look for different patterns of activity across a range of detectors. In this case there are two detector sets.

The first set of detectors uses quartz crystals, similar to the devices used in digital watches. We measure time to high accuracy, by monitoring the vibrations of the crystal electronically. When quartz watches first came out they were so far superior to common clocks and watches that the variations in accuracy from watch to watch were hardly noticed. But purity and surface contamination do effect the frequency and this property is exploited here.

The first part of the nose consists of polymers coated onto a set of eight quartz crystals. Select molecules stick to these polymers and change the frequency at which the crystal vibrates. The pattern of vibrations across the crystal set is the first detector signal.

The second uses a set of semi-conductor metal-oxide gas detectors. Incoming gases react with oxygen on the detectors and change the conductivity. This provides the second pattern. The nose is sufficiently sensitive to detect different variations of coffee (Copley, 2001).

Gardner at the University of Warwick is developing a combined artificial tongue and nose. Myron Krueger in DARPA (Defense Advanced Research Projects Agency) is pursuing transmission of olfactory data for quite a different reason – telesurgery. Smell is extremely useful to a surgeon as he cuts into the body. Krueger's approach is to build a backpack containing phials of different liquid odours. These are delivered as bursts of vapour by forced evaporation using ultrasound. The time lag with this method is down around 0.1 s. The problem, as the reader might guess by now, is defining the component odours to use, still a difficult problem (Pescovitz, 1999).

Ambryx is a company specifically trying to exploit developments in neuroscience and molecular biology for sensory applications (Brown, 1999). Recently they have patented work on the so-called sentinel molecules, TR1 and TR2, which sense bitter and sweet flavours on the surface of the taste cells.

Electronic noses are pushing into territory outside conscious human experience. One new device, the Cyranose (after Cyrano de Bergerac), is achieving success rates at detection of lung cancer from a person's breath of around 85%. It has considerable potential as a screening tool (News track, 2005). It is interesting to speculate whether some seemingly paranormal abilities might rest on an ability in some people to recognise or classify such odours subconsciously.

9.7   Envoi

Chemical senses are amongst the oldest of evolution. On the one hand they are the simplest, basically an associative memory, but on the other the least well understood. The human genome project, and our ever increasingly sophisticated genetic tools, are helping to unravel how they work. But from a virtual reality or computer storage and transmission perspective they lag behind the other senses. The next few years will be very interesting indeed.

10 The somatosensory system

10.1 Introduction

This chapter collects together two quite different systems. The first is the sensor system distributed throughout the body which monitors what is going on inside or outside, the temperature, chemical and tactile (pressure and vibration).

The second is the vestibular system which monitors the position and movement of the body in space. It interacts quite strongly with the tactile system. So, for example, the receptors in the skin in the feet contribute to maintaining our balance and can take over if the vestibular system gets damaged. The vestibular system is also tightly coupled to the visual system, through the opto-kinetic control of eye movements and gaze, a topic which §12.1 takes up in detail. Yet it is located inside the ear. Unsurprisingly, these systems share common principles and some transduction mechanisms with the senses considered earlier in the book.

These two systems are at the forefront of developments in virtual reality and computer games. Haptic interfaces are fast growing in importance (§10.8.4) and monitoring gaze has numerous evolving applications in human computer interface design The vestibular system appears obliquely in its role in physically interactive games and total immersion (§12.1.8).

Collectively these sensors comprise the the *somatosensory system*, from the Greek word *soma* for body. The sensors embedded in the skin monitor the impact of the external world and gather information about it. Hearing and vision are active senses in that we turn our heads or move our eyes to attend to a stimulus, and we work from hypotheses of what we might be seeing or hearing. Touch is both passive and active: the skin constantly monitors passively environmental influences upon it. Yet, we also use our hands actively for exploring the shapes and surfaces of objects (*stereognosis*) while a cat uses its whiskers for modelling the immediate external environment, sometimes in total darkness (§10.6.1). This active tactile exploration is usually referred to as *haptics*.

But the brain and body also need to know what is going on inside the internal organs, the muscles and the skeletal system. The *proprioceptive* mechanisms monitor where the body is, the strength with which we grasp things and all the complex interactions of our body with the environment. In keeping with the central theme of the book, of collecting information from the environment, emphasis in this chapter will be the sensing of external influences (§1.1).

Nevertheless, there is an interesting interrelationship between external and internal. Sometimes the internal monitoring interacts with the perception of sensory data. Examples include suppression of an animal's own acoustic output (§5.4.4.1), being insensitive to tickling (§10.4.9) and suppression of visual information during saccades (§12.1.6).

Because touch involves direct contact with the body, it subserves all sorts of passive warning mechanisms, from fast reflexes to chronic irritation and pain. So, it is not surprising that it has direct pathways to the emotional areas of the brain, some linked to specific sensors and some more sublimal. Studies seem to show, for example, that people have more positive interactions with others if they receive light, even unnoticed, gentle touches. This direct access to emotional centres has obvious potential in the games industry, not yet fully exploited.

At the same time, touch helps us to actively build up a representation of the three-dimensional (3D) world. It furnishes information on the 3D shape of objects and on the nature of surfaces, their smoothness, roughness, regularity, stickiness and conductivity/temperature.

The somatosensory system exhibits the same, common information processing principles as the other senses:

- Each receptor does not necessarily contribute an independent piece of information; extensive pooling occurs.
- There is streaming into separate pathways; whether this is an evolutionary phenomenon or a computational efficiency strategy is not fully known.
- It is very sensitive, probably close to the theoretical limits, but less is known in the somatosensory system and such judgements are less clear-cut.

The tactile system is very diverse, and varies along a number of dimensions:

- The *functionality varies.* Sensors report: threatening stimuli (pain); position of body parts; things touching the surface of the skin or moving across it; and properties of what goes on inside the body.
- The *stimulus type varies* and may be mechanical, thermal or chemical.
- The *spatial resolution varies* from pain, which is often hard to localise, to the fine textural sensitivity of the fingertips.
- The *temporal properties vary.* The dynamics may be tonic (where the signal continues while the stimulus is present) or transient (where the signal onset and/or end is signalled) as was evident in both hearing and vision. Some cells respond quickly, others slowly and may form independent stimulus maps.

Touch is especially interesting. On the one hand it is a relatively primitive sense in evolutionary terms; on the other it is the progenitor of much other sensory apparatus:

- Sensitivity to air or water currents is present even in the cilia of bacteria.
- Touch sensitive hair cells become the transduction mechanisms of the ear and the vestibular organs.
- Specialised touch receptors in the lateral lines of fish become tuned to water waves.

- The elephant's trunk has exquisite sensitivity, enabling the elephant to pick up tiny objects (§10.6.3); the whiskers of the cat, or *vibrissae*, go beyond mere passive sensation to the construction of a 3D world view (§10.6.1).
- The texture and the surface properties of things we put into our mouths is also as important as the taste; the tongue has close integration of touch (texture) and chemical (taste) receptors, probably tighter peripheral coupling than almost any sense, leading to the complex mixture of sensations in the things we eat.

10.1.1   Overview of the chapter

Section 10.3 describes the primary somatic detectors which subserve the sense of touch, focusing mainly on mechano-receptors (§10.3.1), but considering briefly thermal (§10.3.4), pain (§10.3.3), proprioception (§10.3.2) and visceral (§10.3.5) receptors. Section 10.4 discusses the neural pathways and §10.5 links the receptors and pathways to some of the limits determined by psychophysics.

Section 10.5.4 covers unusual phenomena of current research interest, phantom limbs, where there is discord between the brain's map of the body and the sensory signals coming in. Closely related to phantom limbs (discussed in Chapter 12) are out-of-body experiences where somatosensory and visual information conflict. Section 10.6.2 considers some unusual tactile mechanisms found elsewhere in the animal kingdom.

Two lengthy sections conclude the main content of the chapter. Section 10.7 considers the vestibular system and §10.8 sketches the developments in the exciting field of haptics – artificial touch systems for the interfaces of computer games, telework and robotics.

10.2   The physical limits to touch

Air molecules constantly bombard the skin of an animal on land, creating background noise which sets a lower limit to the sensitivity to pressure receptors in the skin and to the information available. There would be little value in being much more sensitive than this. But it is obviously advantageous to get down to this level. Detecting a slight rustle of wind and its direction, and the possible direction from which predators might appear, could be of survival value. Inhabitants of warm climates might prefer our skin to be a litle *more* sensitive to the injection of a mosquito's proboscis!

Just as with an image on the retina, the spatio-temporal pattern on the skin, and its Fourier Transform determine the sampling requirements. But whereas in vision diffraction imposes an exact spatial-frequency cut-off there is no such *theoretical* limit for touch. This is somewhat similar to hearing, where there is no exact frequency cut-off, but the tuning of the ear canal and the mechanical properties of the basilar membrane limit the ultimate frequency response[1].

---

[1] One would expect that the tuning and frequency limit of the neural system of the ear and the elements of mechanical transduction would have evolved to approximately match.

The elastic/mechanical properties of the skin determine whether a point indentation will excite more than one receptor, somewhat analogous to the spreading of light through the lens of the eye, so that in a photoreceptor array with sampling matched to a finite pupil, a point of light covers more than one receptor. (§6.5.2). Another implicit cut-off is the fluid sac of the Pacinian corpuscle (§10.3.1.2).

The sensitivity of mechano-receptors falls into a number of distinct categories:

- Detection of a slight pressure, the mosquito bite test.
- Detection of slight vibration, the threshold determined by the frequency and the amplitude of the vibration.
- Spatial resolution of two adjacent pressure points – similar to two-point resolution of points of light on the retina.

The mechano-receptors vary widely in spatial density and do not form arrays of crystalline regularity like the cones in the fovea of the eye. Nevertheless, the sampling theorem will still be useful in thinking about tactile resolution.

The detection of vibration probably reaches its most sensitive in hearing. Chapter 5 described how hair cells are responsive to amplitude vibrations of mere nanometres, with the sensitivity to vibration of the basilar membrane close to the physical limit set by Brownian motion (§5.2.1.4). But the neural response to individual mechano-receptors is close to perfect as with rods: just as we can detect a single photon event in the rods, so we can detect a single mechano-receptor signal, despite the considerable convergence (Gardner & Kandel, 2000).

The optimisation of proprioception, the monitoring of the muscles, tendons and skeleton is determined more by the accuracy of motor and limb control than by physical variables. Motor control systems and sensors are co-optimised.

Thermal and chemical sensors need to be sufficiently sensitive to detect damage. It is interesting that evolution has not yet managed to sense ultra-violet light, or ultra-violet damage to the skin, as witnessed by the alarming rates of skin cancer in countries such as Australia.

10.3   Primary somatic detectors

As with the other senses, somatic information is streamed in a number of pathways and is almost like several senses all rolled into one. But the bandwidth and information processing are less well understood. They comprise four *modalities*:

- *Thermal* sensations, hot and cold and also chemicals (like chilli and mint) which induce thermal sensations.
- *Pain* produced by damaging stimuli in the so-called *nociceptors* (from the Latin *nocere* to injure).
- *Proprioceptive* signals[2], which monitor the skeletal framework and its supporting tissues, the positions of bones, tendons and muscles.

---

[2] From the Latin *proprius* – own.

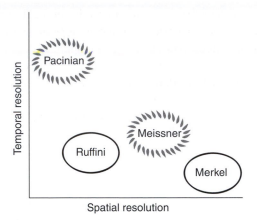

Figure 10.1 Spatial and temporal resolution of the major mechano-receptors. Starbursts denote fast adapting receptors. (Data from Johnson and Yoshioka, 2002.)

- *Exteroceptive signals*, or touch signals, which indicate stimuli external to the body acting on the skin.

These receptors are distributed throughout the body, but, unlike the other senses, their anatomical morphology is quite diverse. In some cases the sensory nerves are just loose in the skin. At other times there is some sort of terminator on the ending, subserving a protective or filtering role.

10.3.1   Mechano-receptors

There are several different types of mechano-receptor, but they fall into two categories based on their dynamics, numbering some 6–10 million overall (Grunwald, 2004). They may be fast adapting (FA) and thus maximally sensitive to things which move or vibrate. They may be slow adapting (SA) and hence more responsive to displacement of the skin. The nomenclature varies somewhat between humans and other mammals.

Figure 10.1 shows the main categories. Similar to the situation in vision, the X/Y ganglion cells of the cat, and the M/P ganglion cells in primates, the faster responding cells have lower spatial resolution (§6.6).

SA1 fibres are the primary transducer of static resolution information. Goodwin (1998) reports that only 50% of FA1 fibres respond to static stimuli, FA2 do not respond at all and SA1 respond the most vigorously[3].

Given the varying dynamics of the receptors, it is not surprising that the thresholds for moving stimuli are lower, as dicussed in §10.5. So, in general more afferents will contribute a signal from a moving stimulus. In general the firing of any given mechano-afferent is stronger for moving stimuli.

The fingers stroked over a surface give a clear impression of surface roughness. Connor and Johnson (1992) provide a nice illustration of this with a rotating drum

---

[3] These results are for monkey, which does not have SA2 receptors.

covered with a braille pattern (a collection of raised dots on the surface). As the spacing between the dots is varied the Merkel discs give sharp responses localised over each dot, with the strongest response for dots around 3 mm apart. The Meissner corpuscles give slightly less precise responses, while the Pacini corpuscles respond independently of dot spacing, sensitive only to things moving within their wide receptive field.

Sections 10.3.1.1, 10.3.1.3, 10.3.1.2 and 10.3.1.4 discuss the mechano-receptors in detail while §10.4 and Table 10.2 elaborate on the innervation and pathways.

10.3.1.1   Meissner corpuscles – fast adapting type 1 (FA1)

The Meissner corpuscles (MCs) are oval in shape, around $150 \times 50\,\mu$ (Goodwin, 1998) and filled with fluid. They are fast-acting sensors with a temporal frequency response of 10–200 Hz. They occur in areas of high tactile discrimination such as the fingertips, concentrated in humans in digits I–III (Johansson & Valbo, 1979). Their primary role is sensing surface texture and properties by stroking or touching something which is moving past or vibrating. The high density of MCs facilitates the detection of the softening texture of ripe fruit and this might have been an evolutionary adaptation. Hoffmann *et al.* (2004) find a reasonable, though imperfect, correlation between the density of MCs and percentage of fruit in the diet. MCs have a high displacement sensitivity, from 4 to 400 μ, with a maximum density at the tip of around $1.4\,\text{mm}^{-2}$ (Goodwin & Wheat, 2004; Johansson & Valbo, 1979), about 1.2 mm apart in the fingertip at the highest density and up to 10 mm in the palm. Interestingly this density is greater than the high resolution-form detecting system, the Merkel receptors (§10.3.1.3), but it seems to respond to skin motion only (Johnson & Yoshida, 2002). Despite their higher density, their spatial resolution is lower. Their minimum detectable element size is at least 3 mm, or a spatial frequency of just 0.12 cpmm (Johnson & Yoshida, 2002). There are 30–80 MCs per neuron, yet each may contact with up to six neurons (myelinated) with additional innervation from unmyelinated fibres which may subserve a nocioception (pain sensitive) role (Paré *et al.*, 2001). But as many as 50% of MCs may have just one nerve fibre (Guinard *et al.*, 2000).

The sampling analysis is thus not clear-cut. The convergence ratio varies from one to at least six, so, as a mosaic, the sampling properties are poorly defined.

10.3.1.2   Fast adapting type II: Pacinian corpuscles

The Pacinian corpuscles (PCs) are large ovoid, fluid-filled receptors, up to 1 mm in length (Goodwin, 1998). They are also rapid acting, but are responsive to an even higher frequency range than MCs – 70–1000 Hz – with the peak from 200 to 400 Hz. Their displacement sensitivity is also extremely high at just 3 nm for direct prodding of the corpuscle to 10 nm displacements of the skin (Johnson & Yoshida, 2002).

The PCs convey very little spatial information and may respond to deviations over huge areas, as much as an entire hand or part of the arm. There are about 2500 in the hand, 350 per finger and 800 in the palm, with one neuron per PC. Curiously they are reported to be absent from the face (Macefield, 2005). Their spatial resolution requires stimuli at least 5 mm in size (Johnson & Yoshida, 2002). Although the maximum frequency of detection is up around 1 kHz, this is very high for a single neuron to track. However, in building an artificial system, we need to be able to generate (encode, transmit, decode) at this frequency to provide a realistic stimulus.

Scheibert *et al.* (2009) have recently demonstrated that the ridged structure of finger-prints acts to amplify vibrations and increase the signal picked up by PCs.

10.3.1.3 Merkel receptors – slow adapting type 1 (SA1)

Merkel receptors, clustered into so-called Iggo domes, respond rapidly with a strong signal and adapt slowly, usually labelled as SA1 (Johnson & Yoshida, 2002). These domes comprise clusters of Merkel disc receptors, most numerous in whiskers (Halata *et al.*, 2003). The Merkel epithelial cells wrap around the unmyelinated nerve ending, but it is thought that it is the nerve ending itself that is mechano-sensitive and not a transduction process within the Merkel cells. The nerve itself is myelinated (Halata *et al.*, 2003). The SA1 cells have centre-surround organisation, so they have little sensitivity to uniform stimuli (cf. retinal ganglion cells in §6.6). But again, this seems to be generated by the mechanical sensitivity rather than by an inhibitory sound, such as we find in retinal ganglion cells.

SA1s vary considerably in sensitivity, by as much as a factor of four (Goodwin, 1998). Their response is a complex mixture of pressure and curvature of surface. Goodwin gives the following expression for SA1 response in monkey (1998):

$$R = s\phi[1.91 - 1.63e^{-0.00243\kappa}] \qquad (10.1)$$

where s gives the sensitivity of the fibre, ϕ is a function of pressure and κ is the curvature in m^{-1}. s is around 64 and responses, r, as spikes in the first second after contact, range up to around 150. Thus for spheres down to around 2 mm. the response is approximately linear with curvature. ϕ varies approximately linearly with a value of around 1 for a force of 15 g (see Figure 10.2). The spatial resolution is about the same in monkey and humans (Johnson & Yoshida, 2002).

The response is maximal when the sphere is placed in the centre of the receptive field as would be expected. To get unambiguous information about curvature requires separating out the components of force and curvature. The profile of a *population* of SA1 units varies with curvature, with force scaling the intensity of the response. Although the Merkel receptors would seem to be the source of high spatial resolution, their density is lower than that of the MCs – about 0.7 mm^{-2} at the maximum density, corresponding to a distance between centres of 1.2 mm and a sampling frequency of 0.83 cpmm (Johansson & Valbo, 1979), higher than for monkey (Goodwin & Wheat, 2004).

Johnson and Yoshida (2002) suggest that the difference between the FA (Meissner) and SA (Merkel) receptors is rather like the difference between the scotopic and photopic (rod/cone) visual systems. The RAs are very numerous, individually very sensitive, but have low overall resolution. Also, like the rods, they do not effectively form one big receptor, but are interdigitated with many receptor neuron mapping.

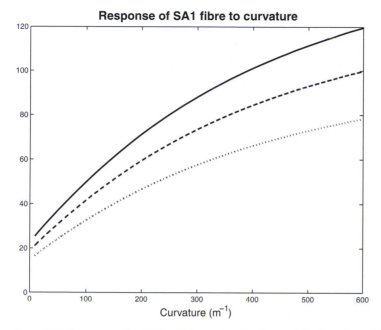

Figure 10.2 Response of an SA1 unit in monkey for force of 10 (dotted), 15 (dashed) and 20 g (solid). (From Johnson, 2002.)

10.3.1.4 Ruffini receptors – slow adapting type 2 (SA2)

Ruffini's end organs lie deep in the dermis, the inner layer of the skin; they do not cease firing while a stimulus is present, and provide continuous (annoying) response. Again, the Ruffini label refers to the complex of cells terminating in the nerve ending. These are the SA2 receptors. Their receptive fields are almost an order of magnitude larger and the sensitivity to indentation an order of magnitude smaller. Their sensitivity to skin stretch is double that of SA1 and detecting skin stretch, particularly during hand flexing, is thought to be one of their main roles (Johnson & Yoshida, 2002). They have little role in forming perception.

10.3.1.5 Hair cells

The upper frequency psychophysical threshold closely parallels the sensitivity of the PCs. Just as in vision, the cells with the highest temporal resolution are not the cells with the highest spatial resolution. That accolade goes to the Merkel disc receptors. But one of the most exquisite displacement sensitive receptors is the hair cell with a trigger nerve at the base of the hair. This piece of biotechnology gets refined and reused in special touch systems such as whiskers (§10.6.1) and in other senses such as the hair cells of the ear.

10.3.1.6 Location in the skin

The different types of receptor appear in different layers of the skin. Structurally the organisation is quite complicated, with the receptors esconced in the fabric or ridges and folds that make up the fingerprint. The Merkel and Meissner receptors, those with the highest spatial resolution, appear near the surface at the junction of the top layer, the epidermis and the inner layer, the dermis, giving a spatial resolution in the fingertip of around 2 μm (Maheshwari & Saraf, 2006). This is not surprising. The mechanical properties of the skin act like a positional blur the deeper one goes. But this argument does not apply to temporal properties. The receptors with the highest temporal sensitivity, the PCs, are buried deep in the dermis, along with the Ruffini endings, both connected to myelinated nerve fibres (but see §10.6.3 on the elephant trunk). The hair cells also have their endings buried deep in the dermis.

10.3.2   Proprioception

With the exception of the Ruffini receptors, which have a significant role in monitoring skin stretch, all the mechano-receptors monitor impact or pressure on the skin. But there is a whole set of receptors for monitoring where the skeleton and muscles are, the proprioceptors, and detailed maps of body location and posture exist too (Graziano *et al.*, 2002). There are at least four types of proprioceptor (Goodwin, 1998):

1. Golgi tendon organs which monitor tension in the muscles.
2. Muscle spindles which monitor length and velocity of the muscles.
3. Joint receptors in the joints themselves.
4. Cutaneous receptors near the joints and the hairy skin of the hand which contribute to measurement of the position and velocity of the fingers.

There is of course huge variation in these proprioceptive systems across the animal kingdom.

The muscle spindles are the most numerous with about 25–30 000 throughout the body. Each muscle is made up of a number of spindles, each containing muscle fibres and neural innervation. The latter consist of one primary fibre, known as Group Ia and up to five secondary fibres or Group II. Group I are faster (72–120 ms^{-1} compared with 20–72 ms^{-1}) and more sensitive to velocity and acceleration in the muscle. There are about 4000 in each arm and 7000 in each leg (Prochazka & Yakovenko, 2002). The number of spindles in a muscle, M_s, is given approximately by:

$$N_s = 38\sqrt[3]{M} \tag{10.2}$$

where M is the muscle mass in grams. So, a whale has a lot of muscle spindles – and needs a large brain to match!

The Golgi tendon organs, 0.2–1 mm long capsules, are less numerous, about 2500 in the arm, innervated by Group 1b fibres.

The last category, the joint receptors, are the least numerous with only a few hundred in the whole arm. Whether there are proprioceptors in bone is still open to question. There are lots of fine nerve fibres entering bone but these are more likely to be linked to pain or control than mechanical 'osseoreceptors'. Numerous PCs often lie alongside bones but their precise role is uncertain (Rowe *et al.*, 2005).

The action of the proprioceptive neurons is quite non-linear and tightly coupled with the motor control units. Prochazka estimates that a million impulses per second come in to the spinal cord (Prochazka & Yakovenko, 2002) from each leg. Thus the information rate from the proprioceptive system is at least 1 Mbps.

10.3.3 Nocioreceptors and pain

Pain arises from two sources: the first, nocioceptive pain, arises from specialised threat or damage detectors; the second, neuropathic pain, arises from damage to pain-transmitting nerve cells themselves. Only the first is considered here.

Pain is a relatively low bandwidth channel. It uses thin, weakly myelintated and slow $A\delta$ fibres or even slower C fibres, with response latencies 100–300 ms and 1000 ms respectively (Gross *et al.*, 2007) (see Table 10.1). At the same time it is *immensely complicated*. Simple linear thinking has very little place here. We get the most acute confusion of where the painful stimulus is, and all sorts of sensitisation and blocking phenomena occur at the biochemical level.

Nocioreceptors are all free nerve endings which fall into three categories (Gardner & Kandel, 2000):

1. Mechano-receptors, which are sensitive to sharp objects and use $A\delta$ neurons (see Table 10.1).
2. Thermal receptors, which are responsive to hot or cold, such as temperatures above 50°C or below 5°C. The non-painful discriminations of warmth and cold shut down when these limits are exceeded. These also use $A\delta$ fibres.
3. Polymodal receptors responsive to mechanical, heat and chemical stimuli. These use the unmyelinated C fibres. Thus their signals arrive centrally after the thermal and mechano-receptor signals. They are found in, for example, tooth pulp and can generate prolonged nagging pain.

The subjectivity of pain is complex with pain suppressed in some situations involving major injury, yet remaining chronic in the absence of ongoing damage or even in phantom limbs (§10.5.4). It now appears that oscillatory behaviour may relate to the actual perception of pain. Gross *et al.* (2007) show that gamma oscillations, in this case in the range 60–95 Hz, are induced by painful stimuli in S1 along the $A\delta$ (Table 10.1) first-pain fibres. But the oscillations are much stronger when the pain is just perceived near threshold than when a stimulus is just below the subjective threshold of pain. It is as if the oscillations subserve a salience function, indicating action is necessary – see §4.6.3.

The chemical response may be direct or through chemicals released by traumatised cells. Nocioreceptors tend to be slower conducting than touch in general as one might expect for unmyelinated fibres and do not tend to adapt rapidly.

One of the disturbing trends in our increasing knowledge of neuroscience is the way we find more and more human behaviours and characteristics in lower animals. Those of us who consider fish to be an honorary vegetable might find the latest results on their pain sensitivity disturbing. Fish with cartilaginous (as opposed to bony) skeletons, such as sharks, do not seem to have nociceptors. Not so the delectable trout, unfortunately. Sneddon and colleagues (cited in Holden, 2003c) have found 58 receptors on the trout's head which operate suspiciously like pain sensors. The trout's heart rate goes up when

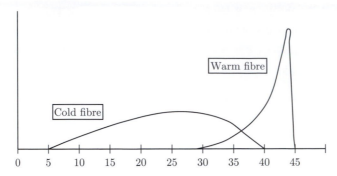

Figure 10.3 Response of cold or warm thermal receptors. (Redrawn from Gardner *et al.*, 2000.)

they are stimulated (as does ours when we experience pain) and the trout's subsequent behaviour (rubbing its nose) is typical of the way we would respond to a painful stimulus.

10.3.4   Thermal receptors

Thermo-receptors respond to cold (25–30 °C) or falling temperature or to warm (−40–42 °C) or rising temperature, i.e. variations either side of normal body temperature (around 37 °C) (Figure 10.3) (Gardner *et al.*, 2000). Thermo-receptors have bare endings in the skin, like the nocioreceptors for pain (§10.3.3). There are separate sets of receptors to signal pain from temperature, which cut in at the extreme limits (less than 5 °C and greater than 50 °C) of hot and cold sensations.

Thermal receptors do not need to be very plentiful. Since heat is conducted across the skin and inwards, there would not be much advantage in having a particularly high spatial resolution thermal map of the body. However, a fast temporal response is desirable, so that we can immediately avoid a hot or cold object.

There is a familiar link with smell and taste. Chillis taste hot and menthol (the active ingredient of mint) tastes cool. A few years ago, a receptor which is sensitive both to capsaicin (active ingredient of chilli) and heat was discovered. Now genetic techniques have revealed a receptor protein sensitive to cold and menthol (Seydel, 2002). However, these receptor proteins are not the only ways in which heat and cold are sensed; other ion channel mechanisms have now also been established. In fact, injury to or malfunction of some of these ion channels may make cells unduly sensitive to cold. Taste is similar with some modalities (e.g. salty) operating via ion channels and some (e.g. bitter) via receptor proteins (§9.5.2).

Interactions also occur between mechano-receptors and heat. The skin temperature affects the perceived magnitude of vibration in the PCs (Verrillo & Bolanowski, 2003) but has little or no effect on the other types.

10.3.5   Keeping the body functioning

The last set of receptors barely impinge upon our conscious activity. They monitor the activities of the internal organs, the *viscera,* and thus do not have any specific

Table 10.1 Afferent fibres of the somatosensory system. Numerals classify fibre diameter. Roman and Greek letters denote conduction velocities. The unmyelinated fibres have the smallest diameter and lower conduction velocity even for comparable diameter. (After Gardner *et al.*, 2000.)

Myelination and size	Muscle nerve	Skin nerve	Diameter (μm)	Conduction velocity (ms^{-1})
Yes, large	1	Aα	12–20	72–120
Yes, medium	2	Aβ	6–12	36–72
Yes, small	3	Aδ	1–6	4–36
No	4	C	0.2–1.5	0.4–2.0

relevance to human computer interaction, robotics, animats or avatars in virtual worlds. Detectors exist for organ deformation (similar to, for example, PCs in the skin), chemical behaviour (glucoreceptors for sugar and pH receptors), flow and pressure and other bodily maintenance activities. The afferents have their own viscera, sympathetic and para-sympathetic which end up in the dorsal root (§10.4.3). They are usually slow and unmyelinated.

10.4 Pathways

The somatosensory pathways stream into a number of major nerves and then into either the spinal cord for the body or the cranial nerves for the head. The spinal cord terminates in the brainstem, from whence information travels to the thalamus and finally the somatosensory cortex. There is some evidence for distinct pathways from different receptor classes to the cortex as we have seen for other senses.

There are a number of different nerves and nerve fibre types, with different labels according to whether they innervate muscles or the skin (cutaneous). Table 10.1 shows the fibre names and conduction velocities.

10.4.1 Receptor afferents

Two nerves transport information from the mechano-receptors: the *median* and *ulner* with conduction velocities around 20–80 m^{-2}. The type 1 afferents have small circular or ovoid receptive fields. They show some degree of convergence (Goodwin, 1998; Macefield, 2005) with hot spots corresponding to the individual receptors. The type 2 afferents, with bigger receptive fields with ill-defined borders, are the other way round with one or more afferents per receptor.

- The Merkel receptors, which feed slow-adapting fibres (SA1), have a convergence of around 3:1. These fibres have a density around 1.4 mm^2 and represent about 25% of the fibres in the median nerve which supplies the hand.
- The Meissner corpuscles, which feed fast-adapting fibres (FA1) have a similar convergence to Merkel, but each neuron now innervates several Meissner receptors at

Table 10.2 Summary of properties of cutaneous mechano receptors in the hand. Figures from Macefield and Gardbner, see text.)

Receptor	Fibre		Median nerve %	Convergence	max freq.
FA I (Meissner)	Aα, β	43		12–17	100
SA I (Merkel)	Aα, β	25		4–7	100
FA II (Pacinian)	Aα, β	13			200
SA II (Ruffini)	Aα, β	19			20

an afferent density of around $0.7 \, \text{mm}^2$. These represent about 43% of the median nerve.

- The Ruffini receptor feeds one or more nerves at a density of $0.09 \, \text{mm}^2$ and 19% of the median nerve.
- The Pacinian corpuscle has its own individual nerve at a density of $0.8 \, \text{mm}^2$ and about 13% of the median nerve.

From Table 10.2 we can now estimate the average information capacity per fibre in the median nerve, assuming 1 bit per spike as about 100 bps, or around 400 kbps from the hand.

From the above densities, the sampling theorem could now give us a maximum grating resolution without aliasing. But the assumption that the density of afferents gives the cut-off frequency is erroneous, very similar to the situation with the cat retina, where there are 50% more cones than ganglion cells. It is the receptor density which determines the cut-off. The convergence acts like a low-pass filter. Given a convergence rate of around 4:1 and the density of 0.7 afferents mm^2, the sampling theorem would predict a maximum spatial frequency with a period of 0.6 mm. The measured receptive fields have a width of around 2 mm, suggesting that there is some degree of undersampling. If we fit a Gaussian to the receptive field, we get, very approximately, $e^{-\pi r^2}$, where r is the distance from the centre. The Fourier Transform of this is simply $e^{-\pi v^2}$, where v is the spatial frequency in cycles per mm (cpmm). If we take the sampling rate of the nerve cell fibres, the separation is about 1.2 mm giving a Nyquist rate of 0.42 cpmm. At this frequency we find the power of the receptive field is still close to 60%, suggesting some undersampling, albeit mitigated by the mechanical properties of the skin.

This might be an oversimplification, however. If we think of the individual receptors as just creating one big receptor, then the argument holds. But if we consider the sampling array to be the receptors themselves, the story is more complicated. Now we would need to know the filtering imparted by the skin. There is some cross-wiring of receptors – each Merkel cell feeds more than one nerve cell, suggesting this picture might be the more accurate one.

The psychophysical resolution is smaller, implying analogous to hyperacuity, but there is no evidence of aliasing type effects, perhaps as a result of the filtering emposed by the convergence ratio (§10.5.1.2).

As noted above, no physical mechanism imposes a rigid cut-off frequency, as does diffraction in vision. Yet the mechanical properties of the skin impose a softer limit. If we ignore the filtering effects of the skin, we can calculate the effective undersampling assuming that all prefiltering is carried out just by the receptive field of the SA1 receptor. Assuming the field is Gaussian, then we get about a factor of two undersampling, very similar to the situation in vision.

The cut-off for discrimination of curvature is about a radius of 1.5 mm (§10.5), which is close to the afferent spacing. Thus some level of receptive field overlap must occur, suggesting that there is no significant undersampling.

10.4.2 En route to the brain

The structure and organisation of the brain varies radically even amongst mammals. Obviously cat needs rather different wiring than whale. Whereas in vision and hearing there is considerable variation in scaling and resolution and maybe anisotropy, this is nothing compared to the differences imposed by different body shapes, different limb usages and the many other ways cat's body is different to man. Even within monkeys, variations are considerable, both in the brain areas within the somatosensory cortex and the way these areas are connected together.

The somatosensory system in the detection and transmission phases is very diverse across all three modalities, time, space (in this case over the surface and through the volume of the body) and streams. Unlike all the other senses two primary neural pathways (dorsal root ganglia and cranial nerves) transmit information to the brain.

10.4.3 The spine

The spine is the big somatosensory transmission channel for most of the body, the equivalent of the transoceanic optical fibres which keep the internet alive. It has many different sorts of input and output, making the present account a little simplistic. The cross-section of the spine has a central area a bit like an X or H and is known as the 'horn' with ventral and dorsal, left and right symmetric components. The horn contains ten laminae (layers) of grey matter (containing cell bodies), surrounded by columns of white matter. The *dorsal horn* primarily takes input from the peripheral nerves, the *dorsal root ganglia* (DRGs), destined for the brain and comprises laminae I–VI (Gardner & Kandel, 2000). The *ventral horn's* primary role is the transmission of signals from the brain back to the muscles and internal organs. So, this chapter deals with the dorsal horn. Figure 10.4 shows the mapping of different somatosensory receptors to the dorsal horn laminae, emphasising the streaming of different types of information en route to the brain.

The 31 nerves which enter the dorsal horn, the *dorsal root ganglia*, fall into several classes:

the **5 sacral nerves** for the lower limbs;
the **5 lumbar nerves** for the lower back;

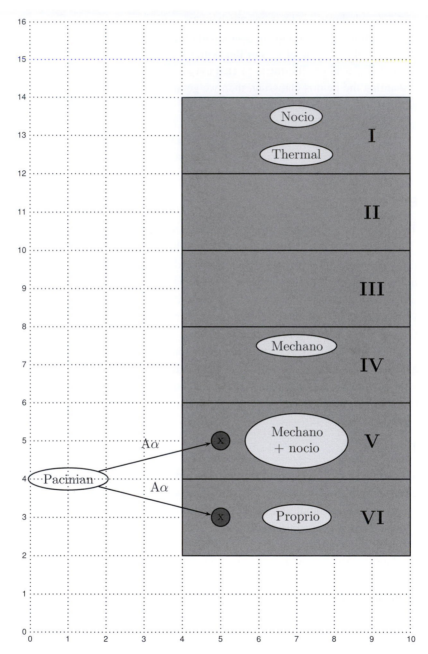

Figure 10.4 Streaming of information from different types of receptor into the dorsal horn.

the **12 thoracic nerves** for the chest and upper body;
the **8 cervical nerves** for the neck;
the **coccygeal nerve** for the coccyx.

Each of these 31 nerves corresponds to a particular part of the body surface, known as a *dermatome*. So, for example, the first sacral nerve, S1, innervates the ankle and back

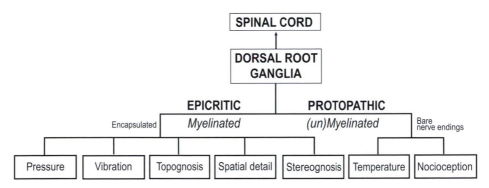

Figure 10.5 Spinal cord pathways.

and outer edge of the foot. Two distinct pathways operate: the dorsal column-medial leminiscal system transfers touch and limb proprioception, proceeding on the ipsolateral side and crossing over in the medulla/pons area; the anterior-lateral system crosses over to the contralateral side lower down on entry to the spinal cord and transfers chemical and thermal information. Figure 10.5 shows the pathways feeding into the spinal cord from a functional perspective.

Figure 10.5 shows an overarching distribution of fibres within the dorsal root ganglia. There are two categories: *epicritic* neurons are encapsulated, i.e. they end in some sort of transduction entity like a Merkel receptor; *protopathic* neurons have just bare nerve endings embedded in tissue. The epicritic handles spatial and temporal information, essentially the sense of touch. The protopathic handles pain and temperature sensitivity and proprioception.

10.4.4 The trigeminal pathway

Input from the face and skull does not travel through the spine, but though the *trigeminal pathway* in the fifth cranial nerve. Like the spinal cord it has a laminar structure with different mixes of receptor types feeding into it. It has two distinct nuclei (Iggo, 1982):

caudal nucleus, which handles a mixture of receptor types and projects contralateraly to the ventral posterior medial (VPM) nucleus in the thalamus;
principal sensory nucleus, which handles fast mechano-receptors and projects to the ventrobasal (VBM) nucleus of the thalamus.

One might guess that the main source of information from the neural matter of the inside of the teeth is nocioceptive, pain in other words!

10.4.5 Somatosensory cortical pathways

The pathways described here are broadly based on primates, and reasonably close to what we expect in humans. The interested reader might consult the detailed review by Kaas and the recent studies of humans by brain imaging (Burton, 2002; Kaas *et al.*, 2002).

There are two distinct generic somatosensory areas in mammals, SI and SII, but unlike V1 and V2 in vision, where V2 is fed primarily by V1, SII receives a lot of direct input. But like V2, SII carries out more complex processing operations with upwards of a third of its cells being bilateral (Eickhoff *et al.*, 2006a; Eickhoff *et al.*, 2006b), with the ispilateral side receiving input through the corpus collosum (Fabri *et al.*, 1999). Phase locking of neural activity in contralateral SI and SII has been observed using MEG (Simões *et al.*, 2003).

In human, SI is usually considered to be Brodmann areas 3b, 3a, 2 and 1. Area 3b is strongly myelinated, with columnar organisation separating body areas, such as face from hands, or just the fingers of the hand. It is critical for tactile form perception. Although other areas may also have a columnar organisation, it has not been fully demonstrated and the strong myelination is absent. The other areas are fed by 3b as well as from direct input as we shall see below. Area 3b processes surface roughness, based mostly on inputs from SA1 fibres. Vibrational information does not contribute to the perception of surface roughness (Lederman & Klatzky, 2002). Each area contains its own topographic map of the body, making four distinct maps in SI.

SII, although defined long ago (Adrian, 1940), is more complicated and is approximately equivalent to Brodmann areas 40 and 43. It lies ventral to SI and has three distinct areas, in primates referred to as SII (a subsidiary component of overall SII), PV and VS. Recently Eickhoff *et al.* (2006a) have identified analogues in an area, the *parietal operculum*, labelled OP1 to OP4, running caudal to rostral. OP1 corresponds to SII, OP4 to PV and OP3 to VS. OP2 is not uniquely somatosensory. Each of these areas contains an invidual map, with OP1 and OP4 mirror images as shown in Figure 10.6 for non-human primates. The cells in these areas are distinguishable by staining methods (Eickhoff *et al.*, 2006a), suggesting different functions for the different maps, but further work is needed to fully define them. These different maps are also identifiable by fMRI imaging as the subject's skin is brushed with a sponge (Eickhoff *et al.*, 2006b). These areas just outside SII may have specific pain functionality (Treede *et al.*, 2000).

Whereas in SI individual fingers can be seen by fMRI, in SII the maps are more diffuse and the extent of overlap of the individual fingers is uncertain (Ruben *et al.*, 2001). Also distinct from SI is the bilateral sensitivity of a significant fraction of cells.

Two other Brodmann areas, 5 and 7, located in the *posterior parietal cortex* are also components of the somatosensory system. Area 5 receives input from SI and integrates left and right across the corpus collosum, combining proprioceptive and mechano-receptor information. Area 7 goes even further combining visual with somatosensory information.

Somatosensory information has quite a number of streams but there are three essential pathways, labelled here as SP1, SP2 and SP3. Each of these separate pathways has its own topographic maps, so the body surface is represented numerous times throughout the cortex as discussed above:

SP1, the mechano-receptors; it is thought that the four mechano-receptor types, RA, SA1, SA2 and PC form four independent subpathways throughout, carrying out different processing tasks (Johnson & Yoshida, 2002);

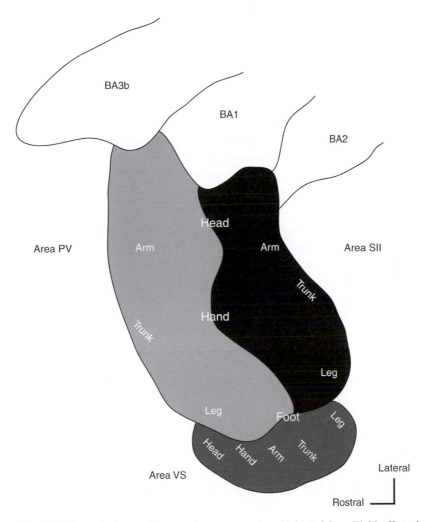

Figure 10.6 Layout of secondary somatosensory cortex. (Adapted from Eickhoff *et al.*, 2006b.)

SP2, the muscle spindles (proprioception and kinaesthetics);
SP3, the thermal/chemical sensors.

There is an additional set of fibres running in the peripheral nerves, the *C-fibres*, which are unmyelinated. One role they have, certainly in the hand, is for thermal and noxious stimuli and they also seem to transmit mechno-receptor activity in the glabrous skin of the forehead (Macefield, 1998). But they may have an additional inhibitory role. A selective drug such as capaicin, the active ingredient of chilli, which suppresses these fibres, causes cortical receptive fields to *expand*. In other words, the receptive field's boundaries are *defined by inhibition* as occurs throughout sensory processing. If this inhibition is removed, then the receptive fields get bigger (Calford *et al.*, 1998).

The first stage of entry to the brain in the somatosensory cortex is the medulla. The external cuneate nucleus (ECN) receives information from SP2, the muscle spindles, i.e. the proprioceptive and kinaesthetic information. Naturally, the ECN projects to the cerebellum as well as to the thalamus, the common relay station for sensory information. The RA and SA mechano-receptors project to three other nuclei in the medulla, the gracile, cuneate and trigeminal nuclei (GCTN), which also project to the thalamus.

The thalamus has three principal tactile modules of its 50 or so subdivisions. The ventroposterior superior nucleus (VPS) receives input from the ECN and thus carries the muscle information. The middle module, the ventroposterior nucleus (VPN) receives the mechano-receptor information from the GCTN, which in turn splits into a module for the face (the ventroposterior medial nucleus, VPM) and the limbs (ventroposterior lateral, VPL). The lower module, the ventroposterior inferior nucleus (VPI) receives thermal/chemoreceptor data.

From here things get really complicated! There are at least four primary areas (Brodmann 1, 2, 3a, 3b) in SI and three in SII with direct projections from the thalamus to both areas. As noted earlier, 3b is the main area for mechano-reception and thus receives its input from VP. It projects to 1, then on to 2 and then on to 5/7, which also receives visual input. It also projects to SII. So pathway SP1 travels through GCTN to VP to 3b and hence to 1, 2 and SII. Areas 1 and 2 also project to SII.

The area for muscle information is 3a, so pathway SP2 travels through ECN, VPS to 3a, and 3b also projects to 3b (recall how skin-stretch can contribute to monitoring hand position).

Finally, with SP3, the thermo/chemical information bypasses SI completely and goes straight in to SII.

Complex though it may seem, this is still something of a schematic oversimplification, since many of these areas have other, weaker projections! For example, VPS and VP to area 2 and VPI to many areas. Another nucleus, VMpo, may be specific to pain/temperature information and possibly be responsible for the affective aspects of pain stimuli (Willis, 1995) (see Figure 10.7), and there appear to be complex interactions between pain and temperature (see §10.5).

Most of these areas contain full body maps, i.e. they are topographically organised. We will leave our exploration of the somatosensory system at this point, before the abstraction of shape and form. Nelson *et al.* (1980) provide a detailed map of mammalian SI showing multiple topographic maps of the body surface. Also evident is the same strategy found in vision (fovea) and hearing (speech frequencies) non-uniform allocation of cortical space (§10.4.8).

10.4.6   Affective and discriminative distinct pathways

The pathways to the somatosensory cortex handle discriminative aspects of touch. They provide spatial awareness of where stimuli are on the body and the nature of the stimuli, deformation and vibration. Separating out the distinctive functions of each pathway is difficult, but sometimes a genetic or other malfunction offers extra insight. In one such case, Olausson *et al.* (2002) report a subject who was lacking the fast myelinated nerve fibres characterising much of the tactile system. They were thus able to identify the role

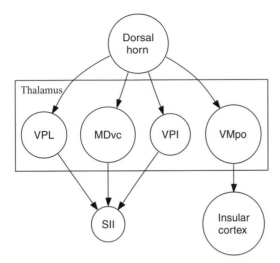

Figure 10.7 Pain/thermal pathways after Willis (1995). Middle level nodes are in the thalamus: VPL, ventral posterior lateral nucleus; MDvc, medial dorsal (ventral caudal part); VPI, ventral posterior inferior nucleus; VMpo, posterior ventral medial nucleus.

of another stream to the tactile system, consisting of slow unmyelinated fibres. These respond to soft stroking of the skin (in the hairy, rather than glabrous areas), but not to vibration. The subject reported very little positional information associated with the stimulus. Concurrent fMRI showed that the signals were going directly to the insular cortex and *not* through the somatosensory cortex. Thus there seems to be an *affective* (emotional) stream of tactile information as well as the discriminative stream.

But the signals arising from a textured surface are complicated, dependent on pressure and speed of movement. But, at the cortical level, some of these detecting factors seem to be normalised out, much in the manner of colour constancy (Chapman, 1998). The visual areas V1 and V2, which receive most of the input from the thalamus, do not exhibit colour constancy, which first appears in V4 – there is some hierarchical structure. More cells are speed sensitive in 3b than in area 1. This is suggestive of a gradual extraction of object characteristics (surface roughness) over details of the data collection (speed of movement of the hand and pressure on the surface).

There is also evidence that the affective properties of touch are streamed separately. Rolls *et al.* (2003) found that pleasant and painful, as opposed to neutral, stimuli activate specific and different areas in two brain areas, the orbitofrontal cortex and the anterior cingulate cortex.

10.4.7 The cortical control of attention

Attention seems to have little effect on the firing of neurons in the thalamus, but possibly some effect in the cortex (Hsiao & Vega-Bermudez, 2002). The primary somatosensory cortex, SI, exhibits some effects, but the suppression effects are large in SII. The control of attention seems to originate in the posterior parietal cortex.

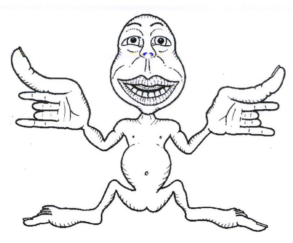

Figure 10.8 A somewhat unpleasant humanoid figure showing the relative proportions of the somatosensory cortex devoted to different parts of the body.

10.4.8   The cortical map of touch

The layout of the body in the somatosensory cortex is uneven. Some areas get more cortical space than others, as displayed graphically by the gargoyle in Figure 10.8. The shading on the figure shows the two-point resolution we discussed in §10.5. In effect this is like a resolution map, reflecting the density of receptors in different parts of the body, where the allocation of cortical processing power neatly parallels the peripheral receptor density.

As one might expect the hands get a lot of processing power. A bit more surprising is the amount of cortex devoted to the lips which get the largest component of all. Facial expressions are very important for communication. The eyes are frequently assumed to be the 'windows of the soul', and recent work on autism suggests understanding the emotions conveyed by the eyes is very important. But, as computer animators will endorse, the mouth rather carries a great deal of expression of emotion (Fleming & Dobbs, 1999), sometimes more so than the eyes. Speech is of course strongly influenced by the lips and thus we would expect strong proprioceptive control in fine adjustments of the mouth.

The 'use-it-or-lose-it' maxim applies to the somatosensory cortext just as to other parts of the brain. The area is very malleable and adaptable and §10.5.4 describes how, following an injury, rewiring occurs quite rapidly.

With vision and hearing the eye and the inner ear are pretty much the same size across the human race. But people vary enormously in height and girth. So one might ask whether bigger people have more touch receptors or just lower resolution. At present not much is known about variations across individuals or changes during the lifespan, although there are some indications of loss of nerve density with age (Goodwin & Wheat, 2004). At the level of the nerves feeding back to the brain up the spinal cord, the density is roughly constant, so the cortical representation of the body has roughly the same information requirements. Across animal species the story is different. The brain size of animals does not map directly to their intelligence. Bigger animals, most noticeably whales and pachyderms, require larger brains to manage the body. Measures of intelligence (without asking too closely what that exactly means) tend to be found in the size of the neocortex, which, of course, is largest of all in humans. The complexity of social interaction is thought to be one driver of neocortical size (Dunbar, 1996)[4].

10.4.9   Modification of somatosensory cortical activity

Although information in the cortex is a somewhat distorted map of the sensory areas over the body, the cortex does not necessarily represent what is coming in from the sensors, but more what is actually being perceived. Feedback from higher levels modifies the extent and position of activity. One common somatic illusion is so-called *tactile funnelling*. Two stimuli which are in distinct places seem to occupy a single central position. Recently Chen *et al.* (2003) have shown in squirrel monkey that in conditions in which humans suffer this illusion, in area 3b (responsible for form perception) of the somatosensory cortex, there is indeed just a single representation in the middle of the two stimuli. Thus at this first level of processing in the brain, the activity reflects what we perceive, not what we sense. Similar things happen in vision (§6.6.2.1). This is another example of the paradigm shift which has been taking place over the last few years in the role of top-down processing in sensory systems (§2.3.4). Another example is the strange phenomenon of being unable to tickle oneself. Blakemore *et al.* (1998) found that the cerebellum is instrumental in suppression of sensory activity brought about by motor actions of the body – i.e. it predicts the sensory outcomes and cancels this prediction against the actual outcome, somewhat analogous to predictive coding (§4.2.5). Their fMRI results show less activity from self-generated to external stimulation in the somatosensory cortex and the anterior cingulate cortex (which might be associated with the specific affective properties of being tickled), resulting from these cerebellar signals.

10.4.9.1   Touch information

One of the earliest attempts to look in detail at the spike train was by Werner and Mountcastle in the 1960s, reanalysed by Rieke *et al.* (1996; §3.2.2) They measured the firing rate in mechano-receptor response to varying static deflections of the skin in monkey. It turns out that, consistent with the other results we have seen, the information rate is about 2.5 bits per spike.

In the hand at least each individual fibre can generate activity elsewhere. McNulty and colleagues (2001; 1999) showed that muscle activity is detectable from stimulation of single FAI, FAII or SAII fibres.

If we add up the receptor counts across the body, we find about 300 receptors in each fingertip giving around 1400 per hand with about 4000 for the palm, say around 5000 per hand (§10.3.1). Given the huge fall-off of density on other parts of the body, the

---

[4] This is possibly controversial. A finding just published indicates that the frontal cortex is not *disproportionately larger*, but simply scales with overall brain size. See Dunbar and Schultz (2007).

total receptor count will be around ten times this or about a quarter of a million across the body. Thus the total bandwidth for touch falls somewhat below that of vision and hearing and is fairly similar to that of the chemical senses. Although the bandwidth for touch is well below that of vision, the temporal sampling demands are quite high. At least a 2 kHz sampling rate is necessary to keep pace with the fastest mechano-receptors.

10.5   Psychophysics of the somatosensory system

Figure 10.5, which showed the pathways leading into the spinal cord (§10.4.3), illustrates the different types of information coming into the somatosensory system. The temperature and pain pathways are unmyelinated, therefore transmit more slowly and are thus lower bandwidth channels. Two of these, stereognosis and topognosis require integration of other senses and are deferred to §7.2.3.

Many of the early somatosensory measurements date from the 1950s, particularly from the seminal work of Nobel Laureate Vernon Mountcastle (Iggo, 1982). The same core ideas of *where* and *what* hold for mechano-receptor and touch as they do for vision and hearing. In the first case, the spatial location measurements are resolution and detection of gratings (Fourier-like components), but there is also an analogue to depth perception in the use of touch to get information about object curvature, i.e. 3D shape (*stereognosis*).

10.5.1   Spatial analysis

The first measurement of interest is two-point resolution. In this case, if a sharp instrument is used to gently prick the skin simultaneously in two different places, then there will be a minimum separation of the points of contact required before it is possible to distinguish two distinct positions. The two-point resolution is shown on the grotesque cartoon in Figure 10.8. As you might expect the fingertips have excellent resolution at around 1–2 mm. The worst is on the broad expanses of the body such as the torso where it is at least ten times worse (see also §10.6.2). The discrimination of position in the fingertips is close to the limit set by the MC density.

The tongue has perhaps the best resolution (Johnson & Yoshida, 2002; van Boven & Johnson, 1994) and its tactile performance is currently being harnessed to help in surgery (§10.8.3).

10.5.1.1   Temporal and spatial frequency analysis

The Fourier perspective requires measuring the temporal and spatial frequency limits. To measure *temporal* resolution requires the application of a vibrating stimulus to the skin and increasing and decreasing the frequency until the vibration is no longer detectable. Frequencies in excess of 1 kHz are detectable in line with the peak sensitivity of the PCs. But the minimum amplitude vibration detectable occurs at a frequency of around 250–300 Hz.

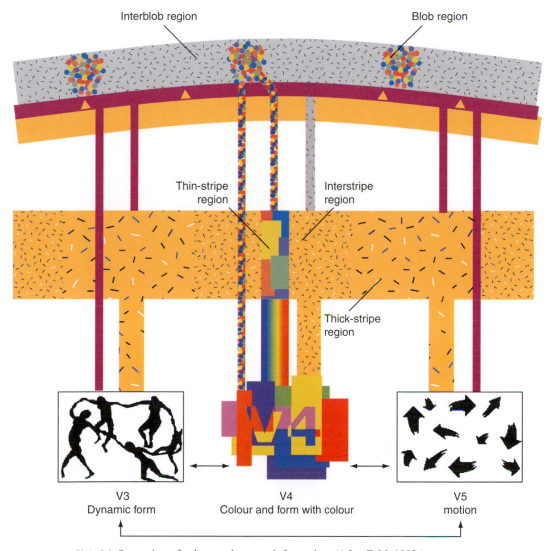

Plate 8.1 Streaming of colour and texture information. (After Zeki, 1993.)

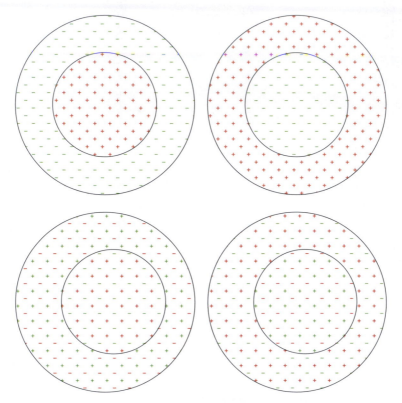

Plate 8.2 Single (top) and double (bottom) opponent cells. Plus signs denote excitation and minus signs inhibition. These cells are sensitive to the *wavelength* of the incident light.

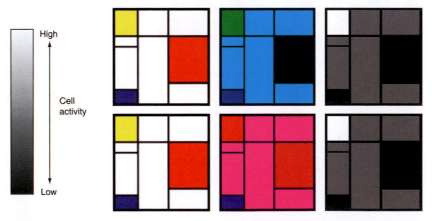

Plate 8.3 Schematic *illustration* of the activity of a hypothetical yellow V4 cell. The Mondrian on the left is illuminated by green light (middle, top) and magenta light (middle, bottom), generating green and magenta casts respectively. Magenta is minus green, so yellow under magenta light appears red. The diagram on the right shows the activity of the colour sensitive cell which responds to all three patterns even though the wavelengths reflected are quite different. Since colour constancy is actually operating in the viewing of this diagram, the reflectances are deliberately exaggerated. Redrawn from Zek, see text.

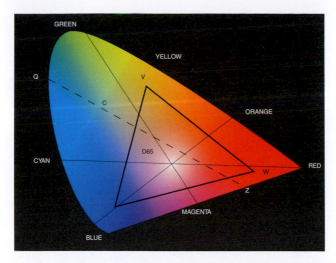

Plate 8.4 The chromaticity diagram. D65 is the white point discussed in the text. The line defining the the curved part of the horseshoe are the chromaticity coordinates of single wavelengths, the straight line has the non-spectral colours. The triangle illustrates the gamut possible with just three sources, typical of most CRTs and almost all LCD and plasma screens at the time of writing. It is impossible to cover the whole human gamut with just three phosphors or LEDs.

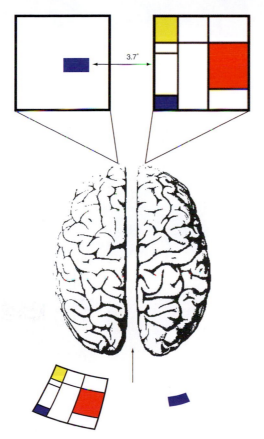

Plate 8.5 Colour constancy works across the hemispheres, i.e. the illumination estimated from the Mondrian on the right allows a good estimation of the colour on the left. The surround will influence the perceived colour for up to 10 degrees in one visual field, and in this experiment up to 3.7 degrees. (Redrawn from Zeki, 1999, with permission; based on experiments by Land, 1990, see text.)

Plate 10.1 Elephants use their trunk for a variety of tasks, including social interaction (§10.6.3).

Plate 10.2 The star-faced mole. Reprinted with permission from Macmillan Publishers Ltd: Catania, 'Olfaction underwater 'sniffing' by semi-aquatic mammals', Nature, **444**, 1024–1025 © 2006.

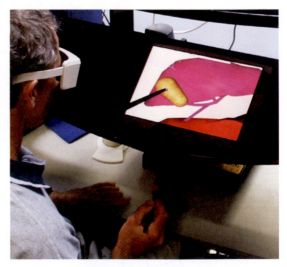

Plate 10.3 Haptic workbench. (Courtesy of Dr Dean Economu, CSIRO.)

The other Fourier parameter is the minimum and maximum *spatial* frequencies of patterns on the skin. But, whereas with vision the resolution of two spots of light does not reliably indicate the separation of photoreceptors, in touch the two-point separation does seem to indicate the separation of the Merkel disc receptors. In such a case, the two-point receptor distance becomes the sampling frequency. The minimum detectable spatial frequency is around 1 cycle mm^{-1}, but whether we should really call this a spatial frequency or element resolution is open to doubt, since only a small number of cycles would fit on the fingertip.

There are strong information processing analogies to static and varying stimuli in vision (§7.5). Using a similar paradigm of detecting the orientation of a grating (in this case a set of bars created by a microarray of tactile probes), Bensmaïa *et al.* (2006) report that the threshold spatial frequency goes down as the temporal spatial frequency goes up. They argue that it arises from interference between FA1 and SA1 channels.

Also, dynamic stimuli tend to give lower thresholds than static ones. Pressure sensitivity is about 2–3 times higher for a static stimulus than for a moving stimulus where a surface ridge of 1–3 μ is detectable. Similarly, for a 1 mm spatial frequency grating, the minimum detectable frequency changes is around 5%, corresponding to a change in period of 52 μ (Chapman, 1998).

10.5.1.2 Curvature

Curvature is another important object characteristic, independent of shape, size and movement. For example a cube and sphere of the same size will require the same hand position, and proprioceptive information about the object. Curvature, however, clearly distinguishes the two (Goodwin, 1998).

The detection of curvature varies approximately linearly, with about 75% correct responses occurring at a sphere radius of around 3 mm. A 10% difference is detectable.

The positional accuracy of location of a sphere depends on its curvature: at a small radius of 1.92 mm the minimum detectable shift in position is 0.38 mm, rising to 0.55 mm for a sphere of radius 5.80 mm (Goodwin, 1998). These distances are smaller than the distances between afferents and even smaller than the distances between the receptors themselves. Hence in vision terms, this would class as *hyperacuity* (§6.7.1.2).

10.5.1.3 Texture and surface properties

Two of the object characteristics which have driven a lot of investigation are texture and shape. The characteristics of surface texture, which are often highly indicative of object identity, may be categorised as:

- *roughness*;
- *hard/softness*;
- *sticky/slippery*, which has only recently become recognised as an independent surface dimension.

10.5.2   Attention in the somatosensory system

Again, as in vision, attention drives how the cortex processes tactile information. Amongst these similarities, there are preattentive properties, windows of attention and the modification of somatosensory activity according to attention and perception.

10.5.2.1   Preattentive stimuli

The first parallel with vision is in preattentive stimuli (§6.7.2, §8.8). Some tactile stimuli are perceived fast, in parallel, others seem to require serial processing.

Parallel in the somatosensory context means parallel across the surface of the body, or even just between fingers. Preattentive stimuli can be picked out from distractors elsewhere, say a stimulus on one finger with different stimuli on the others. Hot/cold, rough/smooth, hard/soft are all preattentive (Hsiao & Vega-Bermudez, 2002). Texture and vibration are possibly preattentive, perhaps dependent on the precise nature of the stimulus.

If the system has to piece together information to arrive at a percept, preattentiveness is less likely. So, spatial analysis tasks, stereognosis and other complex shape-form percepts are not preattentive.

10.5.2.2   The window of attention

Attention is like a searchlight, picking out features of the world for detailed examination. In vision, the fovea normally is the recipient of attention. If some distracting feature in the periphery appears, the eyes shift to bring it into the foveal region. The high resolution of the fovea sets a natural spatial window to attention.

In the somatosensory system, this strategy of bringing high resolution to bear is not available. The areas of high resolution such as the lips and hands are fixed in particular parts of the body. Of course a tactile interrupt anywhere on the body may bring the hands or other senses into play as we swat (hopefully) a biting mosquito.

But confined to the local tactile sense alone, the window of attention can be as small as a single fingerpad. When two stimuli fall within this window, confusion can occur, slowing down perception. The window is not only spatial, but temporal, allowing easier distinct perception, when stimuli occur at different times.

10.5.3   The perception of pain

Pain is surprisingly illusive anatomically and physiologically. Although the pain receptors within the body itself are easily identified, the cognitive contributions to painful stimuli are very complicated. Of perhaps all sensory activity, the relationship between stimulus, peripheral nerve activity and cortical processing is most complex.

From a psychophysical perspective pain is exceptionally diverse, but the parameters which have characterised the other senses, spatial mappings and temporal dynamics are less useful. The information transmitted and utilised about pain is less about spatio-temporal properties of something happening to, or inside, the body, but more about needs and priorities for action. Thus we shall not consider pain in any great detail.

There are some perceptual characteristics worth noting:

- Pain is often referred. Not only is it not spatially precise, but it may appear in a completely different part of the body!
- The effect of a damaging stimulus is often two phase – a subjectively sharp pain (first pain), followed by a more prolonged burning or throbbing sensation (second pain), reflecting the multiple neural mechanisms (§10.3.3).
- Pain can be turned off, at a peripheral level, by vibration, rubbing, shaking, or at a cognitive level by natural opiates or other means unknown.
- Subjectively pain can equally be turned on at a cortical level, arising from no clear peripheral activity. The most striking and distressing example of this is pain in phantom limbs following amputation (Ramachandran & Blakeslee, 1998) (§10.5.4).
- Painful stimuli are very persistent! Yet at the same time pain can sometimes be suppressed at a cognitive level. New research is exploring the possibility of pain relief using virtual reality, literally taking the patient's mind off pain by involving audiovisual stimuli.

Pain is one area where we probably do not need to do any research in the human–computer interface, unless the pursuit of realism in computer games extends to new levels!

10.5.3.1　Temperature and pain

As discussed in physiology and pathways above, pain and temperature are communicated along low bandwidth channels. These are often diffuse experiences, not precisely localised. They often interact in surprising ways too. One such example is the *thermal grill* (Craig *et al.*, 1996). Here a pattern of alternating bars of warm and cool produces a surprising sensation of intense, painful cold. Furthermore, although cool and warm do not significantly activate the anterior cingulage cortex, an area actively involved in emotional processing, this combination does.

This effect seems to be related to the way cold inhibits pain. Craig *et al.* (1996) suggest it arises through *disinhibition* of the pain pathway in thalamo-cortical pathways. This underlines the extremely complex nature of pain perception and its often confusing relationship to external stimuli.

10.5.4　Phantom limbs

One of the most curious aspects of the somatosensory system is the phantom limb phenomenon. A person who has lost a limb may still feel it from time to time, as anything from an itch to severe pain. This is an activity within the brain, since the input is of course no longer there.

Ramachandran and Blakeslee (1998) give a highly readable account of how, because the somatosensory cortex covers such a wide range of stimulus sources, if one area fails to receive input other adjacent areas may take over. Thus in one patient, who had lost a hand, the index finger no longer provided input, but an adjacent input area, the upper lip, would now produce a sensation in both the lip and the 'phantom' finger.

This rewiring happens really fast. Kelso (1995) describes experiments on owl monkey which demonstrate this. When a finger is amputated, the regions of the adjacent fingers move in over a period of a few weeks, even in the adult monkey. In these experiments Merzenich and colleagues found remapping of cortex three weeks after a nerve has been cut (Holloway, 2003). In other examples, women who have had masectomies experience mapping of the phantom nipple to the sternum or even the earlobe within a few days of surgery.

But higher pathways which take input from this finger area may not rewire properly (or at all) and thus they effectively do not realise what has happened. So stimulating the nose might activate an area which now includes the lost finger area so creating a sensation in that finger.

A less tragic example of this is Ramachandran's explanation (ABC Radio) for why toe-sucking is often found to be erotic. Forget Freudian phallic symbolism and just note where things are distributed in the cortex!

This makes me wonder about the basis of foot fetishes in normal people . . . I suggest that the reason is quite simply that in the brain the foot lies right next to the genitalia. Maybe even many of us so-called normal people have a bit of cross-wiring which would explain why we like to have our toes sucked.

An even more extraordinary idea, reported by Ramachandran in the media (Williams, 2007), occurs with people who have undergone gender reassignment. Following injury or pathology related surgery in which a man loses his penis, a phantom may be experienced. But for people who undergo surgery following a sex change, this does not always occur. The brain's map of the body seemed, in these cases, to never have included the male genitalia.

10.6   Special tactile adaptations

Tactile sensors, just like eyes or ears, come in lots of different forms. Some are unique to particular animals, such as the elephant's trunk (§10.6.3), while the star-faced mole takes the prize for most unusual (§10.6.2).

10.6.1   Whiskers

Many small mammals, such as rodents and cats, have whiskers which help them navigate their environment. In the serious fires in the hills above Berkeley in the 1990s, many pet cats survived the inferno by digging in underground. Yet many got their whiskers burnt and were seen for some weeks afterwards bumping into things!

The whiskers form a distinctive spatial map. In the rat, for example, there are about 30 or so whiskers arranged in a grid, or *vibrissae*, each with its own vibrissal nerve (Gardner & Kandel, 2000). Each nerve carries about 100 neurons, conveying the displacement in time and space of the whisker. Each vibrissal nerve projects to a patch of cortex shaped like a barrel, with similar centre-surround properties to those in the other senses (Fox *et al.*, 2003). The whiskers have a fairly neat topograhic array and a similar topographic organisation is found in the *barrel* cortex with five horizontal rows, just as we find topographical mapping of the retina photoreceptors or the hair cells of the ear.

Just as with other sensory systems, early deprivation of inputs leads to distortion of the maps and loss of plasticity in the adult (Petersen *et al.*, 2004; Rema *et al.*, 2003) and the very precision of the barrel map and its mapping to the whiskers has made it very suitable for such studies.

The directional tuning of whisker movement is achieved by inhibitory mechanisms mediated by the neurotransmitter GABA. This is analogous to orientation tuning in the visual system, and, if GABA antagonists are introduced into SI, directional tuning decreases (Calford *et al.*, 1998).

But, just like other areas of cortex, the whisker barrels area can turn out to be something else. On the other hand, if the incoming nerves from the whiskers are transplanted into prospective visual cortex, the barrels form in this new place.

10.6.1.1 Information transmission by whiskers

A rat uses its whiskers actively for rapid identification. It will palpate or *whisk* – briefly brush against an object – and make a decision about where to go or what to do within a few hundred milliseconds. Thus it cannot average over a train of spikes. The whole operation generates around just one spike in the cells which receive the greatest stimulus (Rieke *et al.*, 1996, p. 55), while a whisker deflection generates around 0–2 spikes from layer 4 (which receives thalamic input) (Finnerty *et al.*, 1999). Panzeri *et al.* (2001) find that the first spike carries 83% of the information in the first 40 ms and that the average bits per spike is around 1.6, a similar figure reported for insects (§5.4.3.1). Arabzadeh *et al.* (2006) obtain similar figures, comparing information capacity from spike timing data versus spike counts, achieving around 1.5 bits per spike with a spike timing window of 4 ms.

Jones *et al.* (2004) show how precise this spike encoding is using exactly the same linear convolution kernel approach developed by Rieke and others (§4.6.2). Exposing a whisker to white noise stimuli in its preferred orientation and recording the spike trains leads to a kernel independent of stimulus frequency and highly accurate for acceleration, and reasonably so for velocity and position. Optimum performance was around 125 Hz suggesting that whiskers perform best for sharp transitions, i.e. as they brush against an object.

Each whisker has a preferred direction of movement. A strategy from de Ruyter computes kernels which perform even better. If the kernel is computed using stimuli in the preferred direction and mirror imaged to the reverse direction and the composite spike train, the computed kernels perform even better. This suggests that integration of different directions in the somatosensory cortex gives highly accurate deflection information.

Whiskers do far more than sense when they brush against an object. The dynamics of their function is still a subject of ongoing research. For example, Williams and Kramer (2010) show that there are computational advantages to having a tapered whisker even though this might mean that the tips break off more frequently. Whiskers are widespread in mammals (but not monotremes) and amongst numerous distinct species, including cat, rat, fox, badger and others, there is a varying ratio of base to tip diameter which is often an order of magnitude.

Going beyond the shape of a surface with which the whiskers are in contact is the characterisation and identification of this surface. Thus the 3D surface texture is important as we saw in §7.2.5. In relatively recent work it transpires that whiskers have narrow tuning frequencies, just like the hair cells of the ear or the bandpass cells of the visual cortex. Hartmann (2003; Neimark *et al.*, 2003) argue that this tuning subserves texture discrimination, analogous to the braille example above (§10.3.1).

Mounting evidence suggests that the process of whisking has a strong top-down component. Whisker twitching starts with periodic neuronal pulses at around 7–12 Hz in the barrel fields which spread downwards. As larger exploratory movements begin, the frequency drops to 4–6 Hz (Nicolelis *et al.*, 2002). Somatosensory cortical activity is radically different depending upon whether the rat is actively exploring or whether it is being exposed to a stimulus passively (Krupa *et al.*, 2004). The frequency of whisking depends too on the task. The frequencies for sensing a texture are much higher than those for determining shape and positions of objects (Kleinfeld *et al.*, 2006).

10.6.1.2 Robot whiskers

The wide variety of animal senses and specialisations within them frequently follow adaptations to lifestyle and environment. Moles, which spend most of their time underground, or bats, which hunt at night, find other senses serve them better than vision. So with man-made systems mimicking animal senses is often a good strategy. Robot whiskers are now a reality, although with relatively small 4 × 1 arrays at the time of writing (Solomon & Hartmann, 2006). By computing the bending moments at the base of each whisker made of spring steel wire and fitted with a strain gauge, the whisker array could whisk a complicated 3D object and produce an accurate description of the 3D surface.

10.6.2 The strangest tactile organ

Moles are often thought to be blind, not surprising when you consider how much time they spend underground in dark burrows. One species of mole has an extreme adaptation of its tactile sense to compensate for its impoverished visual environment, described recently by Catania (2002). The star-faced mole (*Condylura cristata*) has a huge star of touch sensors on its nose! Plate 10.2 shows this grotesque anatomy.

The star is a *tactile*, not an olfactory organ (Catania, 2002). Each of the 11 appendages on each side is touch sensitive, containing 25 000 *Eimer organs*. Each has both a vibration sensitive and continuous pressure sensitive detector, unique to moles, plus some additional highly sensitive detectors right at the tip (Catania, 2002). The innervation is impressive with around 100 000 nerve fibres, around six times the number of the human hand!

The detectors number 11 are the most sensitive and Catania (2002) draws two interesting parallels to visual strategies:

1. The star can rotate to bring these most sensitive elements into contact with an object of interest, analogous to bringing an object into foveal vision.
2. The area of brain devoted to the star is relatively larger for these receptors than for the others, also analogous to the fovea (and of course to the area of somatosensory cortex devoted to hand and tongue (§10.4.8)).

This similarity to vision illuminates the way computational strategies get reused in different contexts. The relative allocation of cortical space is also analogous to the cortical magnification factor (§6.6.2).

10.6.3 The elephant's trunk

From the dark underground world of the small, secretive star-faced mole, the elephant is something of a contrast. Elephants see, hear and smell comparable to many large herbivorous mammals, but they also have one of the most sensitive and versatile of tactile organs, the trunk. They use their trunk for foraging, for drinking water (and spraying it everywhere), but it also subserves vibration detection in the ground and chemoreception.

The elephant trunk has a super-sensitive area on the dorsal tip, sometimes called the finger, analogous to the fovea in vision. Anecdotal evidence suggests it is sufficiently sensitive to be able to pick up a single grain of rice (Rasmussen & Munger, 1996) or a coin from the floor (Hoffmann *et al.*, 2004), although the African elephant, *Loxodonta africana*, appears more highly developed than the Asian elephant, *Elephas maximus* (Hoffmann *et al.*, 2004).

As with the human hand, there are a range of receptors, similar, although not identical, to the hand: free nerve endings; PCs; simple multi-branched corpuscles receptors, similar to receptors found in the mouths of dogs; and many hair cells with Merkel-like terminations. The PCs are smaller in the trunk finger (0.2 mm compared to up to 1 mm elsewhere), suggesting higher resolution, consistent also with them relatively closer to the surface than in human skin. Neither elephant seems to have the MCs found in humans (Hoffmann *et al.*, 2004).

Elephants also use their trunk for chemoreception. They also have a flehmen (§9.1) behaviour where they collect animal markings with the trunk and then insert them directly into the vomeronasal organ (Rasmussen & Munger, 1996) (§9.1) and sniff. When they meet a trunk shake will often occur (Plate 10.1) transferring some mixture of tactile and chemosensory cues.

10.7 Vestibular system

The vestibular system monitors the position and motion of the head. Many of the building blocks occur elsewhere in other senses and the system is physically co-located within the ear. Information streaming occurs in two ways: the principal streams relate to the type of information such as direction and nature of movement (sensitivity to velocity or acceleration, linear or rotational); the other is dynamic, common across the senses, the streaming of fast action and adaptation versus slower tonic response.

We can walk perfectly well with our eyes closed. We can touch our fingertips together at arms length. We can do this because we have an excellent awareness of not just where our own limbs and body parts are, but also a strong sense of our position and orientation relative to gravity.

A large alcohol intake seriously affects balance and the ability to walk in a straight line. Although the effect of alcohol on the brain is responsible for slurred speech and, at the limit, visual disturbances, the loss of balance is more peripheral. Alcohol changes the relative density of fluids within the canals and consequently disrupts the functioning on the vestibular system.

Some people function quite well being colour blind or having no stereopsis, sometimes without even being aware of the limitation (§7.2, §8.3.1.4). It is similarly possible to compensate to some extent with the senses, such as vision, if the vestibular system malfunctions through injury or disease, but serious problems remain[5]. The vestibular system responds rapidly and provides far faster balance control than is possible through senses such as vision. It also mediates correct posture.

The vestibular system is functionally separate from the ear but shares with the ear space in the petrous bone cavity and the fluid systems are interconnected (Benson, 1982). Like the proprioceptors of the somatosensory system, its role is in providing information about the head, its orientation and movement. It consists of two independent sensors:

> **the otolith organs** monitor linear acceleration of the head relative to gravity;
> **three semicircular canals**, at right angles which detect angular motion.

10.7.1   Functional components: otolith organs

There are two otolith organs, the *sacculus* and the *uricle*. Each operates in the same way, but one handles vertical movement, the other horizontal. Just as in the ear, the movement of hair cells transduces sound, so movements of hair cells in the otolith organs transduce the effects of head velocity and acceleration.

In each organ, movement of the head leads to movements of the fluid within, causing hair cell movement. As the head moves, so does the internal fluid, but the presence of heavy mineral particles, the otolith dust (calcium sulfate) adds inertia to the fluid[6]. So its movement lags behind that of the head, producing a shear on the hair cells. The hair cells cover an area inside the otolith organ, referred to as the *macula*[7]. In the case of the

---

[5] With the minor compensation that one would never get nauseous from travel or virtual reality.

[6] Otolith comes from the Greek words, *ous*, ear, and *lithos*, stone.

[7] Macular occurs often in anatomical descriptions. It comes from the Latin spot, and gives us via Italian the description of coffee, with just a dash of milk, *caffè macchiato*.

sacculus, the macula is vertical, in the case of the uricle it is horizontal, corresponding to the direction of sensitivity.

10.7.2   The semicircular canals

There are three semicircular canals, providing a complete 3D description of angular acceleration. One is horizontal, the other two are vertical, each at 45 degrees to the straight-ahead direction, diverging outwards, as if both sides of the head together formed the tips of an X pattern.

Each canal functions in the same way. It is fluid filled and has a bulge at one end, the *ampulla*, where hair cells are located. The canal is completely sealed, by a water tight diaphragm, *the caulla*, so fluid cannot go round in circles. So angular acceleration in the fluid exerts a pressure on the membrane, the *cupulla* at the end. The hair cells attached to it thus move and generate neural signals.

As with the other senses, hair cells come in *tonic* and *phasic* forms, sensitive to steady-state and transient stimuli.

Anatomically the hair cells come in type I and type II. Anatomically, they look different: the former have a bulb-like ending, the latter more cylindrical (Benson, 1982). The packing of cilia is approximately hexagonal, like photoreceptors, fitting as many as possible into a given area for a given diameter of fibre.

10.7.3   Vestibular pathways

The vestibular system shares the same nerve, the 8th cranial nerve, as the ear, to transmit to the brain, and has around 20 000 myelinated axons. There is some convergence since the utricle has about 30 000 hair cells and the saccule, 16 000 (Goldberg & Huspeth, 2000). The vestibular system goes straight to the brainstem, with a pathway bifurcating off to the cerebellum, responsible for much movement control. A population of Purkinje cells in the cerebellar *vermis* computes the translation from the head-centre (egocentric) reference frame of the otolith organs and semicircular canals to a world-centred (allocentric) reference frame (Yakusheva *et al.*, 2007).

There is also a thalamic pathway, projecting to the ventral posterior and lateral nuclei (Benson, 1982) in layers 2V and 3a.

From the thalamus, pathways project to the somatosensory and parietal cortex[8].

The mid-brain contains a module known as the interstitial nucleus of cajal (INC) which integrates eye movements but also controls head orientation. If damage occurs to either the right or left side, or there is an imbalance in input due to damage downstream, the head ends up on its side. This condition is known as *torticollis* (Klier *et al.*, 2002).

The eyes interact with the visual system in the control of gaze, an issue discussed further under cross-sensory integration (§12.1).

---

[8] The sacculus is now thought to have some direct emotional pathways operating at high sound levels in excess of 90 dBA. Maybe this is why rave parties are so popular.

10.8   Artificial tactile systems

The use of artificial touch information has grown rapidly over the last decade, from expensive haptic systems in laboratories to vibrational alerts on mobile phones. Space constraints allow only the briefest of overviews.

10.8.1   The JACK system

The JACK system was a development of the Centre for Human Modelling and Simulation at the University of Pennsylvania (Foley *et al.*, 1996). It aimed to provide accurate kinematics for complex walking, bending and gripping operations.

To build such a system requires very detailed information about the human body itself. It was already to hand, from the anthropomorphic survey of the US Army carried out in 1988. The JACK system modelled 73 joints with 136 degrees of freedom.

Typical applications include seating people in virtual enclosures. In the automotive industry such an enclosure would be a car, allowing assessment of comfort and accessibility of controls. In computer games giving characters smooth flexible body action is still a major challenge. Walking and running still (in 2011) looks stilted.

10.8.2   Robot hands

Hands are very popular robotic devices serving different functions, for input and output. The flexible and complex manipulation of the hand make it an excellent input device to a game or virtual reality (VR) system. Such technology is well advanced and becoming rapidly cheaper. Conversely, its value as an input device to a human user holds a lot of potential to give feedback about virtual objects.

Data gloves which track human hand movements have been around for a long time, but are still quite expensive. They are sophisticated pieces of microelectronic and sensor technology. A typical data glove (Foley *et al.*, 1996, p. 302) uses fibre optics to measure deflections. At one end of the fibre is an LED, at the other a phototransistor. Bending causes a loss of light and concomitant change in signal from the phototransistor.

The glove may also include a sensor which gives the overall position and orientation in 3D space. An example is the Polhemus sensor. It consists of three transmitter and three receiver coils oriented at right angles (somewhat *similar to the semicircular canals* of the vestibular system). Each transmitter sends out a pulse in succession, creating nine numbers for each cycle through the transmitter coils, enough to orientate and position. The range is restricted to a couple of metres, but the temporal response, at around 10 ms, is good. The spatial resolution is about 2.5 mm translation and 0.5 °C angular accuracy.

Paulson (2002) discusses the haptic interface which enables parents to feel the growing foetus. A haptic stylus provides feedback on skin surface and is pressure sensitive, a bit like a mechanical whisker. Clinical applications are foreseen.

Force feedback has reached a high level of development in the games industry. Steering wheels in driving games provide torque feedback, and in some of the most recent games

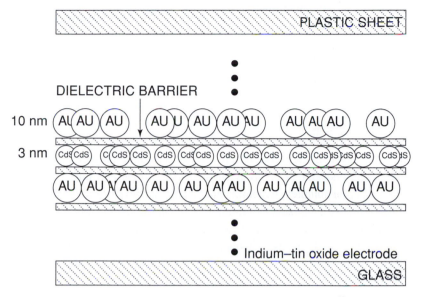

Figure 10.9 Touch sensor design using nanoparticles, redrawn from Maheshawi and Saraf (2006). Gold (AU) layers are separated from cadmium sulfide (CdS) by organic dielectric layers with a plastic sheet at the top and a glass cover to an electrode of indium–tin oxide at the bottom.

vibrations of the ground as monsters walk upon it are felt through the controller. In 2005 Siemens launched a mobile phone with a variety of vibratory options, aimed principally at the mobile phone games market.

The Tacta Vest (Lindeman *et al.*, 2006) is a neoprene upper body garment, made out of several segments with embedded tactile stimulators, controllable for vibrational frequency and intensity. Currently around eight sensors require a bandwidth of around 10 bit samples at 100 Hz, making around 8 kbps. This is well within the capacity of Bluetooth (400 kbps), allowing wireless control of several panels at the same time.

Building robot hands or other special touch sensors, say for applications in robot-assisted surgery, is a different matter. Touch sensors still lag some way behind the capabilities of the human hand (Crowder, 2006), particularly in very sensitive areas, as, say, the human fingertip. However, experimental demonstrations of new sensors based on nanotechnology have already taken place.

Maheshwari and Saraf (2006) developed a sensor, 100 nm thick and 2.5 cm² long composed of alternating layers of nanoparticles of gold 10 nm in diameter and cadmium sulfide, 3 nm in diameter, each layer separated by a dielectric as shown in Figure 10.9. The layer emits light in response to stress and this is captured by a 512 × 512 pixel CCD. This sensor has a spatial resolution of 40 μm and a height resolution of better than 5 μm. Although at a very early stage of development, this would promise tactile performance of a robot hand comparable with human. But, unlike the animal tactile sense, one sensor will subserve both static texture and dynamic slip and movement (Crowder, 2006).

10.8.3   Sensors on and in the body

The usual focus on enhancing the body's sensory systems is outwards. How can we see in the dark (infrared)? How do we sniff dangerous chemicals? How do we taste things which might be poisonous? But there is another interesting area of development. How do we enhance the monitoring of our own body? Knowing whether somebody else is under stress can be quite useful. With the rise in global terrorism, rapid and effective truth validation becomes ever more important. But individuals might wish to monitor and control their own stress levels by sensing skin conduction, heart rate or even stress hormone levels in the blood. External sensors for internal states, particularly stress, are already around, from skin conductance to EEG. One expanding research field is that of wearable computers, which promise to go far beyond the Dick Tracy watch. For a start, wearable computers can monitor body states, akin to an extension of the internal somatosensory system. There is a big need to manage type 2 diabetes in the western world, where incidence is growing almost as rapidly as obesity, to which it is linked.

One of the most surprising developments features the tongue, which as noted above has very good tactile resolution. The tongue display unit under development (Holden, 2003d) is an array of electrodes on the tongue which can provide a 3D representation of the surgical site hooked up to the surgeon's scalpel.

But sensors can potentially go deeper, actually inside the body. Type 2 diabetes, rising faster and faster in the developed world alongside rising obesity, requires careful monitoring of blood sugar levels. Wearable blood sugar monitors already exist and play a role in the plot of the recent film *Panic Room*. Heart rate monitors are also current technology. But it might also be possible to develop sensors which look for other signs of pathology, monitoring arthritis, abnormal blood cell counts, maybe even tumour growth.

Internal sensors are still some way off, because of problems of bodily rejection, but are an active area of research (Service, 2002). But the possibilities are endless: sensing of brain hormone levels for managing mental illness; monitoring stress on muscles, ligaments, bones, following injury or during execessive training. Another fledgling technology monitors eyelid movement to detect (and stop) somebody falling asleep.

Military applications abound, too. Apart from augmenting the management of the soldiers' physiology, detectors of weaponry beyond our current sensory systems are possible, such as odourless, toxic gases or smart sensor dusts.

There are clear opportunities for computer games. Making a game adaptive, sensitive to the player's state can make it more or less challenging, or in other ways enhance the player's enjoyment. One company active in this area is Emotiv which makes EEG headsets for capturing emotional states in computer games (Emotiv, accessed 2010). But there is even an entire genre of games, albeit a minor one at present, which is concerned just with relaxation and stress relief. Maybe in our frenetic world, these games are set to become mainstream.

10.8.4   Haptic interfaces

Haptic interfaces are becoming quite widespread from the simple feedback controllers of computer games to devices for feeling and prodding virtual objects. Plate 10.3 shows a state-of-the-art device (mid 2002!) for virtual surgery. LCD glasses provide full stereo vision, while the scalpel or syringe feels the back pressure as it enters the skin. The virtual patient even lets out a virtual scream if the procedure is too rough! Virtual sword fights could take on a whole new realism.

The system shown is driven by a Silicon Graphics Onyx workstation with two processors and huge memory. But with the current rate of growth of microprocessors, this performance should be available in the personal computer domain within a few years.

Grasping virtual objects requires feedback about the nature of the object. Thus appropriate pressures and tensions have to be applied to the hand. Modern gloves look like, well, gloves, but there is still some way to go in building convenient devices.

10.8.5 Using stochastic resonance for enhancing touch

As people age, their senses diminish in numerous ways: hearing loss at high frequencies and less discrimination in the voice frequency range; glaucoma and junk floating around in the eye. The somatosensory system is no exception and, for example, balance may be affected by deteriorating proprioceptive sensors in the feet. A novel solution to this uses stochastic resonance (§4.6.4). Many small vibrating rods are embedded in a special sole, stimulating the feet. The level of vibration is set at below detectability, but its effect is to increase proprioceptive sensitivity and overall to improve balance (Choi, 2003; Harry et al., 2005).

10.8.6 Fixing the spinal cord

One of the most devastating injuries a person can suffer is damage to the spinal cord, resulting in paraplegia or quadraplegia. The spinal cord is immensely complicated and the outlook for repair is still not very good. Unlike vision and hearing, repair strategies focus on biological repair, such as inducing neural growth, rather than man-made implants (Schwab, 2002).

10.8.7 Adding artificial limbs

Really exciting advances in bioengineering feature artificial limbs of all kinds. The mechanical properties of artificial hands are reasonably well resolved. The big problems lie in the control systems, i.e. getting the brain's intentions to the device. The performance artist Stelarc (http://www.stelarc.va.com.au) has actually succeeded in adding a third arm.

When somebody has suffered an injury, there may be intact neural pathways into which implants can tap, similar to retinal (§6.10.2) or cochlear implants (§5.5.5). But some illnesses or large-scale damage may have destroyed these pathways and it may be necessary to go directly to the brain (Craelius, 2002). To get motion control one possibility is to try and pick the instructions up from the scalp using EEG. But the information from the brain's surface obtained via EEG is quite limited and nowhere near enough to control sophisticated movement as yet. However, success in getting subjects to spell out letters at a few characters per minute has already been achieved. A standard BCI (brain–computer interface) is under development (Wickelgren, 2003) which will enable different brain signals to be accessed under programmer control.

EEG is attractive because it is non-invasive. Yet ultimately it is unlikely to ever have the resolution of putting sensors deep inside the brain. With a billion cells, how could we ever get enough wires in and out? Fortunately it seems that there is sufficient redundancy to get by with a *much* smaller number.

At least 1000 electrodes are needed to get reasonable movement control (Craelius, 2002). This is possible. What is a little inconvenient is the massive electronics required to do the analysis! Thus the technological hurdle is in miniaturisation. There are biological hurdles too. Just as with organ transplants, the body reacts adversely to things implanted within it and will destroy or reject whatever it can. At present 100 electrodes is achievable, and one exciting new development is the provision of tactile feedback via actuators onto the skin for a robot limb (Wickelgren, 2003).

Using monkeys as subjects, Nicolelis and Chaplin (2002) implanted very fine stainless steel wires, 50 μ thick, directly into the brain, the wires lying next to and sampling the activity of neurons at around 10–20 Hz. Surprisingly they found that around just 100 neurons would achieve a hand movement accuracy of around 70% with less than 1000 predicted to achieve 95%. They now have implanted 704 microwires across eight brain centres and these seem to remain viable for months or even years. Thus the future for prosthetic devices for people who have suffered catastrophic body injury is promising.

10.8.8   Haptic devices for games

Computer games have had various sorts of haptic devices since the late 1990s. At first the focus was on providing a realistic feel to controllers, where one could actually feel the forces of cornering in a car or fast manoevering in a plane.

Probably the first device which became really mass-market was Microsoft's *Sidewinder* joy stick. It sold over a million units. Around the same time steering wheels, providing feedback for car-racing games appeared. Sony introduced the dual shock force-feedback controller for the Playstation.

More recently Nintendo introduced the *Rumble Pack*, one of the first developments to add environmental vibration to the experience of a game. One neat idea, which didn't sell very well, was the *backpack thumper* – a device which would thump you in the back whenever you got shot!

Immersion formed in the USA in 1993 has been developing vibration/tactile enhancements for the computer games industry for over a decade. The latest technology, *VibeTonz*, is now available in over 20 mobile phones. Developer kits allow control of vibration along multiple dimensions, similar to those of the human somatosensory system itself and to synchronise tactile signals with audio and visual. There seeems to be huge potential here. At present entertainment effects are largely heuristic and the opportunities for transmitting structured information in a human–computer interface are wide open.

10.9   Envoi

The sense of touch is a powerful one and it has mysterious qualities still not fully understood. Unlike vision and hearing, where artificial systems have been around for decades and are quite well understood, haptics is still a rapidly growing area.

From prosthetic limbs to cybersex, research in haptic devices holds great promise (Strauss, 1999).

11 Non-human sensory systems

As human beings we are aware of just five senses – vision, hearing, touch, taste and smell, which we have looked at in some detail in the previous chapters. Are there any others? As we noted in Chapter 9 we do have a subsidiary olfactory sense, the vomeronasal organ, known to be sensitive to pheromones in other mammals. There is still argument about whether human beings do indeed detect pheromones, but the evidence is mounting that we do (§9.3.9). Thus, as we come to the end of the book, we look at other senses which human beings either do not have or senses which have atrophied to the extent that they are no longer accessible to consciousness.

But are there other senses of which we are not aware? If we leave out paranormal phenomena, the only possibility might be some sort of innate navigational system. But in other animals there are documented electrical, magnetic and infrared image sensors. Compared to the other senses discussed previously, these are far less studied. Thus the information estimates in this chapter are considerably less accurate and more speculative than before.

11.1 Electrical sense

If you have the misfortune to be chased by a shark, don't bleed! Sharks have an exceptionally acute sense of smell. But even if, wrapped up tight in a wet suit, from whence you omit no detectable odour, the worst may still not be over. Sharks have an electrical sense, able to pick up the electrical signals from animal neural and other activity (Kalmijn, 1971). It gets worse if you bleed – injuries release electrolytes into the water enhancing the electrical activity and the shark uses its electrical sense to home in on the victim at close range (Fields, 2007b).

Numerous aquatic animals have such an electrical sense. The platypus, with its vacuum cleaner-like bill, has electrical sensors within it. This makes evolutionary sense. River water may be murky, with all sorts of ambient sound and a variety of smells. Electrical sense may be the most precise way of locating the small animals in the water which the platypus likes to eat.

Electrical senses are generally confined to aquatic animals. This makes sense too. The electrical conductivity of water is so much higher than that of the air, making the signals stronger and more usable. But the fields fall off rapidly with distance, so the electrical sense is very much short-range.

The echidna (*Tachyglossus aculeatus*), one of the few monotreme species, has an electrical sense, but it is totally terrestrial (Gregory *et al.*, 1989). It has much lower behavioural sensitivity than platypus and it is possible that its electrical sense is a sort of leftover from an aquatic ancestor in the 100 million year lineage of monotremes, although it still functions. Alternatively it might be useful in the humid environment of termite mounds when hunting or in finding prey just below the surface of the soil (Gregory *et al.*, 1989).

But animals can also use electric fields for communication. One such example is the knife fish (*gymnotiformes* which lives in murky river water. However, using electrical signals to communicate increases the risk of predation by animals such as sharks and electric eels which use them to hunt. However, some knife fish have evolved a more complex electrical signal, with a higher frequency component, which makes it less detectable by predators (Stoddard, 1999). This coevolution of sensory systems in predator–prey networks extends to other modalities (Ryan, 1999).

11.1.1 Electrical transduction

In shark electrical transduction takes place in sacks, called *ampullae of Lorenzin*. They contain an electrosensitive gel which deforms in the presence of electrical fields. Within this gel hair cells sense this movement and generate a neural signal. Thus the machinery is rather similar to that for hearing and some of the mechano-reception.

11.1.2 Sensitivity

Pettigrew and colleagues (Fjällbrant *et al.*, 1998) have studied platypus electrosensitivity in some detail. The bill has around 40 000 receptors arranged in parasaggital stripes along the bill, with a distinct axis of maximum sensitivity point 60 degrees downwards and to the side with a 60-fold variation in sensitivity. The bill acts as an antenna, optimised for weak signals close up rather than more distant stronger signals. Some controversy surrounded the sensitivity of its electro-sense. The individual receptors have quite a low sensitivity, around $2\,\mathrm{mV\,cm^{-1}}$. But both behavioural measurements and recording from cells in the cortex give sensitivities two orders of magnitude higher. The explanation lies in integration across the bill, making the sensitivity at least $25\,\mu\mathrm{V\,cm^{-1}}$. This is close to optimal, since the electromygenic signals (the electrical signals emitted by prey) are in the range $12\text{--}1800\,\mu\mathrm{V\,cm^{-1}}$.

The echidna only has around 100 receptors (Proske *et al.*, 1998). But platypus has not by any means the most sensitive of electrodetectors. Catfish are much lower and sharks come in at an astonishing $25\text{--}30\,\mathrm{nanoV\,cm^{-1}}$ (Kajiura & Holland, 2002). This could be thought of as an illustration of *scaling*. Sharks swim much faster, so they need to detect prey further away, perhaps analogous to the telephoto lens found in birds of prey (§2.2). The sensitivity of shark is extraordinary with voltage sensitivity comparable with the voltages created in the ion channels of nerve cells themselves (Kalmijn, 1982).

11.2   Heat sensors

Infrared image sensors and detectors are a relatively new innovation on the night vision front, but are now a standard tool used by naturalists, soldiers and others out in the depth of the night. But animals got there first.

Warm-blooded animals have a bit of a problem in attempting to use thermal detection. Their own body temperature produces a very high noise level, making discrimination of temperature differences in the environment very difficult. Dusenbury (1992) has estimated the theoretical sensitivity limits to thermal detectors, but no animal comes close, mainly because of environmental noise. Temperature gradients within thermal detectors limit sensitivity and temporal response (they take time to heat up and cool down). Cold-blooded animals, such as nematodes, seem to be sensitive to about 0.001 °C whereas warm-blooded creatures such as humans are around 0.05–0.08 °C.

Thermal sensors serve a variety of unanticipated uses, such as detecting current flows in water. The jewel beetle, *Melanophila acuminata*, has sensors under its wings for infrared radiation from forest fires. This enables it to seek out smouldering bark where it lays its eggs (Grimm, 2004). But the most developed mechanism is that found in snakes, which have a basic eye sensitive to infrared wavelengths.

11.2.1   The rattlesnake pit

The western diamondback rattlesnake (*Crotalus atrox*) has one of the best developed thermal sensors, an infrared eye. The sensor is a pit just behind the head. The detector of infrared radiation of around 5–30 μ is a very thin membrane suspended in air, essentially inside the pit cavity. For a thermal detector to be reasonably rapid in response it must have a low thermal capacity, hence the emphasis on minimal thickness. Figure 11.1 shows the black body radiation at 37 °C (human body temperature), which neatly matches the rattlesnake sensitivity.

The membrane has an area of about 3–4 mm^2, an approximately 60 × 60 square array of detector elements innervated by 3500 nerve cells (Dusenbury, 1992). Each is about 40 μ in diameter, creating some overlap in their sensitivity direction. Overall the organ covers about 45–60 degrees and the rattlesnake can strike in total darkness to an accuracy of about 5 degrees based on signals from the pit, but there is also tight integration with vision (Hartline *et al.*, 1978).

Although functionally the detector is akin to an eye, in fact it has much more in common with the somatosensory system, particularly the thermal and nocioceptor channels. In fact the heat sensitive transducer is the TRPA1 receptor, known as the wasabi receptor. In humans it is activated by allyl isothiocyante, the active ingredient of wasabi and mustards (Gracheva *et al.*, 2010). The thermal sensitivity is five orders of magnitude below the potential resolution of the detectors, but has clearly evolved to just the level of sensitivity required. So, a temperature difference of 0.003 °C in 0.1 s is detectable at a threshold of around 28 °C, corresponding to object temperature variations of 0.1 °C or a mouse at 30 cm.

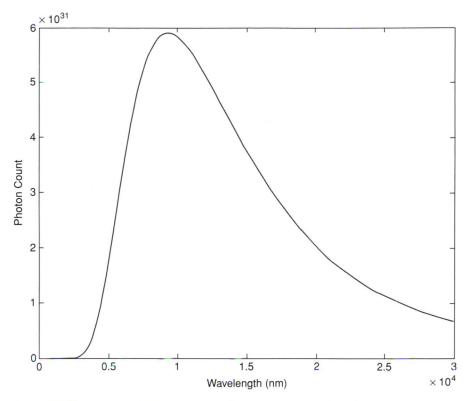

Figure 11.1 Black body radiation at human body temperature, cf. Figure 6.1 for the electromagnetic radiation from the sun.

Despite the efficacy of the pit, it is still a relatively low information sensor. Using the integration time of 0.1 s and assuming 10 bits per time interval, we would end up with an information rate of around 0.35 M bits per second, well down on vision or hearing for mammals but probably not dissimilar to the other senses of the snake.

The California ground squirrel, *Spermophilus beechey*, exploits this lack of spatial resolution in a curious way. Rundus *et al.* (2007) show that the squirrels channel heat and wag their tails to deter rattlesnakes, but do not use this heat defence for other snakes which do not have thermal detection.

11.3   Magnetic sense and navigation

Birds and other animals which migrate over huge distances, such as sea turtles, have long been thought to use the earth's magnetic field (Brown, 2001a). But the mechanisms have been elusive and there is still some doubt as to what the actual detectors really are.

There are still many details to unravel in the magnetic sense, which is quite widespread in at least birds (Mouritsen & Ritz, 2005), turtles (Walker *et al.*, 2002), fish (Walker *et al.*, 1997), honey bees (Walker *et al.*, 2002) and cetaceans (Kirschvink *et al.*, 1986).

Experimental studies demonstrated that the leatherback sea turtle, *Dermochelys cori-acea*, makes use of the earth's magnetic field. These turtles hatch in Florida but then spend years swimming around the Atlantic between North America and Europe. They cannot risk straying too far north, for strong currents will carry them into very cold waters where they die from hypothermia. Similarly, currents will carry them too far south, if they stay outside a region known as the North Atlantic gyre.

Researchers (Lohmann *et al.*, 2001) observed the directions turtles swam in large tanks, where the magnetic field could be controlled. Leatherhead turtles are sensitive to the intensity and inclination of the field and swim in the tanks in exactly the directions they swim in the wild in the same magnetic field conditions.

Instead of looking for the transduction mechanisms, Lohmann *et al.* (2001) looked at the behaviour of cells in the superior colliculus of the mole rat. This area is implicated in various sorts of spatial functioning. The cells showed a distinct preference for polarity of the magnetic field and it appears that the mole rat has a topographic map of the magnetic field variations.

Claims for a human magnetic sense go back some way (Baker, 1981) but recently new evidence for some residual magentic sense in humans has emerged (Baker, 1989). Thoss and others (Phillips, 2002) at the University of Leipzig discovered that the threshold for detection of a small spot of light is lower along the direction of a magnetic field. This would seem to operate in low light level conditions, while Carruba *et al.* (2007) found that magnetic fields evoked potentials similar to other senses. This is clearly a fairly controversial area.

11.3.1   Tranduction of magnetic signals

Despite experimental demonstrations of the need for magnetic fields, and despite the existence of magnetic sensitive materials, the mechanisms of this magnetic sense have been a complete mystery. Magnetite (a magnetic oxide of iron) particles were discovered, for example, in the skulls of pigeons, but this does not actually prove that they are used for a magnetic sense; in fact it now transpires that some birds do *not* use magnetite exclusively (Nemec *et al.*, 2001).

Experiments with robins suggests that the magnetic fields are not operating like a traditional compass. The current theory is that magnetic fields interact with the energy levels in photoreceptors in the eye, rather than with ferrite particles. How the detection occurs is still a mystery. But in newts it seems that there may be special photoreceptors for magnetic detection, by changing the sensitivity of the photopigment according to Ritz and Phillips (2002).

Thus in birds there are at least two transducers (Mouritsen & Ritz, 2005): in pigeons magnetite (a ferromagnetic material) is present in the beak and innervated by the trigeminal nerve; in songbirds cryptochromes' flavon couplings are the transducers. The latter story was a complicated assembly of several components and the final piece fell into place in 2005. Many birds probably use both (Mouritsen & Ritz, 2005).

The sensitivity relative to physical limits is unknown, but sensitivities of four orders of magnitude below the earth's magnetic field have been measured (Walker *et al.*, 2002). The earth's magnetic field is quite noisy, dependent upon the distribution of magnetic rocks, for example. Various other factors cause random variations from day to day, including variations in particle movements in the atmosphere to solar activity. These variations can reach a few per cent (Davis, 2004). Thus one would guess that the performance of animal magnetic sensors is close to the limits imposed by external noise.

The precise navigational strategies and their neural architecture await elucidation.

11.4   Sonar

As we have seen, animals develop signals for communicating with one another, across hearing, vision and the chemical senses. But there is a different kind of signal where animals generate the stimulus they sense, where the stimulus is an environmental probe. The phosphorescence of deep ocean fishes probably falls into this category. Deep down in the ocean there is no natural light, so animals create their own.

Creating our own light is very nineteenth century but a powerful technique, which man is still keen to emulate, is biological sonar. Two general classes of animals make extensive use of sonar – bats, water-dwelling mammals such as whales and dolphins and some species of birds, such as swiftlets and oilbirds. In fact sonar works much better underwater (and in fact it is in the ocean context it has the most man-made uses, such as in submarines), because there is an impedance matching problem between the animal (water) and the air, making power transfer difficult (Holland *et al.*, 2004).

11.4.1   Bats

The sounds the bat makes for sonar are a mixture of two components, a whistle followed by a chirp. The following discussion is based on work by Suga (1990). The first component, the whistle, or continuous tone, provides Doppler information about the relative velocity of the objects off which it bounces. This is essentially a hunting tool. Prey, such as moths, can be detected and tracked both from their flight and the beating of their wings.

The second component is a mixture of frequencies (FM), from which the bat deduces range. Neural circuitry introduces delays between the initial signal sent out and the echoes coming back in. Thus cells fed by the initial frequency delayed by some amount and an echo arriving at the same time fire to indicate an object at a particular distance.

The bat has a lot of specialised wetware to do all of this computation. There is an extensive thickening of the basilar membrane around the frequency bands of the signals it generates and there are correspondingly many more neurons devoted to these bands. The working of the delay lines is not fully understood, although it seems possible that extensive chains of neurons may be used.

The bat has evolved a neat way of handling the huge confusion of sounds which arise in a colony in a cave, where thousands of bats are all sending out echo-location signals. The second harmonic in the FM signal is very weak, only about 1% of the total energy of the signal. Thus other bats do not hear it. But the bat hears its own signal through bone conduction. Thus it can do comparisons against its own sonar pulses, while its coincidence detecting cells ignore all the signals from other animals.

There is much to be learned on how bat sonar operates. Yovel *et al.* (2008) report that different plant species give distinctively different reflection spectra to bat ultrasound emissions, thus enabling the bat to differentiate vegetation type. Behavioural evidence shows that a wide range of bat species are able to extract and use such information.

11.4.2   Whales and dolphins

Odontocetes (dolphins, porpoises and other toothed whales) also use sonar for navigation and for spatial orientation to complement vision (Verfuss *et al.*, 2005). Just as we saw that elephants, the largest of terrestrial mammals, communicate with very low frequencies, commensurate with their size, so we find the same very low frequencies in the leviathans of the ocean, whales. But water is a much better conductor of sound than air, thus whale signals travel for huge distances. In fact one issue of ecological concern is the extent to which the great increase in ocean noise from shipping has an impact on the well-being of whales.

11.4.3   Brain implications of sonar for brain architecture

The perception of timing differences from emitted pulses may have led to a relative increase in the size of the cerebellum. Clark *et al.* (2001) found in two mammalian taxa, *cetacea* (whales and dolphins) and *microchiroptera* (bats), a fraction of total brain volume of around 20% compared to a fairly constant 13.5% for other mammals including the *megachiroptera* (megabats, which do not echo locate). It also turns out the *mormyrid* fishes which use electrial pulses for location also have the largest cerebellum amongst teleost fishes.

11.4.4   Echo location in games

Time-of-flight cameras have been available commercially for some time. They use lasers to measure the distance of points in a scene from the sensor and build up a three-dimensional depth image by scanning. In 2009 a good quality time-of-flight camera would cost around Australian $10 000. However, Microsoft have just launched, as this book goes to press, the Kinect system, which records depth maps, adding a new level of interactivity to computer games. The technology is still very new, but there is a buzz of activity around developing new games to fully exploit its capabilities.

11.5   Enhancing the senses with implants

William Gibson, often said to have coined the popular term, *cyberspace*, had a deep vision of the future of computer networks. In *Neuromancer*, the characters plug their brain directly into the internet using cyberdecks. Characters such as the razor girls have claws like a cat, made out of tiny razor blades. They are under direct neural control. In Fred Hoyle's *Black Cloud*, the super intelligent cloud downloads directly to a human brain – and unfortunately overloads it! How far are we away from bypassing our native biological senses?

At the European Conference on Artificial Life in Brighton in 1997, Stelarc, one of the world's most extraordinary performance artists, showed just what can already be done. Stelarc's stunts have always been way out. In early days he was hung, full monty, way way above street level on tall cranes, held by steel hooks inserted in his skin.

At the conference he had a third, robot arm. This arm had direct neural implants and he had already made substantial progress in controlling it. He demonstrated that he could write with both right hands simultaneously, each hand writing its own letter! So, we can already do quite interesting things with motor skills.

An alternative to brain or neural implants is to extract information from the outside of the brain. The electrical activity within the brain produces fluctuating electrical fields on the surface of the skull. For many years this activity has been measured and correlated with behaviour. For example, the several different types of sleep display distinct patterns in so-called electroencephalography (EEG). The onset of an epileptic fit is characterised by strong synchronous waves in the EEG.

However, until recently EEG was a tool of very low precision. But improvements in sensors coupled with powerful computer analysis have dramatically increased its resolution. In fact it is now possible to extract around 1 bit per second (not much on the scale of sensory systems!) from the EEG of the motor cortex and use this to carry out simple actions such as dialling a telephone number (König & Verschure, 2002). There is considerable scope for EEG since the equipment is relatively low cost, it is totally non-invasive and the required computing power is getting cheaper all the time.

Emotive Technologies have been developing EEG for use in computer games. Their device has now been demonstrated at a number of big conferences, such as the Game Developers Conference in the USA (in 2009) and looks set for the market.

11.6 Sensing animats

Our virtual worlds at the present time are forgeries. Games may have realistic looking characters in realistic looking worlds, but they are fakes. Movements, world behaviours are scripted, or at least procedural, maybe slightly randomised, but far, far from an ongoing simulation of a world.

The designer of a virtual world for a computer game or other simulation can set this world on earth, the solar system or the known universe, or in some other fanciful universe. George Gamow in his amazing series of Mr Tomkins books imagined a world with different values for the speed of light or Planck's constant, different worlds indeed.

But a lot of the time, games stay within our universe and are constrained by the laws of physics and their constants. At which point the coherence and realism of the game requires its animats to behave in a physically consistent way. We do not want them to see through solid objects, or to navigate through inanimate worlds with vision in total darkness. Thus an understanding of what is and what is not possible is very valuable to the game designer.

Scripting is also brittle. An animat cannot behave in a plausible way if the context goes off script. But it will not always be this way. Ray Kurzweil (1999) estimates that by 2019 an Australian $1000 computer will match the processing power of the human brain. By 2029, the same $1000 computer will have the power of 1000 brains. With such awesome power, the environments we create, and the animats that inhabit them, will be much more like actual creatures in autonomous, real virtual worlds.

The animats of the future will sense their virtual environment, communicate with each other and other species, learn, adapt and think. At some point in the future, maybe within the lifetimes of some of our readers, they may have a consciousness all their own.

12 Sensory integration

The Leatherback turtle had a magnetic compass 100 million years ago.

In the last decade a lot of growth has occurred in our understanding of the integration of sensory systems. In fact entire conferences are now devoted to integration, both biological and robotic. The individual senses usually reinforce one another, enhancing feature detection and object recognition. But in some cases there may be contradictions and all sorts of strange experiences result.

From the perspective of the themes of this book, integration seems contrary to the fundamental strategy of streaming. So what we are really interested in is the way these streams are used to cross-check each other and how they are used to make decisions and guide behaviour. There are some deep questions of information theory, again beyond the scope of this book in any quantitative treatment. The data from one sensory stream can act as a cross-reference or prior, in effect reducing noise. As a rapidly expanding area, this chapter is thus a selection rather than comprehensive overview of integration. The topics it covers are:

- System integration exemplified by the optokinetic system (§12.1). The vestibular system provides feedback to the eye muscles to control gaze direction independently of head movement. This is the most complex integration system discussed and is fundamental to visual processing.
- Cross-calibration, where one sense is used to calibrate or adjust the inputs from another (§12.2). Unlike the opto-kinetic system, which operates continually, calibration systems may operate intermittently, e.g. at dusk.
- Integration areas in the brain where data are collated from more than one sense (§12.4.3). This is obviously a huge topic, where at best a few key ideas can be discussed.
- Unusual effects arising from sensory conflicts, such as various kinds of touch illusion and out-of-body experiences (§12.1.8).
- Consciousness is a philosophical grand challenge. What is it, what animals have it, what are its brain correlates (if any) and many other questions are still earnestly debated. Needless to say this book cannot really enter into these arguments. But our impression of the sensory world is a unitary one, and there are some sensory discoveries which shed a little light on these most difficult of questions (§12.5).

12.1   Eye movements

Movement of the eyes is something of which we are acutely aware. We like people with stong gaze, we are sensitive to eye contact. We trust or believe people who look at us directly and refer to people whose eyes drift away as shifty, with perjorative implications.

Whereas many animals have rather limited eye movement, the human eye roves around constantly. Apart from the dynamic process of building up an integrated picture of the world, the eyes convey a lot of psychological information too. Baron-Cohen (1997) considers the development of eye contact and assessment of eye position in other people as a crucial development in children, and considers poor development to be a causal factor or consequence of autism. The oft-noted extra sensitivity of women to emotions is reflected in a more efficient process of gaze direction (Bayliss *et al.*, 2005). Thus the movement of the eyes and the direction of gaze are fundamental to how we extract information from the world (Land & Tatler, 2009).

The eye movement system is extemely sophisticated employing the fastest muscles in the body (Alpern, 1982) and a complex system of excitatory and inhibitory pathways from brainstem to cortex. As we might expect by now, Helmholtz was interested in and aware of the importance of eye movements, but it was Edwin Landott in 1890 who first recognised the jerky, *saccadic* movements and Raymond Dodge in 1902 who identified five distinct movements (Goldberg, 2000).

A lot of the complex control mechanisms take us outside our primary interest of sensory information per se, so we shall try to extract a few relevant ideas. Section 12.1.1 discusses the muscle systems which control these movements and how different nerves control them and project to different brain areas. Section 12.1.2 describes the types of movement themselves and how they relate to visual information processing.

12.1.1   The eye muscles

The eye which is approximately spherical can rotate in three dimensions, hence it needs three sets of muscles to control rotation around three axes. Each muscle pair acts in a push/pull fashion:

1. The **lateral and medial rectus** muscles rotate the eye around a vertical axis producing *adduction* and *abduction* away from and towards the nose respectively.
2. In the straight-ahead position the **superior and inferior rectal** muscles operate in the horizontal axis joining the two eyes, producing *elevation* and *depression*. Again in the straight-ahead position the **superior and inferior oblique** muscles rotate the eye on a front-back axis through the pupil, approximately the optical axis, producing *extorsion* and *intorsion* away from and towards the nose respectively.

The superior oblique muscle acts on the eye through a pulley known as the **trochlear** and it is controlled throught the **trochlear** nerve, part of cranial nerve IV. The lateral rectus is controlled through the **abducens** nerve in cranial nerve VI and the rest, the

Table 12.1 Different types of eye movement.

Movement	Description	Control	Section
Zitterbewegung	Small, random trembling	Random	§12.1.3
Vergence	Contrary motion for accommodation	Vestibular	§12.1.4
Saccades	Sudden jumps from one fixation point to another	Cognitive	§12.1.6
Pursuit	Smooth tracking of a moving object	Cognitive	§12.1.5
OKN	Optokinetic nystagmus	Vestibular	§12.1.5
Vestibulo-ocular	Head movement to keep gaze constant	Vestibular	§12.1.5

medial rectus, the inferior oblique and the superior and inferior recti are controlled by the *oculomotor nerve*, cranial nerve III (§2.5.1 and Figure 2.2).

12.1.2 Classification of eye movements

There are several different types of eye movements, which we can classify in two different ways. One way is the control mechanism – attentive and conscious or automatic. The other involves the head, i.e. those movements which involve the head and those which do not. The former require close coupling with the vestibular system, which is why eye movements are considered here in this chapter on sensory integration. In the first category lie the processes of fixing and focusing the eyes on some target, notably *fixation* and *vergence* movements, and some of the movement tracking mechanisms, known as *pursuit* movements. The eyes can track quite quickly, at around $100°s^{-1}$, i.e. sweeep from one side to the other in around 1 s. Finally we have the most fundamental movement of all, the *saccade*. Our eyes do not drift endlessly, but stop and fixate in one place, then move rapidly to somewhere new in a saccade movement[1].

In the second category, there are the movements which involve the head too. They are the *optokinetic* and *vestibulo-ocular movements*. As the names imply the first of these is driven by the visual image itself, the second by the vestibular system (Goldberg, 2000). As we have noted earlier, the vestibular system reacts a great deal more quickly in control of balance than the visual system and so it is here. The vestibulo-ocular system corrects for rapid head movements, while the optokinetic system operates for longer, smooth head rotation, as in panning a camera.

Table 12.1 shows the categories and their driving mechanisms.

12.1.3 Zitterbewegung: small random jitter

Eye movement varies from a slight, high frequency trembling, the motion, *zitterbewegung*, to the wholesale shift of gaze. The eye not only moves to look at different things, it *has* to move. The image rapidly fades to grey if the eye is kept perfectly still, the

---

[1] This is a slight simplification. The eye does have some other movements, such as the *zitterbewegung*. Post-processing removes all these movements from our awareness.

Troxler effect. But because of the slight trembling motion, the eye is never completely still. To show this effect requires *retinal stabilised images,* created with an optical mirror system. The eye sees the world through a mirror, the position of which is moved dynamically by an electromechanical control system to exactly counter the movements of the eye.

12.1.4   The vergence movements

In all except one of the different types of movement, the eyes move together towards the same focus point, so-called *conjugate* eye movements. In vergence movements, however, when the eye is moving to focus at different distances, the eyes move inward towards each other, to focus on a closer object, or move outwards to focus on something further away. Such movements are called *disconjugate.*

The vergence movements are part of three linked systems: accommodation, to bring objects at a particular distance into focus (§6.4.2.2); the vergence movements themselves; and constriction of the pupil which affects depth of field (§6.4.2).

12.1.5   Pursuit movements and optokinetic nystagmus

Pursuit movements allow the eye to track a moving target, keeping the target in the same place on the retina. Apart from keeping the high resolution fovea focused on the target, they also serve to effectively increase the integration time for the photoreceptors. At a simple level this increases the signal to noise ratio. Pursuit movements are relatively slow with a maximum speed of around $100\,°\text{s}^{-1}$.

As the head moves around, it implicitly shifts the visual field. If the eyes stayed still within the head, their direction within the world would move around. Thus to avoid blurred vision, or a sensation of the visual field bobbing around (*oscillopsia*), a sophisticated dynamic correction mechanism is required. This is supplied through information from the vestibular system (§12.1). Thus the direction of gaze is a balance between head and eye movement (Goldberg, 2000).

Smooth movement of the eyes is interrupted by occasional rapid jumps backwards. This alternation of slow and fast movements is known as *nystagmus.*

The speed of eye movement adapts similar to the response to other sensory stimuli. Maddess and Ibbotson (1992) found a big adaptation in optokinetic nystagmus, which serves to stabilise images on the retina.

12.1.6   Saccades and attention

When something attracts our attention we usually move our eyes towards it, in a rapid jump, the saccades are really fast, around $900\,°\text{s}^{-1}$. Whereas pursuit movements require a moving stimulus, saccades operate by a different mechanism and do not require movement in the visual field. The full voluntary control of intentions ultimately resides in the executive areas of the prefrontal cortex, but much of the control of eye movements and saccades takes place in the parietal lobe.

The lateral intraparietal area (LIP) efffectively maintains a priority map of attention (Yantis, 2003), determining the competition for visual attention. Attention usually moves to the target of a saccade before the saccade takes place, although LIP is *not* thought to be a saccade map – an area controlling where and when saccades occur.

Bisley and Goldberg (2003) found that the LIP provides a distributed representation of attentional value of features across the visual field, but that it is not possible to determine from looking at absolute LIP activity where attention is or where it is going, but rather it is the region of greatest activity which dictates the attentional focus. At this region of the visual field, there is increased perceptual sensitivity – we are more sensitive to things to which we are paying attention.

Conversely the sensitivity of cells in V1 in *expected* directions may increase. In single cell monkey studies, Sharma *et al.* (2003) found that the response latencies of such cells decreased with increasing predictability of where a visual target would appear. Effectively top-down signals provide an internal representation for location and this representation accords well with a Bayesian prediction of target appearance in a particular place. In fast reaction first-person shooter computer games, where villains can pop up anywhere, there is an interesting opportunity here for games to confound these predictions as difficulty level goes up!

We can plan and make not just a single saccade, but sequences of them. This seems to require information flowing back to higher saccade planning areas of where a saccade is going to end up. This information is known as a corollary discharge and Sommer and Wurtz (2002) were recently able to demonstrate its pathway in the macaque monkey. The discharge begins in the superior colliculus just before a saccade is generated, passes through relay neurons in the mediodorsal thalamus and onto the frontal eye fields. These are located in the premotor cortex in the frontal lobes and are responsbible for voluntary saccades and pursuit (tracking) eye movements (§12.1.5).

12.1.7 Blurring

The speed of movement of the eyeball during a saccade is really quite impressive. One might wonder, then, why the visual image does not appear blurred. There is some sort of suppression of vision, while the movement takes off, the information is blocked off and we just fill in with what we have, a bit like filling in frames of a movie.

Saccadic suppression is a far from new idea. How it works in detail has remained a mystery, but important advances in our understanding of the mechanism appeared in early 2002. The retina is capable of tracking fast moving objects. So, movement across the retina, itself, should not cause suppression. But maybe there could be a way of detecting movement of the entire visual field within the retina which would generate the inhibitory signal.

However, we now know that the suppressive force is extraretinal. Cells in the middle temporal (V5) and middle superior temporal areas fall into different categories according to how they behave during a saccade. One interesting category flips its directional sensitivity during the saccadic movement! So the suppression results from the cancellation of two signals, one in phase and the other out of phase (Thiele *et al.*, 2002). This is really quite a surprising result. Even for peripheral sensory mechanisms we do not yet have all the answers. Finally the saccade is an impulsive movement and an area of medial frontal lobe, the anterior cingulate cortex, monitors for errors in the intended destination of a saccade (Ito *et al.*, 2003).

12.1.8   Motion sickness

The vestibular system, also located inside the ear, uses similar transduction hairs in fluid. The geometrical orientation of the canals picks up movement and acceleration. The vestibular system, based as it is in the physical position and movement of the head and body, experiences *cognitive dissonance* in virtual reality. Our virtual experience from vision and maybe hearing might be one of flying through the air like superman. But the vestibular system is confused, because it is not experiencing the forces and accelerations which it should. This leads to nausea.

The exact nature of this nausea problem is not fully known, or is classified information! It is a serious problem and has possibly held back mass-market virtual reality. There are three sorts of explanation, none of which are firmly established at the time of writing:

- Cognitive dissonance, which is also offered as the explanation for why we are much more likely to get travel sick in a car or a bus if we try to read; the puzzling thing is why cognitive dissonance should make us feel physically sick. Why do we not get a headache, eye strain or some direct sensory feedback. One nice idea is that through evolution, the only situation where we might have encountered it is following poisoning. Botulism, for example, leads to visual disturbances. Other poisons produce balance problems. Of course the best immediate survival reaction to poisoning is to vomit!

- A much vaguer idea is that the control systems for eye movements get taken outside of their operational frequency. It is possible to develop visual stimuli on a computer screen which confound the eye's tracking mechanism and produce rapid sickness (Maddess, 1990, pers. comm.).

- Maybe related to the above is some sort of global oscillation which sets in throughout the brain. Monitoring the electrical activity of the waking brain shows very noisy electrical activity. But when an epileptic fit occurs, this noisy activity is replaced by strong low frequency oscillations. This sounds extremely strange, but there is some anecdotal evidence that anti-epilepsy drugs suppress motion sickness very effectively.

Some antibiotics, such as streptomycin, can have adverse side-effects and damage the vestibular system. Victims can still function well in light, where vision can make up for the lack of direct bodily feedback. A small compensation is that they do not get motion sickness. Thus although the oscillatory phenomena may be linked to motion sickness, some conflict with the vestibular system is a prerequisite for sickness.

12.2   Cross-calibration

The spatial maps from the different senses need to line up. So some degree of checking that they match is necessary.

12.2.1   Links between visual and auditory cortex

Until very recently the individual primary cortical areas, such as V1 for vision, were seen as unimodal, receiving input from just one sensor, in this case the eye. But recent work suggests this view is oversimplified. At least in the case of auditory and visual cortices there are cross-interactions.

Using event-related fMRI Martuzzi *et al.* (2007) found synergetic functioning of visual and auditory cortices (faster reaction times), greater than would be expected from probability summation of each. Furthermore each primary cortex showed some response to the alternate stimulus (vision for auditory and vice versa). The evidence suggests that the effects arise from direct connections between the areas. This research in cross-modality activation is quite new, but it strongly influences how we think of the early stages of sensory processing.

12.2.2   Alignment of visual and auditory maps

The alignment of spatial maps formed by vision as well as hearing presents a challenge for the owl. In vision we get a spatial map from each eye independently: the two eyes give us depth, for which we have additional non-stereoscopic cues. But the owl gets its spatial map from its *two* ears; so, changes in the size of the head, an ear infection, an injury could all throw its map out of alignment.

Knudsen and his team, the world leaders in barn owl sensory information processing, have now demonstrated how the visual system adjusts the auditory map to keep the two maps aligned (Barinaga, 2002b). The optic tectum of the owl feeds into the cells in the external nucleus of the inferior colliculus, which has an auditory map of the world. Cells in this auditory area will respond to visual stimuli mediated by the optic tectum. When there is a misalignment, the response gets stronger, indicative of an error-correcting mechanism.

But obviously the cells in this auditory map should not fire all the time in response to visual stimuli. In fact to see this interaction between visual and auditory maps Knudsen's team had to block the inhibitory transmitter, GABA (Gutfreund *et al.*, 2002). Thus the visual signal is dynamically suppressed most of the time. What turns it on is an open question at the time of writing. It is consistent with the bigger picture, though, which we discussed in §2.3.4, of a hierarchy of inhibitory networks within the animal brain.

12.2.3   Navigation of songbirds

Another interesting example of cross-calibration of senses has recently been demonstrated in songbirds. The *Catharus* thrush migrates huge distances, but it has been unclear as to whether they use the stars, the earth's magnetic field, polarised light, dead reckoning or other cues to cover these great distances with amazing accuracy. Now Cochran *et al*. (2004) have shown that they use a magnetic compass while they are flying at night but reset it against the sun at twilight.

By subjecting the birds to a wrongly oriented field before they left at sunset to fly through the night, they found that the birds would fly in the wrong direction all night. Next day, they would reset at twilight and fly off in the correct direction. The cue the birds use from the twilight sun is still not known for certain. It could be the position of the sun itself, or the pattern of polarised light in the overhead sky, itself affected by sun position. The latter is considered most likely at present (Cochran *et al.*, 2004; Stokstad, 2004).

12.3   Conflict and priority

Sometimes sensory information may conflict in the real world. We saw this with motion sickness, where visual disturbances are assumed to be the result of poisoning with inevitable consequences. But what of other illusory situations?

With movies in the cinema or on television, vision easily overrides sound. We hear the voices coming from people, rather than from some loudspeaker somewhere else. Ventriloquism is an extreme form, where a dummy, whom we know cannot speak, nevertheless seems to be the source of speech (Botvinick, 2004). But what happens when there is no immediate sensory conflict?

The Pinocchio illusion is one of the strangest sensations. The subject holds onto her nose while a vibrator is applied to muscles in the arm. The vibrations stimulate the proprioception system and create the illusion that the arm is extending. But this is inconsistent with holding onto the nose, unless the nose gets longer. This is exactly what the subject experiences with the eyes closed (Lackner & DiZio, 2002).

Another bizarre illusion is the sense of owning a rubber hand! If a subject's hand is hidden from their view, a rubber hand they can see in front of them can be made to feel like their hand. Their actual hand is stroked gently (out of sight) while the rubber hand is stroked in synchrony. The convergence of tactile and visual stimuli gives the impression that the rubber hand is the subject's own hand (Botvinick, 2004). Even more bizarre, if the rubber hand is substituted for some other object, such as the surface of a table, the illusion still occurs (Ramachandran & Rogers-Ramachandran, 2004).

Vision overrides touch. The sensor integration occurs in the right parietal cortex[2] but Ehrsson *et al.* (2004) show, using fMRI imaging, that the sense of ownership is associated with sensory integration areas in the premotor cortex in the frontal lobes.

12.3.1 Cross-modal interactions

Until the last decade the senses tended to be studied largely independently. There were seemingly aberrant phenonena, such as synaesthesia (§12.3.2), but the widespread cross-modal interactions now becoming apparent were scarcely mentioned.

This is too large and too new a field to go into very much detail, but the phenomena are quite diverse. For example, Dematte *et al.* (2006) find a convincing link between olfaction and touch. Pleasant smells will make fabrics seem softer (with obvious potential for marketing laundry detergents!).

Cross-modal interactions can exist at a very low level in the brainstem itself. Musacchia *et al.* (2007) found enhanced visual/auditory interactions in the *brainstem* of musicians (perhaps related to the links between reading music and the encoding of pitch relationships).

12.3.2 Synaesthesia: when things screw up?

Synaesthesia is a strange phenomenon whereby input to one sensory modality produces an effect in another. So somebody might experience different notes of the musical scale as having individual colours. Association of numbers with colours also occurs.

Ackerman (1990) quotes the fraction of synaesthetes in the population as around 1 in 500. Countless authors, painters and composers have reported synthaesthesia. In some cases it has infused their art, beyond it being a mere part of their everyday existence. Kandinsky, one of the prominent abstract painters of the early twentieth century, made colour and its meaning his mission in art. Albert Dove produced evocative paintings, such as *foghorns* which capture the distinctive sound on canvas (Hughes, 1997). The composer Scriabin wrote pieces with specific colour references and auxillary scores for the colour organ (a curious device built by Rimington at the end of the nineteenth century having 14 controllable coloured lights, but no sound) (Campen von, 1999).

The most common form of synaesthesia seems to link colour and sounds. Although there seem to be commonalities across cultures, the relationships are in no way fixed. Rimsky-Korsakov linked the key of C major to white, whereas for Scriabin the same key was red. But there do seem to be shape–colour linkages. Nabokov associated the shapes of letters with colours.

The neurophysiology of synaesthesia is still obscure, but Ramachandran's toe-sucking argument of §10.5.4 looks like the same sort of mechanism. On the other hand, Snyder and Mitchell (1999) have argued recently that synaesthesia is a latent tendency in

---

[2] If this area is damaged, then patients suffer the disorder known as *somatoparaphrenia*, in which they feel that their own limbs on the left-hand side belong to somebody else or are artificial (Botvinick, 2004).

everybody. By increasing the number of links and associations to sensory phenomena we can heighten our pattern recogntion and speed of processing.

Ramachandran and Hubbard (2003) now show that synaesthesia has some properties of real sensation. In one experiment they embed a triangle of twos inside a matrix of fives. Non-synaesthetes have difficulty in picking out the triangle. It does not pop out preattentively (§6.7.2) and they have to track the twos one by one. However, for a person without synaesthesia, if the twos are coloured red, say, and the fives, something different, say, yellow, the triangle pops out. But synaesthetes see the numbers as different colours anyway. So in fact they experience pop-out of the triangle, even when the numbers are all black. They still know that the numerals are black, but have this sensation of a colour being associated with them.

There are also candidate areas for where the crossover is occurring. One is in the fusiform area of the temporal lobe where V4 (§8.4.2) and some object recognition occurs (§6.11). In fact in some synaesthetes it is possible to observe using brain imaging activity in V4 even when the stimuli are monochrome (Ramachandran & Hubbard, 2003). These people are cross-linking the grapheme (the Arabic numeral 5, not, say, the Roman numeral) and the colour.

Other synaesthetes cross-link the more abstract, ordinality property of numbers and see, say, days of the week in different colours. In this case the interaction is occurring higher up in an area referred to as the TPO (the conjunction off temporal, parietal and occipital lobes). Located here is the angular gyrus, an area which integrates different sensory cues associated with an object or animal.

At first aberrant cross-wiring was proposed as the cause of synaesthesia, but now a failure of inhibition between areas is espoused. This fits in with the Snyder and Mitchell (1999; Snyder *et al.* 2004) view discussed above. As human infants, synaesthesia is present which disappears as inhibitory processes develop. It is also consistent with the links between creativity and the removal of inhibition and with the statistical finding that creative people are seven times more likely to be synaesthetes.

12.4   Sensory integration

Imaging techniques and ever more sophisticated neurophsyiology and concomitant signal processing have led to great advances in understanding how the senses integrate within the brain. Most of the discussion in the previous chapters has concerned areas which are part of the pathway of a single sense. Yet, at some point everything has to come together for a uniform view of the world. Integration is not only exciting as the summation of sensory information processing, but it is also a key discipline for robotics and the subject of entire conferences in its own right.

Two kinds of integration are possible. One is direct space driven – sound vision from the same point in space, for example. The other is where cells are driven by some higher order concept which has a distinct multi-sensory representation.

12.4.1   Sensory synergy

Frequently an object or event will have multiple sensory cues, visual, auditory, chemical associated with it. As one might expect, these add together to push recognition over threshold, when each individual sense does not have quite enough information. In the fruit fly, *Drosophila*, stimuli learned with both a visual and olfactory cue, both themselves below threshold, are nevertheless learned. Furthermore, there seems to be cross-modal transfer of memory and subsequent retrieval by just one of the subthreshold stimuli (Guo & Guo, 2005). The way these cross-modal mechanisms work is still largely unknown.

Touch provides spatial and surface cues which complement vision. For a cat the surface information he gets from his whiskers is on a par with vision in very low light. Recent studies have shown that in humans haptic input can influence the perceived orientation of surfaces where the visual cues are not very strong (Ernst *et al.*, 2000).

Birds use both visual and magnetic cues for navigation. Guilford *et al.* (2004) showed that pigeons following motorways add as much as 20% to their distance. But the extent to which, if any, the mechanisms augment rather than complement each other is not clear.

Another interesting synergy arises between the vestibular system and other senses. People are not particularly good at judging acceleration and, noting that astronauts take some time to learn to catch things in reduced gravity, Indovina *et al.* (2005) investigated the role of the vestibular system in judging time to collision of accelerated objects. When acceleration corresponds to earth's gravity, g, parts of the vestibular system become active in processing g-like visual motion. They argue that there is a built-in model for g in human processing of movement.

12.4.2   Audio and visual mirror neurons

In the monkey ventral premotor (frontal) cortex, there is an area, F5, which is similar to Broca's area in man, the area considered to be the prime language area. Some cells here respond to abstract concepts of an action. A group at the Universitá di Parma identified *mirror* neurons in F5 which respond whether the monkey performs an action or sees the action performed. Now Kohler *et al.* (2002) from this group have found mirror neurons responding to sound, vision and action in rhesus monkeys. These neurons are activated by common actions the monkey performs, breaking peanuts, tearing, ripping things and so on. Single neurons can respond to the monkey performing the action, seeing somebody else perform the action or hearing the associated sound.

Thus in F5 there are neurons which respond when actions occur (or maybe are thought about) and integrate sensory data associated with the action. It is easy to see how these neurons could be progenitors of abstract thought and perhaps contribute to the debate as to which comes first, language or thought.

12.4.3   Integration areas

Some brain areas, both cortical and non-cortical, integrate information from the senses. Damasio, amongst others, argues that these integration areas are crucial to awareness and consciousness itself (1999). One such area is the *superior colliculus* which sits in the back of the mid-brain, in other words in the older, non-cerebral parts of the brain.

12.4.4   Cross-modal priming

Because the body is movable and deformable, attention frequently has an external, spatial reference. Priming can occcur in cross-modalities, between vision and touch, relative to positions in the external world. So, a light on the left-hand side will prime the left hand to perceive a vibration. But if the subject crosses her arms, the priming now occurs for the *right* hand (Hsiao & Vega-Bermudez, 2002).

12.4.5   Integration of what and where streams

A repeated theme throughout the book has been the streaming of sensory information. Sometimes injury or illness may reveal this in striking selective deficiencies. But the integration of streams in early childhood occurs synchronously with the development of the streams themselves.

The dorsal/ventral where/what streams exist across the different senses. In vision, in very early children, DeLoache *et al.* (2004) show that immature inhibitory mechanisms in children aged 18–30 months may cause scale errors: the child interacts with a miniature object (like a toy car) as if it were the larger one they can get into and pedal – they try to get into the miniature car.

The ventral stream carries the information about the identity of the object (car) and action plans associated with it. The dorsal stream carries spatial and size information. The inhibitory mechanism from the dorsal stream, blocking the action on the miniature car, is not yet fully in place.

12.4.6   Do you see what I see?

One of the classic philosophical conundrums is whether our individual sensations are the same, whether the sensation of red is the same for you and for me. This is at present outside the domain of science, possibly a question forever outside scientific enquiry. But what is becoming apparent is that our brains function in very similar ways.

In an innovative series of experiments using fMRI, Hasson *et al.* (2004) showed five subjects a 30 minute clip of the movie *The Good, the Bad and the Ugly*. They used a pairwise comparison using the activity in each subject's brain to predict the response of another subject, ten comparisons in all. An amazing 30% of activitation was predicted of the brain activation of one subject by another!

They identified two components of synchronisation. The first was not selective for brain region and was strongly affected by emotional points in the movie such as gunshots. Perhaps a good quality film will grasp and direct the attention of the whole of the audience. Just think of the focus of the audience in the scene in *Alien* where the creature bursts out of John Hurt's chest. The second was region specific, with faces, indoor and outdoor scenes and hand movements all showing synchronised activity in localised brain regions. The hand synchronisations were noticeable in somatosensory areas (Brodmann area 5, §10.4). But the synchronisation extended to numerous other areas beyond what would normally be considered the sensory areas. However, the prefrontal cortex, site of executive function, was only weakly predictive from one subject to another. So something of our individuality remains!

12.4.7   Where is consciousness?

Consciousness is one of the great enigmas. We find it difficult to define it, yet it seems to be the intrinsic feature of human life. We do not even know the extent to which animals are conscious, or whether we need the power of human language. Many theories and models appear day by day and the rate of appearance of books is ever increasing. Perhaps it is because brain imaging and techniques such as TMS are taking us closer and closer to a map of brain function, that the last days to make informed guesses are upon us.

Closely related to consciousness in some way is awareness, where some real neurophysiological progress is happening. For example, area V1 of the visual cortex is necessary for awareness, yet strangely, successive areas such as V2 are not (or at least in monkeys). Pascual-Leone and Walsh (2001) use a very interesting TMS experiment to demonstrate the importance of V1.

Using TMS pulses for exciting clusters of neurons, they stimulated the motion area V5. TMS activation of this kind produces phosphenes, flashes of light, in this case moving flashes. At the same time they stimulated V1 in the same visual location, but at a lower level, so that there would be no perceptible effect from the V1 stimulation. When the activation of V1 was delayed by around 25 ms, the phosphenes disappeared, just the time frame for feedback from V5 to V1 to occur. So, it seems that if there is interference of the feedback from V5 to V1, the perceptual experience disappears!

This is very reminiscent of Jackendoff's intermediate theory of consciousness: the brain areas which generate consciousness are not the higher level executive areas of the very low level sensory areas but somewhere in-between (1987).

12.4.8   Out-of-body experiences

One of the more spooky, near-death experiences occasionally reported is the out-of-body experience (OBE). Although like other paranormal phenomena OBEs may be dismissed as lacking scientific reproducibility under controlled conditions, intriguing experiments by Blanke *et al.* (2002) now offer a neurophysiological insight.

In trying to determine the focus of epileptic seizures in somebody with temporal lobe epilepsy, they had to directly stimulate parts of the brain. Stimulating part of the angular gyrus created strange *visual* perceptions in limb position, length and movement. Other effects were a sensation of floating above the bed, a definite OBE experience. These strange effects result from conflicting tactile/proprioceptive, visual and vestibular information in ways not yet fully understood (Blanke *et al.*, 2004).

The angular gyrus contains areas for both somatosensory and vestibular information processing and is thought to involve integration of multi-sensory data for own-body perception. It is interesting though that these experiences are created by atypical stimulation of sensory areas and override cognitive or executive control. In some ways they support Jackendoff's intermediate theory of consciousness rather well (1987).

12.4.9   Do we perceive what we know?

Several examples of previous chapters have shown how we often perceive what we know – we project our previous experience out onto the world. This has advantages for speed, but it can obviously lead to missing some important differences of detail.

But the straight feed-forward pathway may be more important than usually thought. It takes time for feedback to occur from higher centres back down to sensory areas to 'negotiate' the percept. In difficult instances this has to happen. Yet it takes around 10 ms per neuron at least, so, for some decision processes, there is only time for one pass from eye to brain. Fabre-Thorpe *et al.* (2001) show that the feed-forward process may be used quite extensively, in fact for ultra-rapid classification of scenes containing or not containing an animal, despite these images being complex (containing many features) and totally new to subjects.

12.4.10   Competition between senses: consequences of a finite brain

Ripe fruit may give off characteristic odours, but it also changes colour. In fact, in early 2004, new results appeared about the way we humans have a weaker sense of smell than say our cat or dog. It turns out that this is the result of colour vision! Careful comparisons of Old World and New World primates show that as colour develops, so the sense of smell deteriorates and more and more olfaction genes become (non-functional) pseudogenes (Holden, 2004)

The brain is finite, but the idea that areas might compete with one another for space takes a very specific form in this colour-olfaction trade-off. Pääbo and colleagues (Gilad *et al.*, 2004; Holden, 2004) demonstrated that the number of olfactory pseudogenes (§9.3.3) has increased as colour vision evolved, rising from 20% in New World monkeys, which are usually dichromatic, to 30% in Old World monkeys, which are trichromatic.

12.4.11   Synchronisation

In any large scale multi-component computational system there is a considerable problem of keeping all the pieces synchronised. We are some way from understanding all the details, but some of the broader picture is starting to take shape.

From a theoretical point of view, computer scientists are interested in the building of very large parallel computers, such as the SETI@home project for analysing radio telescope data (Kirkpatrick, 2003). To avoid different components getting seriously out of step requires more than local synchronisation. Some small world or scale-free is necessary, but the brain does seem to have this property (§2.4.3).

Finally there is Benjamin Libet's (2004) intriguing work. Through some clever experiments he demonstrated that consciousness or awareness of percepts or decisions made to carry out actions takes place up to half a second *after* the event. This seems at first to challenge the notion of free will, but the philosophical issues are not really relevant here. However, synchronisation across the brain has to take time. Since it takes around 10 ms for synaptic transmission between neurons, then this is time for about 50 signals. This is about the order of the diameter of the brain. Thus we would not expect whole-of-brain synchronisation to occur in less time. Why, or even whether, consciousness and awareness require such synchronisation is an open question.

12.4.12 Sensory inhibition and conscious awareness

We shut out the information we don't need after we have processed it . . .
von Békésy (1967)

In sensory domains such as vision, very precise measurements and understanding of inhibition have been gained. But very recent work following a theoretical paper by Snyder and Mitchell (1999) implies that inhibition may occur on a much larger scale (Bossomaier & Snyder, 2004; Snyder *et al.*, 2004). It is well known that autistic children or savants may have extraordinary arithmetic or drawing skills. Less well known, but very surprising, is that sometimes people following brain damage, say from a stroke, find an *increase* in drawing skills.

It is now looking very much, and at the time of writing these ideas are *new and controversial*, like these capacities exhibited by savants are present in us all but are dynamically suppressed. *Trans-cranial magnetic stimulation* (§2.6.2.1) is capable of turning off the brain modules which suppress, liberating these skills.

12.5 Consciousness

What is it like to be a bat? Thomas Nagel's well-known rhetorical question about the experience of other animals relates to their conscious experience, one of the hot topics in philosophy and neuroscience at the time of writing. There are many views on the mechanisms and locations of consciousness. One such is Jackendoff's intermediate view (1987), which stresses the elements of perceptual awareness (not sensory data – we perceive *red* not 650 nm).

The recurrent theme throughout the book of sensory streaming should have left no doubt that consciousness is nothing like a homunculus sitting somewhere in the brain. Sensory information is taken apart, integrated in various ways at various timescales. The seeming unity of personal experience remains an enigma made more intriguing by evidence that consciousness is delayed (Libet, 2004). Of consciousness itself, there is an increasing number of provocative books. Some, such as Antonio Damasio (1999), take the view that consciousness, and awareness are of, and only of, biological life:

The knowledge would not be expressed in the manner we encounter in human beings and is probably present in many other species . . . The looks of emotion can be simulated but what feelings feel like cannot be duplicated in silicon. Feelings cannot be duplicated unless flesh is duplicated, unless the brain's action on flesh are duplicated, unless the brain's sensing of flesh after it has been acted upon by the brain is duplicated.

Damasio (1999)

But another theme of the book has been the building of senses for robots and animats. As these creatures get ever more complex, and their onboard computational power exceeds that of the human brain, no matter which way you measure it, it seems unlikely that they will not ultimately develop a form of consciousness of their own.

References

Ackerman, D. (1990). *A Natural History of the Senses*. Vintage Books, New York.

Adams, A. (1981). *The Negative*. New York Graphic Society, New York.

Adams, D.L. & Horton, J.C. (2002). Shadows cast by retinal blood vessels mapped in primary visual cortex. *Science*, **298**, 572–576.

Adrian, E.D. (1940). Double representation of the feet in the sensory cortex of the cat. *J. Physiol.*, **98**, 16–18.

Ahnelt, P.K. & Kolb, H. (2000). The mammalian photoreceptor mosaic-adaptive design. *Prog. Retinal and Eye Res.*, **19**, 711–777.

Alpern, M. (1982). Eye movements and strabisxmus. Barlow, H.B. & Mollon, J.D. (eds), *The Senses*. Cambridge University Press, Cambridge, UK.

Alpert, M. (2002). Getting real. *Scientific American*, **287**(6), 124–126.

Amoore, J.W. (1971). Stereochemical and vibrational theories of odour. *Nature*, **233**, 270–271.

Anderson, J.S., Lampl, I., Gillespie, D.C., & Ferster, D. (2000). The contribution of noise to contrast invariance of orientation tuning in cat visual cortex. *Science*, **290**(5498), 1968–1972.

Angioy, A.M., Desogus, A., Barbarossa, I.T., Anderson, P., & Hansson, B.S. (2003). Extreme sensitivity in an olfactory system. *Chemical Senses*, **28**, 279–284.

Anon (2003). Nose in a million. *New Scientist*, **178**(2399), 28.

Anon (2005). How fatty food tickles the tongue. *New Scientist*, **188**(2574), 17.

Arabzadeh, E., Panzeri, S., & Diamond, M.E. (2006). Deciphering the spike train of a sensory neuron: counts and temporal patterns in the rat whisker pathway. *J. Neurosci.*, **26**(3), 9216–9226.

Arnold, K.E., Owens, I.P.FF., & Marshall, N.J. (2002). Fluorescent signalling in parrots. *Science*, **295**(5552), 92.

Atzori, M., Saobo Leil, S., Evans, D.I.P., Kanold, P.G., Phillips-Tansey, E., McIntyre, O., & McBain, C.J. (2001). Differential synaptic processing separates stationary from transient inputs to the auditory cortex. *Nature Neuroscience*, **4**(12), 1230–1237.

Axel, R. (1995). The molecular logic of smell. *Scientific American*, **273**(4), 154–159.

Babcock, D.F. (2003). Smelling the roses. *Science*, **299**(5615), 1993–1994.

Baker, R.R. (1981). *Human Navigation and the Sixth Sense*. Hodder Arnold, Manchester, UK.

Baker, R.R. (1989). *Human Navigation and Magnetoreception*. Manchester University Press, Manchester, UK.

Bar, M., Kassam, K.S., & Ghuman, A.S. (2006). Top-down facilitation of visual recognition. *Proc. Nat. Acad. Sci*, **103**, 449–454.

Barabási, A.-L. (2002). *Linked*. Perseus, MA.

Barbour, D.L. & Wang, X. (2003). Contrast tuning in auditory cortex. *Science*, **299**(5609), 1073–1075.

Barinaga, M. (1999). Salmon follow watery odours home. *Science*, **286**, 705–706.

Barinaga, M. (2000a). Family of bitter taste receptors found. *Science*, **287**, 2133–2134.

Barinaga, M. (2000b). New ion channel may yield clues to hearing. *Science*, **287**(5461), 2132–2133.

Barinaga, M. (2002a). How the brain's clock gets daily enlightenment. *Science*, **295**(5557), 255–257.

Barinaga, M. (2002b). Sight, sound converge in owl's mental map. *Science*, **297**(5586), 1462–1463.

Barlow, H.B. (1959). Sensory mechanisms, the reduction of redundancy, and intelligence. *NPL Symposium on the Mechanization of Thought Process*. No. 10, HM Stationery Office, London, pp. 535–539.

Barlow, H.B. (1964). Physical limits of vision. Gieses, A.C. (ed.), *Photophysiology*. Academic Press, New York.

Barlow, H.B. (1982). General principles: the senses considered as physical instruments. Barlow, H.B. & Mollon, J.D. (eds), *The Senses*. Cambridge University Press, Cambridge, UK.

Barlow, H.B. (1986). Why have multiple cortical areas. *Vision Research*, **26**, 81–90.

Barlow, H.B. & Mollon, J.D. (eds) (1982). *The Senses*. Cambridge University Press, Cambridge, UK.

Barlow, R.B., Birge, R.R., Kaplan, E., & Tallent, J.R. (1993). On the molecular origin of photoreceptor noise. *Nature*, **366**, 64–66.

Barnea, G., O'Donnell, S., Mancia, F., Sun, X., Nemes, A., Mendelsohn, M., & Axel, R. (2004). Odorant receptors on axon termini in the brain. *Science*, **304**, 1468.

Baron-Cohen, S. (1997). *Mindblindness*. MIT Press, Cambridge, MA.

Bayliss, A.P., Pellegrino, G. di, & Tipper, S.P. (2005). Sex differences in eye gaze and symbolic cueing of attention. *Quarterly J. Expt. Psychology*, **58A**(4), 631–650.

Baylor, D.A. & Hodgkin, A.L. (1973). Responses of photoreceptors to flashes. *J. Physiol.*, **234**, 163–198.

Békésy, G. von (1967). *Sensory Inhibition*. Princeton University Press, Princeton, NJ.

Bellevill, S. & Wilkinson, F. (1986). Vernier acuity in the cat: its relation to hyperacuity. *Virtual Reality*, **26**(8), 1263–1271.

Bendor, D. & Wang, X. (2005). The neuronal representation of pitch in primate auditor cortex. *Nature*, **436**, 1161–1163.

Bennett, C.H. (1982). Thermodynamics of computation – a review. *Int. J. Theor. Phys.*, **21**, 905–940.

Bensmaïa, S.J., Craig, J.C., & Johnson, K.O. (2006). Temporal factors in tactile spatial acuity: evidence for RA interference in fine spatial processing. *J. Neurophysiol.*, **95**, 1783–1791.

Benson, A.J. (1982). The vestibular system. Barlow, H.B. & Mollon, J.D. (eds), *The Senses*. Cambridge University Press, Cambridge, UK.

Berry, M.J., Warland, D.K., & Meister, M. (1997). The structure and precision of retinal spike trains. *Proc. Nat. Acad. Sci*, **94**, 5411–5416.

Berson, D.M., Dunn, F.A., & Takao, M. (2002). Phototransduction by retinal ganglion cells that set the circadian clock. *Science*, **295**(5557), 1070–1073.

Bialek, W. (2002). Thinking about the brain. Flyvbjerg, H., Jlicher, F., Ormos, P., & David, F. (eds), *Physics of Biomolecules and Cells*. Springer-Verlag, Berlin, pp. 485–577.

Bialek, W., Rieke, F., Steveninck, R.R., de Ruyter van, & Warland, D. (2001). Reading a neural code. *Science*, **252**, 1854–1856.

Bickerstaff, I., Benson, S., & Hocking, M. (2011). Presentation at Develop, Brighton, UK.

Bickham, J. (2008). The whites of their eyes: the evolution of distinctive sclera in humans. *Lambda Alpha Journal*, **38**, 20–29.

Bisley, J.W. & Goldberg, M.E. (2003). Neuronal activity in the lateral intraparietal area and spatial attention. *Science*, **299**(5603), 81–86.

Blahut, R.E. (1985). *Fast Algorithms for Digital Signal Processing*. Addison-Wesley, New York.

Blakemore, C. & Campbell, F. (1969). On the existence of neurones in the human visual system selectively sensitive to the orientation and size of retinal images. *J. Physiol.*, **203**, 237–260.

Blakemore, S.-J., Wolpert, D.M., & Frith, C.D. (1998). Central cancellation of self-produced tickle sensation. *Nature Neuroscience*, **1**(7), 635–640.

Blanke, O., Ortigue, S., Landis, T., & Seeck, M. (2002). Stimulating illusory own-body perceptions. *Nature*, **419**, 260–270.

Blanke, O., Landis, T., Spinelli, L., & Seeck, M. (2004). Out-of-body experience and autoscopy of neurological origin. *Brain*, **127**, 243–258.

Blinn, J.F. (1978). Simulation of wrinkled surfaces. *Computer Graphics*, **12**(3), 286–292.

Blount, J.D., Metcalfe, N.B., Birkhead, T.R., & Surai, P.F. (2003). Carotenoid modulation off immune function and sexual attractiveness in zebra finches. *Science*, **300**(5616), 125–127.

Bonadonna, F. & Nevitt, G.A. (2004). Partner-specific odour recognition in an antarctic seabird. *Science*, **306**(5697), 535.

Born, M. & Wolf, E. (1980). *Principles of Optics*, 6th edn. Pergamon Press, Oxford, UK.

Bossomaier, T.R.J. & Mackerras, P. (1992). Representation of an image by nonorthogonal basis functions with use of the fast fourier transform. *J. Opt. Soc. Am. A*, **11**(11), 1915–1917.

Bossomaier, T.R.J. & Snyder, A.W. (2004). Absolute pitch accessible to everyone by turning off part of the brain? *Organised Sound*, **9**(2), 181–189.

Botvinick, M. (2004). Probing the neural basis of body ownership. *Science*, **305**(5685), 782–783.

Bracewell, R. (1999). *The Fourier Transform and its Applications*, 2nd edn. McGraw Hill, New York.

Braddick, O.J. (1982). Binocular vision. Barlow, H.B. & Mollon, J.D. (eds), *The Senses*. Cambridge University Press, Cambridge, UK, pp. 192–200.

Bradley, D. (2002). Moving through the landscape. *Science*, **295**(5564), 2385–2386.

Branco, T. & Häusser, M. (2010). The single dendritic branch as a fundamental functional unit in the nervous system. *Curr. Opin. Neurobiol.*, **20**, 494–502.

Brand, A., Behrend, O., Marquardt, T., McAlpine, D., & Grothe, B. (2002). Precise inhibition is essential for microsecond interaural time difference coding. *Nature*, **417**, 543–547.

Broadbent, A.G. (2009). Calculation from the original experimental data of the CIE 1931 RGB standard observer spectral chromaticity co-ordinates and color matching functions. http://www.cis.rit.edu/mcsl/research/broadbent/CIE1931_RGB.pdf.

Brodin, M., Laska, M., & Olsson, M.J. (2009). Odour interaction between bourgonal and its antagonist undecal. *Chemical Senses*, **34**, 625–630.

Brown, K. (2001a). Animal magnetism guides migration. *Science*, **294**(5541), 283–284.

Brown, K. (2001b). A discriminating taste for bitter. *Science*, **291**(5508), 1465–1466.

Brown, K.S. (1999). Are you ready for a new sensation? *Scientific American*, **10**, 38–43.

Buchsbaum, G. & Gottschalk, A. (1983). Trichromacy, opponent colour coding and optimum colour information transmission in the retina. *Proc. Roy. Soc. Lon. B*, **220**, 89–113.

Buck, L.B. (2000). Smell and taste: the chemical senses. Kandel, E.R., Schwartz, J.H., & Jessell, T.M. (eds), *Principles of Neural Science*, 4th edn. McGraw Hill, New York.

Bumsted, K. & Hendrickson, A. (1999). Distribution and development of short-wavelength cones differ between macaca monkey and human fovea. *J. Comp. Neurol.*, **403**(4), 502–516.

Burton, H. (2002). Cerebral cortical regions devoted to the somatosensory system: results from brain imaging studies in humans. Nelson, R.J. (ed.), *The Somatosensory System*. CRC Press, Boca Raton, FL, pp. 27–72.

Burton, R. (1976). *The Language of Smell*. Routledge, Kegan Paul, London.

Caicedo, A. & Roper, S.D. (2001). Taste receptor cells that discriminate between bitter stimuli. *Science*, **291**(5508), 1557–1560.

Calford, M.B., Clarey, J.C., & Tweedale, R. (1998). Short-term plasticity in adult somatosensory cortex. Morley, J.W. (ed.), *Neural Aspects of Tactile Sensation. Adv. Psychology*, **127**. North Holland, Amsterdam.

Campbell, F. & Westheimer, G. (1959). Factors influencing accommodation responses of the human eye. *J. Opt. Soc. Am*, **49**, 568–571.

Campbell, F.W. & Gubisch, R.W. (1966). Optical quality of the human eye. *J. Physiol.*, **186**, 558–578.

Campen von, C. (1999). Artistic and psychological experiments with synesthesia. *Leonardo*, **32**(1), 9–14.

Canoon, L. (2008). Testing a taste test for depression. *Science*, **321**, 1153.

Carr, C.E. & Konishi, M. (1988). Axonal delay lines for time measurement in the owl's brainstem. *Proc. Nat. Acad. Sci*, **85**, 8311–8315.

Carruba, S., Frilot II, C., Chesson Jr, A.L., & Marine, A.A. (2007). Evidence of a nonlinear human magnetic sense. *Neuroscience*, **144**, 356–367.

Catania, K.C. (2002). The nose takes a starring role. *Scientific American*, **287**(1), 38–43.

Catmull, E.E. (1974). A subdivision algorithm for computer display of curved surfaces. Ph.D. thesis, Department of Computer Science, University of Utah, UT.

Cator, L.J., Arthur, B.J., Harrington, L.C., & Hoy, R.R. (2009). Harmonic convergence in the love songs of the dengue vector mosquito. *Science*, **323**, 1077–1079.

Chait, M., Poeppel, D., & Simon, J.Z. (2006). Neural response correlates of detection of monaurally and binaurally created pitch in humans. *Cerebral Cortex*, **16**, 835–848.

Chang, E.F. & Merzenich, M.M. (2003). Environmental noise retards auditory cortical development. *Science*, **300**(5618), 498–502.

Chaparro, A., Stromeyer, C.F., Huang, E.P., & Kronauer, R.E. (1993). Colour is what the eye sees best. *Nature*, **361**, 348–350.

Chapman, C.E. (1998). Constancy in the somatosensory system: central neural mechanisms underlying the appreciation of texture during active touch. Morley, J.W. (ed.), *Neural Aspects of Tactile Sensation. Adv. Psychology*, **127**. North Holland, Amsterdam, pp. 275–298.

Chatternee, S. & Callaway, E.M. (2003). Parallel colour-opponent pathway to primary visual cortex. *Nature*, **426**, 668–671.

Chávez, A.E., Bozinovic, F., Peichl, L., & Palacios, A.G. (2003). Retinal spectral sensitivity, fur coloration, and urine reflectance in the genus *Octodon* (rodentia): implications for visual ecology. *Investigative Ophthal. & Vis. Science*, **44**(5), 2290–2296.

Chen, L.M., Friedman, R.M., & Roe, A.W. (2003). Optical imaging of a tactile illusion in area 3b of the primate somatosensory cortex. *Science*, **302**, 881–885.

Chiao, C.-C., Cronin, T.W., & Osorio, D. (2000). Color signals in natural scenes: characteristics of reflectance spectra and effects of natural illuminants. *J. Opt. Soc. Am. A*, **17**, 218–224.

Choi, C. (2003). Shake, waddle and stroll. *Scientific American*, **288**(1), 16.

Choic, C.Q. (2006). Sense and sensitivity. *Scientific American*, **294**(3), 17.

Clark, D.A., Mitra, P.P., & Wang, S.S.-H. (2001). Scalable architecture in mammalian brains. *Nature*, **411**, 189–193.

Cochran, W.W., Mouritsen, H., & Wikelski, M. (2004). Migrating songbirds recalibrate their magnetic compass daily from twilight cues. *Science*, **304**(5669), 405–408.

Cohen, J. (2002). The confusing mix of hype and hope. *Science*, **295**(5557), 11026.

Cohen, J.D. & Tong, F. (2001). The face of controversy. *Science*, **293**(5539), 2405–2407.

Comon, P. (1994). Independent component analysis: a new concept. *Signal Processing*, **36**, 287–314.

Connor, C.E. (2002). Reconstructing a 3D world. *Science*, **298**(5592), 376–377.

Connor, C.E. & Johnson, K.O. (1992). Neural coding of tactile texture: comparison of spatial temporal mechanisms for roughness perception. *J. Neurosci.*, **12**(9), 3414–3426.

Conway, B.R. (2009). Color vision, cones, and color-coding in the cortex. *Neuroscientist*, **15**(3), 274–290.

Cooley, J.W. & Tukey, J.W. (1965). An algorithm for the machine computation of complex Fourier series. *Math. Comp.*, **19**, 297–301.

Copley, J. (2001). Love is in the air. *New Scientist*, 15.

Corcelli, A., Lobasso, S., Lopalco, P., Dibattista, M., Aranera, R., Peterlin, Z., & Firestein, S. (2010). Detection of explosives by olfactory sensory neurons. *J. Hazardous Materials*, **175**, 1096–1100.

Cox, D., Meyers, E., & Sinha, P. (2004). Contextually evoked object-specific responses in human visual cortex. *Science*, **304**(5667), 115–117.

Craelius, W. (2002). The bionic man: restoring mobility. *Science*, **295**(5557), 1018–1021.

Craig, A.D., Reiman, E.M., Evans, A., & Bushnell, M.C. (1996). Functional imaging of an illusion of pain. *Nature*, **384**, 258–260.

Crowder, R. (2006). Towards robots that can sense texture by touch. *Science*, **312**, 1478–1479.

Curcio, C.A. & Sloan, K.R. (1992). Packing geometry of human cone photoreceptors: variations with eccentricity and evidence for local anisotropy. *Vis. Neurosci.*, **9**, 169–180.

Damak, S., Rong, M., Yasumatsu, K., Kokrahvii, Z., Varadarajan, V., Zou, S., Jiang, P. *et al.* (2003). Detection of sweet and unami taste in the absence of receptor T1r3. *Science*, **301**(5634), 850–853.

Damasio, A. (1999). *The Feeling of What Happens*. Heineman, London.

Davenport, R.J. (2001). New gene may be key to sweet tooth. *Science*, **292**(5517), 620–621.

Davis, J. (2004). *Mathematical Modelling of Earth's Magnetic Field*. Virginia Tech Report, VA.

Davis, R. (1931). A correlated colour temperature for illuminants. *National Bureau of Standards for Research*, **7**, 659–681.

Debevec, P.E. & Malik, J. (1997). Recovering high dynamic range radiance maps from photographs. *SIGGRAPH Proc.*, pp. 369–378.

DeBose, J.L., Lema, S.C., & Nevitt, G.A. (2008). Dimethylsulfoniopropionate as a foraging cue for reef fishes. *Science*, **319**, 1356.

DeLoache, J.S., Uttal, D.H., & Rosengren, K.S. (2004). Scale errors offer evidence for a perception–action dissociation early in life. *Science*, **304**(5673), 1027–1029.

Dematte, M.L., Sanabria, D., Sugarman, R., & Spence, C. (2006). Cross-modal interaction between olfaction and touch. *Chemical Senses*, **31**, 291–300.

Dennett, D.C. (1991). *Consciousness Explained.* Penguin, London.

Derrico, J.B. & Buchsbaum, G. (1991). A computational model of spatiochromatic image coding in early vision. *J. Vis. Communication and Image Representation*, **2**(1), 31–38.

Deutsch, D. (1992). Paradoxes of musical pitch. *Scientific American*, **267**(2), 70–75.

Dominy, N.J. & Lucas, P.W. (2001). Ecological importance of trichromatic vision to primates. *Nature*, **410**, 363–366.

Doujak, F.E. (1985). Can a shore crab see a star? *J. Exp. Biol.*, **116**, 385–393.

Downing, P.E., Jiang, Y., Shuman, M., & Kanwisher, N. (2001). A cortical area selective for visual processing of the human body. *Science*, **293**(5539), 2470–2473.

Drago, F., Myszkowski, K., Annen, T., & Chiba, N. (2003). Adaptive logarithmic mapping for displaying high contrast scenes. *Computer Graphics Forum*, **22**(3), 419–426.

Dulac, C. (2006). Charting olfactory maps. *Science*, **314**(5799), 606–607.

Dunbar, R. (1996). *Grooming, Gossip and the Evolution of Language*. Faber and Faber, London.

Dunbar, R.I.M. & Shultz, S. (2007). Evolution in the social brain. *Science*, **317**(5843), 1344–1347.

Dunkel, M., Scmidt, U., Struck, S., Berger, L., Gruening, B., Hossback, J., Jaeger, I.S. *et al.* (2009). Superscent – a database of flavours and scents. *Nucleic Acid Research*, **37**, D292–D294.

Dunnigan, J.F. (2001). The curse of fast iron. *IEEE Computer*, **34**(4), 100–103.

Dusenbury, A. (1992). *Sensory Ecology*. W.H. Freeman, New York.

Dynamic, B. (n.d.). How do I select my headphones? http://north-america.beyerdynamic.com/service/faqs/kopfhoerer/kaufberatung-fuer-kopfhoerer.html. Accessed 2010.

Editorial (2001). Sock it to the sacculus. *Noise and Vibration Worldwide*, **32**(11).

Ehrsson, H.E., Spence, C., & Passingham, R.E. (2004). That's my hand! Activity in premotor cortex reflects feeling of ownership of a limb. *Science*, **304**(5685), 875–877.

Eickhoff, S.B., Schleicher, A., Zilles, K., & Amunts, K. (2006a). The human parietal operculum. I. Cytoarchitectonic mapping of subdivisions. *Cerebral Cortex*, **16**, 254–267.

Eickhoff, S.B., Grefkes, C., Zilles, K., & Fink, G.R. (2006b). The somatic organisation of cytoarchitectonic areas on the human parietal operculum. *Cerebral Cortex*, **17**(8), 1800–1811.

Emotiv (n.d.). Emotiv. http://www.emotiv.com. Accessed 2010.

Endler, J.A. (1993). The color of light in forests and its implications. *Ecol. Monogr.*, **63**(2), 1–27.

Erichsen, J.T. & May, P.J. (2002). The pupillary and ciliary components of the cat Edinger–Westphal nucleus: a transsynaptic transport investigation. *Vis. Neurosci.*, **19**, 15–29.

Ernst, M.O., Banks, M.S., & Bülthoff, H.H. (2000). Touch can change visual slant perception. *Nature Neuroscience*, **3**(1), 69–73.

Evans, E.F. (1982). Functional anatomy of the auditory system. Barlow, H.B. & Mollon, J.D. (eds), *The Senses*. Cambridge University Press, Cambridge, UK, pp. 251–306.

Eysel, U.T. (2003). Illusions and perceived images in the primate brain. *Science*, **302**, 789–790.

Fabre-Thorpe, M., Delorme, A., Marlot, C., & Thorpe, S. (2001). A limit to the speed of processing in ultra-rapid visual categorization of novel natural scenes. *J. Cog. Neurosci.*, **13**(2), 171–180.

Fabri, M., Polonara, G., Quattrini, A., Salvolini, U., Pesce, M. del, & Manzoni, T. (1999). Role of the corpus callosum in the somatosensory activation of the ipsilateral cerebral cortex and fMRI study of callostomized patients. *Eur. J. Neurosci.*, **11**, 3983–3994.

Fagiolini, M., Fritschy, J.-M., Löw, K., Möhler, H., Rudolph, U., & Hensch, T.K. (2004). Specific GABA$_A$ circuits for visual cortical plasticity. *Science*, **303**, 1681–1683.

Faivre, B., Grégoire, A., Préault, M., Cézally, F., & Sorci, G. (2003). Immune activation rapidly mirrored in a secondary sexual trait. *Science*, **300**(5616), 103.

Fellgett, P.B. & Linfoot, E.H. (1955). On the assessment of optical images. *Phil. Trans. R. Soc. Lond. A*, **247**, 369–407.

Feynman, R.P. (1984). Quantum mechanical computers. *Optics News*, Feb., 11–20.

Feynman, R.P., Leightonm, R.B., & Sands, M. (1963). *The Feynman Lectures on Physics*. Addison-Wesley, Reading, MA.

Field, D.J. (1987). Relations between the statistics of natural images and the response properties of cortical cells. *J. Opt. Soc. Am. A*, 2379–2394.

Fields, R.D. (2007a). Sex and the secret nerve. *Scientific American*, **18**(1), 20–27.

Fields, R.D. (2007b). The shark's electric sense. *Scientific American*, **297**(2), 74–81.

Finger, T.E., Danilova, V., Barrows, J., Bartel, D.L., Vigers, A.J., Stone, L., Helleskamt, G. *et al.* (2005). ATP signalling is crucial for communication from taste buds to gustatory nerves. *Science*, **310**(5753), 1495–1499.

Finnerty, G.T., Roberts, L.S.E., & Connors, B.W. (1999). Sensory experience modifies the short-term dynamics of neocortical synapses. *Nature*, **400**, 567–571.

Fjällbrant, T.T., Manger, P.R., & Pettigrew, J.D. (1998). Some related aspects of platypus electroreception: temporal integration behaviour, electroreceptive thresholds and directionality of the bill acting as an antenna. *Phil. Trans. R. Soc. Lond. B*, **353**, 1211–1219.

Flavornet (n.d.). Flavornet. http://wwww.flavornet.org/flavornet.html. Accessed 2010.

Fleming, B. & Dobbs, D. (1999). *Animating Facial Features and Expressions*. Charles River Media Inc., MA.

Fletcher, H. & Munson, W.A. (1933). Loudness, its definition, measurement and calculation. *J. Acoustic Soc. Am.*, **5**, 82–108.

Foley, J.D., Dam, A. van, Feiner, S.K., Hughes, J.F., & Phillips, R.L. (1996). *Introduction to Computer Graphics*. Addison-Wesley, MA.

Fox, B. (2000). Pump up the volume. *New Scientist*, **168**(2264), 12.

Fox, K., Wright, N., Wallace, H., & Glazewski, S. (2003). The origin of cortical surround receptive fields studied in the barrel cortex. *J. Neurosci.*, **23**, 8380–8391.

Freeman, W.J. (1999). *How Brains Make Up their Minds*. Phoenix, London.

Friedrich, R.W. & Laurent, G. (2001). Dynamic optimisation of odor representations by slow temporal patterning of mitral cell activity. *Science*, **291**, 889–894.

Fries, P., Reynolds, J.H., Rorie, A.E., & Desimone, R. (2001). Modulation of oscillatory neuronal synchronisation by selective visual attention. *Science*, **291**(5508), 1560–1563.

Froehler, M.T. & Duffy, C.J. (2002). Cortical neurons encoding path and place: where you go is where you are. *Science*, **295**(5564), 2462–2465.

Fu, Y., Kefalov, V., Luo, D.-G., Xue, T., & Yau, K.-W. (2008). Quantal noise from human red cone pigment. *Nature Neuroscience*, **11**, 568–571.

Fu, Y.-X., Djupsund, K., Gao, H., Hayden, B., Shen, K., & Dan, Y. (2002). Temporal specificity in the cortical plasticity of visual space representation. *Science*, **296**(5575), 1999–2003.

Fukuchi-Shimogori, T. & Grove, E.A. (2001). Neocortex patterning by the secreted signaling molecule fgf8. *Science*, **294**(5544), 1071–1074.

Gabor, D. (1946). Theory of communication. *J. Inst. Elec. Eng.*, **93**, 429–457.

Gagliardo, A., Loalè, P., Savini, M., & Wild, J.M. (2006). Having the nerve to home: trigeminal magnetoreceptor versus olfactory mediation of homing in pigeons. *J. Exp. Biol.*, **209**, 2888–2892.

Gamow, G. (1944). *Mr Tompkins Explores the Atom*. Macmillan, New York.

Gardiner, J.M. & Atema, J. (2007). Sharks need the lateral line to located odour sources: rheotaxis and eddy chemotaxis. *J. Expt. Biol.*, **210**, 1925–1934.

Gardner, E.P. & Kandel, E.R. (2000). Touch. Kandel, E.R., Schwartz, J.H., & Jessell, T.M. (eds), *Principles of Neural Science*, 4th edn. McGraw Hill, New York.

Gardner, E.P., Martin, J.H., & Jessell, T.M. (2000). The bodily senses. Kandel, E.R., Schwartz, J.H., & Jessell, T.M. (eds), *Principles of Neural Science*, 4th edn. McGraw Hill, New York.

Gaschler, K. (2006). One person, one neuron? *Scientific American*, **17**, 77–82.

Gaskill, J.D. (1978). *Linear Systems, Fourier Transforms and Optics*. Wiley, New York.

Gauthier, I., Curran, T., Curby, K.M., & Collins, D. (2003). Perceptual interference supports a non-modular account of face processing. *Nature*, **6**(4), 428–432.

Gazzaniga, M.S. (1992). *Nature's Mind*. Basic Books, New York.

Gebeshubera, I.C. (2000). The influence of stochastic behavior on the human threshold of hearing. *Chaos, Solutions and Fractals*, **11**(12), 1855–1868.

Genoa (n.d.). Genoa Colour Technologies home page. http://www. genoacolor.com. Accessed 2010.

Georgopoulos, A.P., Schwartz, A.B., & Kettner, R.E. (1986). Neuronal population coding of movement direction. *Science*, **233**, 1416–1419.

Gerzon, M. (1973). Periphony: with-height sound reproduction. *J. Audio Eng. Soc.*, **21**, 2–10.

Gilad, Y., Wiebe, V., Przeworski, M., Lancet, D., & Pääbo, S.J. (2004). Loss of olfactory receptor genes coincides with the acquisition of full trichromatic vision in primates. *PLOS*, **2**, 120–125.

Goldberg, M.E. (2000). The control of gaze. Kandel, E.R., Schwartz, J.H., & Jessell, T.M. (eds), *Principles of Neural Science*, 4th edn. McGraw Hill, New York.

Goldberg, M.E. & Huspeth, A.J. (2000). The vestibular system. Kandel, E.R., Schwartz, J.H., & Jessell, T.M. (eds), *Principles of Neural Science*, 4th edn. McGraw Hill, New York.

Gonzalez, R.C. & Woods, R.E. (1992). *Digital Image Processing*. Addison-Wesley, MA.

Goodwin, A.W. (1998). Extracting the shape of an object from the responses of peripheral nerve fibers. Morley, J.W. (ed.), *Neural Aspects of Tactile Sensation*, Advances in Psychology, vol. 127. Elsevier, Amsterdam, pp. 55–88.

Goodwin, A.W. & Wheat, H.E. (2004). Sensory signals in neural populations underlying tactile perception and manipulation. *Annu. Rev. Neurosci.*, **27**, 53–77.

Gouras, P. (1991). Colour vision. Kandel, E.R., Schwartz, J.H., & Jessell, T.M. (eds), *Principles of Neural Science*, 3rd edn. McGraw Hill, New York, pp. 467–480.

Gracheva, E.O., Ingolia, N.T., Kelly, Y.M., Cordero-Morales, J.F., Hollopeter, G., Chelser, A.T., Sánchez, E.E. *et al.* (2010). Molecular basis of infrared detection by snakes. *Nature*, **464**(7291), 1006–1011.

Grantham, D.W. (1995). Spatial hearing and related phenomena. Moore, B.C.J. (ed.), *Hearing*. Academic Press, San Diego, CA, pp. 297–345.

Graziano, M.S.A., Taylor, C.S.R., & Moore, T. (2002). Complex movements evoked by microstimulation of precentral cortex. *Neuron*, **34**, 841–851.

Gregory, J.E., Iggo, A., McIntyre, A.K., & Proske, U. (1989). Responses of electroreceptors in the snout of the echidna. *J. Physiol.*, **414**, 521–538.

Gregory, R.L. (ed.) (1987). *The Oxford Companion to the Mind*. Oxford University Press, Oxford, UK.

Grimm, D. (ed.) (2004). Beetles may help battle blazes. *Science*, **305**(5686), 940.

Gross, J., Schnitzler, A., Timmermann, L., & Pioner, M. (2007). Gamma oscillations in human primary somatosensory cortex reflect pain perception. *PLOS Biol.*, **5**, 1168–1173.

Grunwald, M. (2004). Worlds of feeling. *Scientific American*, **5**, 56–61.

Guilford, T., Roberts, S., & Biro, D. (2004). Positional entropy during pigeon homing. II: Navigational interpretation of Bayesian latent state models. *Journal of Theoretical Biology*, online.

Guinard, D., Usson, Y., Guillermet, C., & Saxod, R. (2000). PS-100 and NF70-200 double immunolabelling for human digital skin Meissner corpuscle 3D imaging. *J. Histo. Cythochem.*, **48**, 295–302.

Guo, J. & Guo, A. (2005). Crossmodal interactions between olfactory and visual learning in *Drosophila*. *Science*, **309**(5732), 307–310.

Gutfreund, Y., Zheng, W., & Knudsen, E.I. (2002). Gated visual input to the central auditory system. *Science*, **297**(5586), 1556–1559.

Haddock, S.H.D., Dunn, C.W., Pugh, P.R., & Schnitzler, C.E. (2005). Bioluminescent and red-fluorescent lures in a deep-sea siphonophore. *Science*, **309**(5732), 263.

Hagelin, J.C., Jones, I.L., & Rasmussen, L.E.L. (2003). Tangerine-scented social odour in a monogamous seabird. *Proc. Roy. Soc. Lon. B*, **270**, 1323–1329.

Hahnloser, R.H.R., Kozhevnikov, A.A., & Fee, M.S. (202). An ultra-sparse code underlies the generation of neural sequences in a songbird. *Nature*, **419**, 65–69.

Halata, Z., Grim, M., & Bauman, K.I. (2003). Friedrich Sigmund Merkel and his 'Merkel cell' morphology, development and physiology: review and new results. *Anatomical Record*, **271A**, 225–239.

Hamberger, K., Hanser, T., & Gegenfurtner, K.R. (2007). Geometric illusions at isoluminance. *Vision Research*, **47**, 3276–3285.

Hao, S., Sharp, J.W., Ross-Inta, C.M., McDaniel, B.J., Anthony, T.G., Wek, R.C., Cavener, D.R. *et al.* (2005). Uncharged ERNA and sensing of amino acid deficiency in mammalian pyriform cortex. *Science*, **307**(5716), 1776–1778.

Harmon, L.D. & Julesz, B. (1973). Masking in visual recognition: effects of two-dimensional filtered noise. *Science*, **180**, 1194–1197.

Harris, S. (2003). Sweet surround: an interview with DTS CEO, John Kirchner. *Hi-fi News and Record Review*.

Harrison, M.H. (2009). Sivian and white revisited: the role of role of resonant thermal noise pressure on the eardrum in auditory thresholds. arXiv:0910.3170.

Harry, J.D., Niemi, J.B., Priplata, A.A., & Collins, J.J. (2005). Balancing act (noise based sensory enhancement technology). *IEEE Spectrum*, **42**, 36–41.

Hartley, R.V.L. (1928). Transmission of information. *Bell System Tech. J.*, **7**, 535–563.

Hartline, P.H., Kass, L., & Loop, M.S (1978). Merging of modalities in the optic tectum: infrared, and visual integration in rattlesnakes. *Science*, **199**, 1225–1229.

Hartmann, M.J., Johnson, N.J., Towal, R.B., & Assad, C. (2003). Mechanical characteristics of rat vibrissae: resonant frequencies and damping in isolated whiskers and in the awake behaving animal. *J. Neurosci.*, **23**, 6510–6519.

Harvey, F. (2002). Surrounded by sound. *Scientific American*, **286**, 82–83.

Hasson, U., Nir, Y., Levy, I., Fuhrmann, G., & Malach, R. (2004). Intersubject synchronisation of cortical activity during natural vision. *Science*, **303**(56644), 1634–1640.

Hattar, S., Liao, H.-W., Takao, M., Berson, D.M., & Yau, K.-W. (2002). Melanopsin-containing retinal ganglion cells: architecture, projections and intrinsic photosensitivity. *Science*, **295**(5557), 1065–1070.

Hawksford, M. (1998). A closer look at DTS. *Hi-fi News and Record Review*, **43**(2), 60–65.

Haxby, J.V., Gobbini, M.I., Furey, M.L., Ishai, A., Schouten, J.L., & Pietrini, P. (2001). Distributed and overlapping representation of faces and objects in ventral temporal cortex. *Science*, **293**, 2425–2430.

He, S., Dong, W., Deng, Q., Weng, S., & Sun, W. (2003). Seeing more clearly: Recent advances in understanding retinal circuitry. *Science*, **302**(5644), 408–411.

Heath, T.P., Melichar, J.K., Nutt, D.J., & Donaldson, L.F. (2006). Human taste thresholds are modulated by serotonin and noradrenaline. *J. Neurosci.*, **26**(49), 12664–12671.

Heinrich, S.P., Kromeier, M., Bach, M., & Kommerell, G. (2005). Vernier acuity for stereodisparate objects and ocular prevalence. *Vision Research*, **45**, 1321–1328.

Helfert, R.H. & Aschoff, A. (1997). Superior olivary complex and nuclei of the lateral lemniscus. Ehret, G. & Romand, R. (eds), *The Central Auditory System*. Oxford University Press, New York.

Helmholtz, H.L.F. (1877). *On the Sensations of Tone*. Dover, New York (reprinted).

Helmholtz, H.L.F. (1896). *Handbuch der physiologischen optik*, 2nd edn. Voss, Hamburg.

Helmuth, L. (2001a). Location neurons do advanced math. *Science*, **292**, 185.

Helmuth, L. (2001b). Where the brain tells a face from a place. *Science*, **292**(5515), 196–198.

Hensch, T.K. & Stryker, M.P. (2004). Columnar architecture sculpted by GABA circuits in developing cat visual cortex. *Science*, **303**, 1678–1681.

Heron, G. & Charman, W.N. (2004). Accommodation as a function of age and the linearity of the response dynamics. *Vision Research*, **44**, 3119–3130.

Heydt, R. von der, Peterhans, E., & Baumgartner, G. (1984). Illusory contours and cortical neuron responses. *Science*, **224**, 1260–1262.

Hiromádka, T., DeWeese, M.R., & Zador, A.M. (2008). Sparse representation of sounds in the unanaesthetised auditory cortex. *PLOS Biol.*, **6**(1), 124–137.

Hoang, Q.V., Linenmeier, R.A., Chung, C.K., & Curcio, C.A. (2002). Photoreceptor inner segments in monkey and human retina: mitochondrial density, optics and regional variation. *Vis. Neurosci.*, **19**, 395–407.

Hofer, H., Carroll, J., Neitz, M., & Williams, D.R. (2005). Organisation of the human trichromatic cone mosaic. *J. Neurosci.*, **25**, 9669–9679.

Hoffmann, J.N., Montag, A.G., & Dominy, N.J. (2004). Meissner corpuscles and somatosensory acuity: the prehensile appendages of primates and elephants. *Anatomical Record A*, **281A**, 1138–1147.

Holden, C. (ed.) (2001). The t-shirt test tells. *Science*, **292**(5516), 431.

Holden, C. (2002a). Perfumed nests for Corsican birds. *Science*, **297**(5581), 511.

Holden, C. (ed.) (2002b). Breaking the color barrier. *Science*, **298**(5598), 1551.

Holden, C. (2003a). Getting into a cassowary's head. *Science*, **302**(5647), 981.

Holden, C. (2003b). Spiritual bass. *Science*, **301**(5640), 1665.

Holden, C. (2003c). Tender lips. *Science*, **300**(5621), 897.

Holden, C. (ed.) (2003d). Teaching the tongue to see. *Science*, **299**(5613), 1657.

Holden, C. (2004). An eye for a nose. *Science*, **303**(5658), 621.

Holden, C. (ed.) (2005). Voices in the brain. *Science*, **309**(5735), 696.

Holland, R.A., Waters, D.A., & Rayner, J.M.V. (2004). Echolocation signal structure in the Megachiropteran bat *Rousettus aegyptiacus* geoffroy 1810. *J. Exp. Biol.*, **207**, 4361–4369.

Holloway, M. (2003). The mutable brain. *Scientific American*, **289**(3), 58–65.

Hopfield, J.J. (1982). Neural networks and physical systems with emergent collective computational abilities. *Proc. Nat. Acad. Sci.*, **79**, 2554–2558.

Hoyer, P.O. & Hyvärinen, A. (2002). A multi-layer sparse coding network learns contour coding from natural images. *Vision Research*, **42**(12), 1593–1605.

Hsiao, S.S. & Vega-Bermudez, F. (2002). Attention in the somatosensory system. Nelson, R.J. (ed.), *The Somatosensory System*. CRC Press, Boca Raton, FL, pp. 197–218.

Hubel, D.H. & Wiesel, T.N. (1962). Receptive fields, binocular interaction and functional architecture in the cat's visual cortex. *J. Physiol.*, **160**, 106–154.

Huck, F.O. Falesa, C.L., Park, S.K., Speray, D.E., & Self, M.O. (1983). Application of information theory to the design of line-scan and sensor-array imaging system. *Optics and Laser Technology*, **15**(1), 21–34.

Hughes, A. (1977). The topography of vision in mammals of contrasting lifestyle: comparative optics and retinal organisation. Crescitelli, F. (ed.), *Handbook of Sensory Physiology*, vol. 7. Springer, Berlin, Germany, pp. 613–756.

Hughes, A. (1996). Seeing cones in living eyes. *Nature*, **380**, 393–394.

Hughes, R. (1997). *American Visions*. Knopf, New York.

Hunt, D.M., Dulai, K.S., Bowmaker, J.K., & Mollon, J.D. (1995). The chemistry of John Dalton's color blindness. *Science*, **267**, 984–988.

Hyvärinen, A. (1999). Fast and robust fixed-point algorithms for independent component analysis. *IEEE Trans. Neural Networks*.

IEC (2006). *XV Colour Multimedia Systems – Colour Measurement*. IEC 61966–2–4.

Iggo, A. (1982). Cutaneous sensory mechanisms. Barlow, H.B. & Mollon, J.D. (eds), *The Senses*. Cambridge University Press, Cambridge, UK, pp. 369–408.

Imai, T., Suzuki, M., & Sakano, H. (2006). Odorant receptor-derived cAMP signals direct axonal targeting. *Science*, **314**(5799), 657–661.

Indovina, I., Maffei, V., Bosco, G., Zago, M., Macaluso, E., & Lacquaniti, F. (2005). Representation of visual gravitational motion in the human vestibular cortex. *Science*, **308**(5720), 416–419.

Ito, S., Stuphorn, V., Brown, J.W., & Schall, J.D. (2003). Performance monitoring by the anterior cingulate cortex during saccade countermanding. *Science*, **302**(5642), 120–122.

Jackendoff, R. (1987). *Consciousness and the Computational Mind*. MIT Press, Cambridge, MA.

Jacobs, G.H., Neitz, J., & Deegan II, J.F. (1991). Retinal receptors in rodents maximally sensitive to ultraviolet light. *Nature*, **353**, 655–656.

Jacobs, G.H., Williams, G.A., Cahill, H., & Nathans, J. (2007). Emergence of novel colour vision in mice engineered to express a human cone photopigment. *Science*, **315**, 1723–1725.

Janata, P., Birk, J.L., Horn, J.D. Van, Leman, M., Tillman, B., & Bharucha, J.J. (2002). The cortical topography of tonal structures underlying western music. *Science*, **298**(5601), 2167–2173.

Jellinek, J.S. (1992). Perfume classification: a new approach. Toller, S. van & Dodd, G.H. (eds), *Fragrance: The Psychology and Biology of Perfume*. Elsevier, London, pp. 229–242.

Jeong, K.-H., Kim, J., & Lee, L.P. (2006). Biologically inspired artificial compound eyes. *Science*, **312**(5773), 557–561.

Johansson, R.S. & Valbo, A.B. (1979). Tactile sensibility in the human hand: relative and absolute densities of four types of mechanoreceptive units in glabrous skin. *J. Physiol. (Lond.)*, **286**, 283–300.

Johnson, J.I. (1986). Mammalian evolution as seen in visual and other neural systems. Pettigrew, J.D., Sanderson, K.J., & Levick, W.R. (eds), *Visual Neuroscience*. Cambridge University Press, Cambridge, UK, pp. 196–207.

Johnson, K.O. & Yoshida, T. (2002). Neural mechanisms of tactile form and texture perception. Nelson, R.J. (ed.), *The Somatosensory System*. CRC Press, Boca Raton, FL, pp. 73–102.

Jones, L.M., Depireux, D.A., Simons, D.J., & Keller, A. (2004). Robust temporal coding in the trigeminal system. *Science*, **304**(5679), 1986–1989.

Julesz, B. (1971). *Foundations of Cyclopean Perception*. University of Chicago Press.

Julesz, B. (1975). Experiments in the visual perception of texture. *Scientific American*, **232**, 34–43.

Juusola, M. & French, A.S. (1997). The efficiency of sensory information coding by mechanoreceptor neurons. *Neuron*, **18**(6), 959–968.

Kaas, J.H. (1995). Vision without awareness. *Nature*, **373**(6511), 247–249.

Kaas, J.H., Jain, N., & Qi, H.-X. (2002). The organization of the somatosensory system in primates. Nelson, R.J. (ed.), *The Somatosensory System*. CRC Press, Boca Raton, FL, pp. 1–26.

Kajiura, S.M. & Holland, K.N. (2002). Electoreception in juveline scalloped hammerhead and sandbar sharks. *J. Exp. Biol.*, **205**, 2609–3621.

Kalmijn, A.J. (1971). The electric sense of sharks and rays. *J. Exp. Biol.*, **55**, 371–383.

Kalmijn, A.J. (1982). Electric and magnetic field detection in elasmobranch fishes. *Science*, **218**, 916–918.

Kandel, E.R. & Wurtz, R.H. (2000). Constructing the visual image. Kandel, E.R., Schwartz, J.H., & Jessell, T.M. (eds), *Principles of Neural Science*, 4th edn. McGraw Hill, New York.

Kandel, E.R., Schwartz, J.H., & Jessell, T.M. (eds) (1991). *Principles of Neural Science*, 3rd edn. McGraw Hill, New York.

Kandel, E.R., Schwartz, J.H., & Jessell, T.M. (eds) (2000). *Principles of Neural Science*, 4th edn. McGraw Hill, New York.

Kaneko, T., Takahei, T., Inami, M., Kawakami, N., Yanagida, Y., Maeda, T., & Tachi, S. (2001). Detailed shape representation with parallax mapping. *Proc. ICAT*, ICAT, Tokyo, pp. 205–208.

Kanwisher, N. (2006). What's in a face? *Science*, **311**(5761), 617–618.

Katz, B. (1966). *Nerve, Muscle and Synapse*. McGraw Hill, New York.

Kelso, J.A.S. (1995). *Dynamic Patterns*. MIT Press, Cambridge, MA.

Keverne, E.B. (1982). Chemical senses: smell. Barlow, H.B. & Mollon, J.D. (eds), *The Senses*. Cambridge University Press, Cambridge, UK.

Keverne, E.B. (1999). The vomeronasal organ. *Science*, **286**, 715–720.

Kirkpatrick, S. (2003). Rough times ahead. *Science*, **299**(5607), 668–669.

Kirschvink, J.L., Walker, M.M., & Diebel, C.E. (1986). Evidence from strandings for geomagnetic sensitivity in cetaceans. *J. Exp. Biol.*, **120**, 1–24.

Kleinfeld, D., Ahissar, E., & Diamond, M.E. (2006). Active sensation: insights from the rodent vibrissa sensorimotor system. *Curr. Opin. Neurobiol.*, **16**(4), 435–444.

Klier, E.M., Wang, H., Constantin, A.G., & Crawford, J.D. (2002). Midbrain control of three-dimensional head orientation. *Science*, **295**(5558), 1314–1316.

Kohler, E., Keysers, C., Umiltá, M.A., Fogassi, L., Gallese, V., & Rizzolatti, G. (2002). Hearing sounds, understanding actions: action representation in mirror neurons. *Science*, **297**(5582), 846–848.

König, P. & Verschure, F.M.J. (2002). Neurons in action. *Science*, **296**(5574), 1817–1818.

Kram, Y.A., Mantey, S., & Corbo, J.C. (2010). Avian cone photoreceptors tile the retina as five independent, self-organizing mosaics. *PLOS One*, **5**(2), e8992.

Krupa, D.J., Wiest, M.C., Shuler, M.G., Laubach, M., & Nicolelis, M.A.L. (2004). Layer-specific somatosensory cortical activation during active tactile discrimination. *Science*, **304**(5679), 1989–1992.

Kurzweil, R. (1999). The coming merger of mind and machine. *Scientific American*, **10**(3), 56–61.

Lackner, J.R. & DiZio, P. (2002). Somatosensory and proprioceptive contributions to body orientation, sensory localisation and self calibration. Nelson, R.J. (ed.), *The Somatosensory System*. CRC Press, Boca Raton, FL, pp. 121–140.

Land, E.H., Hubel, D.H., Livingstone, M.S., Perry, S.H., & Burns, M.M. (1983). Colour-generating interactions across the corpus callosum. *Nature*, **303**, 616–618.

Land, M. & Tatler, B. (2009). *Looking and Acting: Vision and Eye Movements in Natural Behaviour*. Oxford University Press, Oxford, UK.

Land, M.F. (1990). Vision in other animals. *Images and Understanding*. Cambridge University Press, Cambridge, UK.

Land, M.F. & Nilsson, D.-E. (2002). *Animal Eyes*. Oxford University Press, Oxrord, UK.

Land, M.F. & Snyder, A.W. (1985). Cone mosaic observed directly through natural pupil of live vertebrate. *Vision Research*, **25**(10), 1519–1523.

Landauer, R. (1961). Irreversibility and heat generation in the computing process. *IBM J. Res. Develop.*, **5**, 183–191.

Landisman, C. & Connors, B.W. (2005). Long-term modulation of electrical synapses in the mammalian thalamus. *Science*, **310**, 1809–1813.

Laughlin, S.B. (1981). A simple coding procedure enhances a neuron's information capacity. *Z. Naturforsch.*, **36c**, 910–912.

Laughlin, S.B. & Sejnowski, T.J. (2003). Communication in neural networks. *Science*, **301**(5641), 1870–1874.

Laughlin, S.B., Ruyter van Steveninck, R.R. de, & Anderson, J.C. (1998). The metabolic cost of neural computation. *Nature Neuroscience*, **1**(1), 36–41.

Laurent, G. (1999). A systems perspective on early olfactory coding. *Science*, **286**, 723–728.

Lawton, G. (2003). Armed and dangerous. *New Scientist*, **180**(2420), 34–37.

Lederman, S.J. & Klatzky, R.L. (2002). Feeling surfaces and objects remotely. Nelson, R.J. (ed.), *The Somatosensory System*. CRC Press, Boca Raton, FL, pp. 103–120.

LeDoux, J. (1998). *The Emotional Brain*. Phoenix, London.

Lee, A.B. & Mumford, D. (1999). An occlusion model generating scale-invariant images. *IEEE Workshop on Statistical and Computational Theories of Vision*. Fort Collins, CO.

Lee, B.B., Dacey, D.M., Smith, V.C., & Pokorny, J. (1999). Horizontal cells reveal cone type-specific adaptation in primate retina. *Proc. Nat. Acad. Sci*, **96**, 14611–14616.

Lehky, S.R. & Sejnowski, T.J. (1986). Network model of shape from shading: neural function arises from receptive and projective fields. *Nature*, **333**, 452–454.

Leinders-Zufall, T., Brennan, P., Widlayer, P., Chandramani, S., Maul-Pavicic, S.A., Jäger, M., Li, X.-H. *et al.* (2004). MHC class I peptides as chemosensory signals in the vomeronasal organ. *Science*, **306**(5698), 1033–1037.

Lennie, P. (2000). Color vision. Kandel, E.R., Schwartz, J.H., & Jessell, T.M. (eds), *Principles of Neural Science*, 4th edn. McGraw Hill, New York, pp. 589–591.

Leventhal, A.G., Wang, Y., Pu, M., Zhou, Y., & Ma, Y. (2003). GABA and its agonists improved visual cortical function in senescent monkeys. *Science*, **300**(5620), 812–815.

Levick, W.R. (1986). Sampling of information space by retinal ganglion cells. Pettigrew, J.D., Sanderson, K.J., & Levick, W.R. (eds), *Visual Neuroscience*. Cambridge University Press, Cambridge, UK.

Levitt, J.B. (2001). Function following form. *Science*, **292**(5515), 232–233.

Lewcock, J.W. & Reed, R.R. (2003). ORs rule the roost in the olfactory system. *Science*, **302**(5653), 2078–2079.

Lewicki, M.S. (2002). Efficient coding of natural scences. *Nature Neuroscience*, **5**(4), 356–363.

Li, M. & Vitányi, P. (1993). *An Introduction to Kolmogorov Complexity and its Applications*. Springer-Verlag, New York.

Li, W., Pichl, V., & Gilbert, C.D. (2004). Perceptual learning and top-down influences in primary visual cortex. *Nature Neuroscience*, **7**(6), 651–657.

Liberles, S.D. & Buck, L.B. (2006). A second class of chemosensory receptors in the olfactory epithelium. *Nature*, **442**, 645–650.

Libet, B. (2004). *Mind Time*. Harvard University Press, Cambridge, MA.

Lindeman, R.W., Yanagida, Y., Noma, H., & Hosa, K. (2006). Wearable vibrotactile systems for virtual contact and information display. *Virtual Reality*, **9**, 203–213.

Livermore, A. & Laing, D.G. (1998). The influence of odor type on the discrimination and identification of odorants in multicomponent odor mixtures. *Physiol. Behav.*, **65**, 311–320.

Livingstone, M. (1988). Presentation at Seeing Colour and Contour. Manchester University Press, Manchester, UK.

Livingstone, M. & Hubel, D. (1988). Segregation of form, color, movement and depth: physiology and perception. *Science*, **240**, 740–749.

Lohmann, K.J., Cain, S.D., Dodge, S.A., & Lohmann, C.M.F. (2001). Regional magnetic fields as navigational markers for sea turtles. *Science*, **294**(5541), 364–366.

Loop, M.S. & Bruce, L.A. (1978). Cat colour vision: the effect of stimulus size. *Science*, **199**, 221–222.

Lowe, G. & Gold, G.H. (1995). Olfactory transduction is intrinsically noisy. *Proc. Nat. Acad. Sci*, **92**, 7864–7868.

Lucas, R.J., Hattar, S., Takao, M., Berson, D.M., Foster, R.G., & Yau, K.-W. (2003). Diminished pupillary light reflex at high irradiances in melanopsin-knockout mice. *Science*, **299**(5604), 245–247.

Luo, M., Fee, M.S., & Katz, L.C. (2003). Encoding pheromonal signals in the accessory bulb of behaving mice. *Science*, **299**(5610), 1196–1201.

Lythgoe, J.N. (1979). *The Ecology of Vision*. Clarendon Press, Oxford, UK.

Maas, W. & Bishop, C.M. (eds) (1998). *Pulsed Neural Networks*. MIT Press, Cambridge, MA.

Macefield, V.G. (1998). The signalling of touch, finger movements and manipulation forces by mechanoreceptors in human skin. *Neural Aspects of Tactile Sensation*. Advances in Psychology, vol. 127. North Holland, Amsterdam, pp. 89–130.

Macefield, V.G. (2005). Physiological characteristics of low-threshold mechanoreceptors in joints, muscle and skin in human subjects. *Clin. Expt. Pharmaocol. Physiol.*, **32**, 135–144.

Mackenzie, D. (2003). Tailor-made vision descends to the eye of the beholder. *Science*, **299**(5613), 1654–1655.

Maddess, T. & Ibbotson, M.R. (1992). Human ocular following responses are plastic: evidence for control by temporal frequency-dependent cortical adaptation. *Exp. Brain Res.*, **91**, 525–538.

Maheshwari, V. & Saraf, R.F. (2006). High-resolution thin-film device to sense texture by touch. *Science*, **312**, 1501–1504.

Malakoff, D. (1999). Following the scent of avian olfaction. *Science*, **286**, 704–705.

Maloney, L.T. & Wandell, B.A. (1986). Color constancy: a method for recovering surface spectral reflectance. *J. Opt. Soc. Am. A*, **3**, 29–33.

Manoussaki, D., Dimitriadis, E.K., & Chadwick, R.S. (2006). Cochlea's graded curvature effect on low frequency waves. *Phys. Rev. Letters*, **96**, 088701–088704.

Marcelja, S. (1980). Mathematical description of the responses of simple cortical cells. *J. Opt. Soc. Am.*, **70**, 1297–1300.

Maresh, A, Rodriguez, G.D., Whitman, M.C., & Greer, C.A. (2009). Principles of glomerular organization in the human olfactory bulb implications for odor processing. *PLOS One*, **3**(7), e2640.

Marks, P. (2000). Blast from the past. *New Scientist*, **2226**, 11.

Marr, D. (1982). *Vision: A Computational Investigation into the Human Representation and Processing of Visual Information*. W.H. Freeman & Co., New York.

Martuzzi, R., Murray, M.H., Michel, C.H., Thiran, J.-P, Maeder, P.P., Clarke, S., & Meuli, R.A. (2007). Multisensory interactions within human primary cortices revealed by BOLD dynamics. *Cerebral Cortex*, **17**, 1672–1679.

Masland, R.H. (2001). The fundamental plan of the retina. *Nature Neuroscience*, **6**(9), 877–886.

Mason, B. (2002). Giraffe's elevated view of friendship. *New Scientist*, **175**(2353), 21.

Mast, T.G. & Samuelsen, C.L. (2009). Human pheromone detection by the vomeronasal organ: unnecessary for mate selection? *Chemical Senses*, **34**, 529–531.

Maurer, D., Lewis, T.L., Brent, H.P., & Levin, A.V. (1999). Rapid improvement in the affinity of infants after visual input. *Science*, **286**(5437).

Mazel, C.H., Cronin, T.W., Caldwell, R.L., & Marshall, N.J. (2004). Flourescent enhancement of signaling in a mantis shrimp. *Science*, **303**(5654), 51.

McCann, J.J. (2005). Do humans discount the illuminant? Rogowitz, B.E., Pappas, T.N., & Daly, S.J. (eds), *Human Vision and Electronic Imaging*, Vol. Proc. SPIE 5666, San Jose, CA, pp. 9–16.

McGee, A.W., Yang, Y., Fischer, Q.S., Daw, N.W., & Strittmayer, S.M. (2005). Experience-driven plasticity of visual cortex limited by myelin and nogo receptor. *Science*, **309**, 2222–2226.

McNulty, P.A. & Macefield, V.G. (2001). Modulation of ongoing EMG by different classes of low-threshold mechanoreceptors in the human hand. *J. Physiol.*, **537**, 1021–1032.

McNulty, P.A., Türker, K.S., & Macefield, V.G. (1999). Evidence for strong synaptic coupling between single tactile afferents and motoneurones supplying the human hand. *J. Physiol.*, **518**, 883–893.

Menaker, M. (2003). Circadian photoreception. *Science*, **299**(5604), 213–214.

Mennill, D.J., Ratcliffe, L.M., & Boag, P.T. (2002). Female eavesdropping on male song contests in songbirds. *Science*, **296**(5569), 873.

Merali, Z. (2006). Captured, the sweet smell of happiness. *New Scientist*, **192**(2574), 14.

Mery, F. & Kawecki, T.J. (2005). A cost of long-term memory in *Drosophila*. *Science*, **308**(5725), 1148.

Merzenich, M.M., Kaas, J.H., Wall, J., Nelson, R.J., Sur, M., & Felleman, D. (1983). Topographic reorganization of somatosensory cortical areas 3b and 1 in adult monkeys following restricted deafferentation. *Neuroscience*, **8**, 33–55.

Millar, H. (2006). High dynamic range technology. *Siggraph*.

Miller, G. (2003). Old neurons revisit their youth. *Science*, **300**(5620), 721–722.

Millodot, M. (1982). Accommodation and refraction of the eye. Barlow, H.B. & Mollon, J.D. (eds), *The Senses*. Cambridge University Press, Cambridge, UK, pp. 62–81.

Millodot, M. & Newton, I. (1981). Vep measurement of the amplitude of accommodation. *Brit. J. Ophthamology*, **65**, 294–298.

Mombaerts, P. (1999). Seven-transmembrane proteins as odorant and chemosensory receptors. *Science*, **286**, 707–711.

Mooney, D.M., Zhang, L., Basile, C., Senatorov, V.V., Ngsee, J., Omar, A., & Hu, B. (2004). Distinct forms of cholinergic modulation in parallel thalamic pathways. *Proc. Nat. Acad. Sci*, **101**, 320–324.

Moore, B.C.J. (1995). Frequency analysis and masking. Moore, B.C.J. (ed.), *Hearing*. Academic Press, San Diego, CA, pp. 161–205.

Moore, B.C.J. (1997). *An Introduction to the Psychology of Hearing*, 4th edn. Academic Press, San Diego, CA.

Mouritsen, H. & Ritz, T. (2005). Magnetoreception and its use in bird navigation. *Current Opin. Neurobiol.*, **15**, 406–414.

Müller, H. (1872). *Gesammelte und hinterlassene schriften zur anatomie und physiologie des Auges*. Engelmann, Leipzig, Germany.

Musacchia, G., Sams, M., Skoe, E., & Kraus, N. (2007). Musicians have enhanced subcortical auditory and audiovisual processing of speech and music. *PNAS*, **104**, 15894–15898.

Nagle, M.G. & Osorio, D. (1993). The tuning of human photopigments may minimize red-green chromatic signals in natural conditions. *Proc. R. Soc. Lond. B*, **252**, 209–213.

Nascimento, S.M.C., Foster, D.H., & Amano, K. (2005). Psychophysical estimates of the number of spectral-reflectance basis functions needed to reproduce natural scenes. *J. Opt. Soc. Am. A*, **22**(6), 1017–1022.

Neimark, M.A., Andermann, M.L., Hopfield, J.J., & Moore, C.I. (2003). Vibrissa resonance as a transduction mechanism for tactile encoding. *J. Neurosci.*, **23**, 6499–6509.

Neitz, J. & Jacobs, G.H. (1986). Polymorphism of the long wavelength cone in normal human vision. *Nature*, **323**, 6223–6625.

Neitz, J., Geist, T., & Jacobs, G.H. (1989). Color vision in the dog. *Vis. Neurosci.*, **3**, 119–125.

Nelson, R.J., Sur, M., Felleman, D.J., & Kaas, J.H. (1980). Representations of the body surface in postcentral parietal cortex of *Macaca fascicularis*. *J. Comp. Neurol.*, **192**, 611–643.

Nemec, P., Altmann, J., Marhold, S., Burda, H., & Oelschlaeger, H.H.A. (2001). Neuroanatomy of magnetoreception: the superior colliculus involved in magnetic orientation in a mammal. *Science*, **294**(5541), 366–368.

Neumayer, C. (1998). *Color Vision in Lower Vertebrates*. Walter de Gruyter, Berlin.

News track (2005). Sniffing out cancer. *Comm. ACM*, **48**(7), 9.

Nicolelis, M.A.L. & Chapin, J.K. (2002). Controlling robots with the mind. *Scientific American*, **287**(4), 24–31.

Nicolelis, M.A.L., Fanselow, E., & Henriquez, C. (2002). A critique of the pure feed-forward model of touch. Nelson, R.J. (ed.), *The Somatosensory System*. CRC Press, Boca Raton, FL, pp. 299–334.

Nyquist, H. (1928). Certain topics in telegraph transmission theory. *Trans. AIEE*, **47**, 617–644.

Oka, Y., Omura, M., Kataoka, H., & Touhara, K. (2004). Olfactory receptor antagonism between odorants. *EMBO J.*, **23**, 120–126.

Olausson, H., Lamarre, Y., Backlund, H., Morin, C., Wallin, B.G., Starck, G., Ekholm, S. *et al.* (2002). Unmyelinated tactile afferents signal touch and project to insular cortex. *Nature Neuroscience*, online, 29 July.

Osorio, D. & Bossomaier, T.R.J. (1992). Human cone-pigment spectral sensitivities and the reflectances of natural surfaces. *Biol. Cybern.*, **67**(3), 217–222.

Osorio, D. & Vorobyev, M. (1996). Colour vision as an adaptation to frugivory in primates. *Proc. R. Soc. Lond. B*, **263**, 593–599.

Osorio, D., Ruderman, D.L., & Cronin, T.W. (1998). Estimation of errors in luminance signals encoded by primate retina resulting from sampling of natural images with red and green cones. *J. Opt. Soc. Am.*, **15**(1), 16–22.

Ozaki, M., Wada-Katsumata, A., Fujikawa, K., Iwasaki, M., Yokohari, F., Satoji, Y., Nisimura, T. *et al.* (2005). Ant nestmate and non-nestmate discrimination by a chemosensory sensilium. *Science*, **309**(5732), 311–314.

Pagán, S. (2003). Cardboard can be simply delicious. *New Scientist*, **180**(2420), 11.

Panda, S., Sato, T.K., Castrucci, A.M., Rollag, M.D., DeGrip, W.J., Hogenesch, J.B., Provencio, I. *et al.* (2003). Melanopsin (opn4) requirement for normal light-induced circadian phase shifting. *Science*, **298**(5601), 2213–2216.

Panda, S., Nayak, S.K., Campo, B., Walker, J.R., Hogenesch, J.B., & Jegla, T. (2005). Illumination of the melanopsin signalling pathway. *Science*, **307**(5709), 600–604.

Panzeri, S., Petersen, R.S., Schultz, S.R., Lebedev, M., & Diamond, M.E. (2001). The role of spike timing in the coding of the stimulus location in rat somatosensory cortex. *Neuron*, **29**, 760–777.

Papoulis, A. (1984). *Probability, Random Variables and Stochastic Processes*. McGraw Hill, New York.

Paré, M., Elde, R., Mazurkiewicz, J.E., Smith, A.M., & Rice, F.L. (2001). The Meissner corpuscle revised; a multi-afferented mechanoreceptor with nociceptor immmuno-chemical properties. *J. Neurosci.*, **21**, 7236–7246.

Parker, S. (2006). *Sponges, Jellyfish and Other Simple Animals*. Compass Point Books, MN.

Pascual-Leone, A. & Walsh, V. (2001). Fast backpropagation from the motion to the primary visual area necessary for visual awareness. *Science*, **292**(5516), 510–512.

Pattanaik, S.N., Tumblin, J.E., Yee, H., & Greenberg, D.P. (2000). Time-dependent visual adaptation for realistic image display, 7(7), ACM Press, New York.

Paulson, L.D. (2002). New haptics approach lets parents-to-be 'touch' their unborn children. *IEEE Computer*, **35**(10), 27.

Payne, K. (1989). Elephant talk. *National Geographic*, August, 264–277.

Pelli, D.G. (1999). Close encounters – an artist shows that size effects shape. *Science*, **286** (5429), 844–846.

Peña, J.L. & Konishi, M. (2001). Auditory spatial receptive fields created by multiplication. *Science*, **292**(5515), 249–252.

Perez-Orive, J., Mazor, O., Turner, G.C., Cassenaer, S., Wilson, R.I., & Laurent, G. (2002). Oscillations and sparseing of the odor representations in the mushroom body. *Science*, **297**(5580), 359–365.

Pescovitz, D. (1999). Getting real in cyberspace. *Scientific American*, **10**, 48–51.

Peterlin, Z., Li, Y., Sun, G., Shah, R., Firestein, S., & Ryan, K. (2008). The importance of odorant conformation to the binding and activation of a representative olfactory receptor. *Chemistry and Biology*, **15**, 1317–1327.

Petersen, C.C.H., Brecht, M., Hahn, T.T.G., & Sakmann, B. (2004). Synaptic changes in layer 2/3 underlying map plasticity of developing barrel cortex. *Science*, **304**, 739–743.

Pettigrew, J.D. (1986). The evolution of binocular vision. Pettigrew, J.D., Sanderson, K.J., & Levick, W.R. (eds), *Visual Neuroscience*. Cambridge University Press, Cambridge, UK.

Pettigrew, J.D., Collin, S.P., & Ott, M. (1999). Convergence of specialised behaviour and eye movements and visual optics in the sandlance (*teleostei*) and the chameleon (*reptilia*). *Current Biol.*, **9**(8), 421–424.

Pherobase (n.d.). Pherobase. http://www.pherobase.com.

Phillips, H. (2002). The compass in your eye. *New Scientist*, **174**(2342), 17.

Pidwirny, M. (2010). Solar radiation. Cutlern, J. (ed.), *Encylopedia of the Earth*. Environmental Information Coalition, National Council for Science and the Environment.

Pilcher, H.R.J. (2004). Pigeons take the highway. *Nature*, online, news040209–1.

Popper, K. (1972). *Objective Knowledge: An Evolutionary Approach*. Oxford University Press, Oxford, UK.

Popper, K. & Eccles, J.C. (1977). *The Self and its Brain*. Springer-Verlag, Berlin, Germany.

Poremba, A., Saunders, R.C., Crane, A.M., Cook, M., Sokoloff, L., & Mishkin, M. (2003). Functional mapping of the primate auditory system. *Science*, **299**(5605), 568–572.

Poulet, J.F.A. & Hedwig, B. (2006). The cellular basis of corollary discharge. *Science*, **311**(5760), 518–522.

Prochazka, A. & Yakovenko, S. (2002). Locomotor control from spring-like reactions of muscles to neural prediction. Nelson, R.J. (ed.), *The Somatosensory System*. CRC Press, Boca Raton, FL, pp. 141–181.

Proske, U., Gregory, J.E., & Iggo, A. (1998). Sensory receptors in monotremes. *Philos. Trans. R. Soc. Lond. B*, **353**, 1187–1198.

Quiroga, R.Q., Reddy, L., Kreiman, G., Koch, C., & Fried, L. (2005). Invariant visual representation by single neurons in the human brain. *Nature*, **435**, 1102–1107.

Rajan, R., Clement, J.P., & Bhalla, U.S. (2006). Rats smell in stereo. *Science*, **311**(5761), 666–669.

Rakic, P. (2001). Neurocreationism – making new cortical maps. *Science*, **294**(5544), 1011–1012.

Ramachandran, V.S. & Blakeslee, S. (1998). *Phantoms of the Mind*. Fourth Estate, London.

Ramachandran, V.S. & Hubbard, E.M. (2003). Hearing colours, tasting shapes. *Scientific American*, **288**(5), 42–449.

Ramachandran, V.S. & Rogers-Ramachandran, D. (2004). The phantom hand. *Scientific American*, **5**, 99–100.

Ranganathan, P. (2010). Recipe for efficiency: principles of power-aware computing. *Communications of the ACM*, **53**(4), 60–67.

Rasmussen, L.E.L. & Munger, B.L. (1996). The sensorineural specialisations of the trunk tip (finger) of the asian elephant, *Elephas maximus*. *Anatomical Record*, **246**, 127–134.

Rauschecker, J.P. & Shannon, R.V. (2002). Sending sound to the brain. *Science*, **295**(5557), 1025–1029.

RCA (1968). *Electro-optics Handbook*. Radio Corporation of America.

Reich, D.S., Mechler, F., & Victor, J.D. (2001). Independent and redundant information in nearby cortical neurons. *Science*, **294**(5551), 2566–2568.

Reinhard, E., Stark, M., Shirley, P., & Ferwerda, J. (2002). Photographic tone reproduction for digital images. *Proc. Siggraph.*, **21**(3), 267–276.

Reisert, J. & Restrep, D. (2009). Molecular tuning of odorant receptors and its implications for odor signal processing. *Chem. Senses*, **34**, 534–545.

Rema, V., Armstsrong-James, M., & Ebner, F.F. (2003). Experience-dependent plasticity is impaired in adult rat barrel cortex after whiskers are unused in early postnatal life. *J. Neuroscience*, **23**, 358–366.

Remy, J.-J. & Hobert, O. (2005). An interneuronal chemoreceptor required for olfactory imprinting in *C. elegans*. *Science*, **309**(5735), 787–790.

Ricci, A.J., Kennedy, H.J., Crawford, A.C., & Fettiplace, R. (2005). The transduction channel filter in auditory hair cells. *J. Neurosci.*, **25**, 7831–7839.

Richmond, B. (2001). Information coding. *Science*, **294**(5551), 2493–2494.

Rieke, F., Warland, D., de Ruyter van Steveninck, R., & Bialek, W. (1996). *Spikes. Exploring the Neural Code*. Bradford Book, MIT Press, Cambridge, MA.

Rodieck, R.W. (1973). *The Vertebrate Retina*. W.H. Freeman and Co., San Francisco, CA.

Rolls, E.T., O'Doherty, J., Kringelbach, M.L., Francis, S., Bowtell, R., & McGone, F. (2003). Representations of pleasant and painrful touch in the human orbitofrontal and cingulate cortices. *Cerebral Cortex*, **13**(3), 308–317.

Roorda, A. & Williams, D.R. (1999). The arrangement of the three cone classes in the living human eye. *Nature*, **397**, 520–522.

Rowe, M.J., Mahns, D.A., Sahai, V., & Ivanusic, J.J. (2005). Mechanoreceptory perception: are there contributions from bone-associated receptors? *Clin. Exp. Pharmacol. Physiol.*, **32**, 100–108.

Ruben, J., Schwiemann, J., Deuchert, M., Meyer, R., Krause, T., Curio, G., Villringer *et al.* (2001). Somatotopic organisation of human secondary somatosensory cortex. *Cerebral Cortex*, **11**, 463–473.

Ruderman, D.L. & Bialek, W. (1994). Ruderman, D.L. and Bialek, W. *Phys. Rev. Letters*, **73**(6), 814–817.

Rundus, A.S., Owings, D.H., Joshi, S.S., Chinn, E., & Giannini, N. (2007). Ground squirrels use an infrared signal to deter rattlesnake predation. *Proc. Nat. Acad. Sci.*, **104**, 1433–14376.

Rushton, W.A. (1972). Pigments and signals in colour vision. *J. Physiol.*, **220**, 1P–31P.

Ryan, M.J. (1999). Electrifying diversity. *Nature*, **400**, 211–212.

Sachs, H.G., Schanze, T., Wilms, M., Rentzos, A., Brunner, U., Gekeler, F., & Hesse, L. (2005). Subretinal implantation and testing of polyimide film electrodes in cats. *Graefe's Arch. Clin. Exp. Opthalmol.*, **243**, 464–468.

Sage, C., Huang, M., Karimi, K., Gutierrez, G., Volrath, M.A., Zhang, D.-S., Garcia-Añoveros, J. *et al.* (2005). Proliferation of functional hair cells *in vivo* in the absence of the retinoblastoma protein. *Science*, **305**, 1114–1118.

Sampath, A.P. & Baylor, D.A. (2002). Molecular mechanism of spontaneous pigment activaition in retinal cones. *Biophysical J.*, **83**, 184–193.

Santis, L. de, Clarke, S., & Murray, M.M. (2007). Automatic and intrinsic auditory 'what' and 'where' processing in humans revealed by electrical neuroimaging. *Cerebral Cortex*, **17**(1), 9–17.

Saper, C.B. (2000). Brain stem, reflexive behaviour and the cranial nerves. Kandel, E.R., Schwartz, J.H., & Jessell, T.M. (eds), *Principles of Neural Science*, 4th edn. McGraw Hill, New York.

Savic-Berglund, I. (2010). *Sex Differences in the Human Brain, their Underpinnings and Implications.* Elsevier, New York.

Scentbase (n.d.). Scentbase. http://wwww2.dpss,gu.se/SCENTbase.html.

Scheibert, J., Leurent, S., Prevost, A., and Debrégeas, G. (2009). The role of fingerprints in the coding of tactile information probed with a biometetric sensor. *Science Online*, **323**, 1166467.

Schlaug, G., Jäncke, L., Huang Y., & Steinmetz, H. (1999). *In vivo* evidence of structural brain asymmetry in musicians. *Science*, **267**, 699–701.

Schmiedeskamp, M. (2001). Plenty to sniff at. *Scientific American*, **284**, 23–24.

Schoffelen, J.-M., Oostenveld, R., & Fries, P. (2005). Neuronal coherence as a mechanism of effective corticospinal interaction. *Science*, **308**(5718), 111–113.

Schultze, M. (1866). Zur anatomie und physiologie der retina. *Arch. Mikrosk. Anat.*, **2**, 175–186.

Schwab, M.E. (2002). Repairing the injured spinal cord. *Science*, **295**(5557), 1029–1031.

Schwartz, O. & Simoncelli, E.P. (2001). Natural signal statistics and sensory gain control. *Nature Neuroscience*, **4**(8), 819–825.

Scott, P. (2001). Eye spy. *Scientific American*, **285**(3), 18.

Seifritz, E., Esposito, F., Hennel, F., Mustovic, H., Neuhoff, J.G., Bilecen, D., Tedeschi, G. *et al.* (2002). Spatiotemporal pattern of neural processing in the human auditory cortex. *Science*, **297**(5587), 1706–1708.

Sekiguchi, N., Williams, D.R., & Brainard, D.H. (1993). Aberration-free measurement of the visibility of isoluminant gratings. *J. Opt. Soc. Am.*, **10**(10), 2106–2117.

Serizawa, S., Miyamichi, K., Nakatami, H., Suzuki, M., Saito, M., Yoshihara, Y., & Sakano, H. (2003). Negative feedback regulation ensures the one receptor-one olfactory neuron rule in mouse. *Science*, **302**(5653), 2088–2094.

Service, R.F. (2002). Can sensors make a home in the body. *Science*, **297**(5583), 962–963.

Seydel, C. (2002). How neurons know that it's c-c-c-c-cold outside. *Science*, **295**(5559), 1451–1452.

Shannon, C.E. (1948). A mathematical theory of communication. *Bell System Tech. J.*, **27**, 379–423, 623–656.

Sharma, J., Dragoi, V., Tenenbaum, J.B., Miller, E.K., & Sur, M. (2003). V1 neurons signal aquisition of and internal representation of stimulus location. *Science*, **300**, 1758–1763.

Simões, C., Jensen, O., Parkkonen, L., & Harri, R. (2003). Phase locking between human primary and secondary somatosensory cortices. *Proc. Nat. Acad. Sci.*, **100**(5), 2691–2694.

Simoncelli, E.P. & Olshausen, B.A. (2001). Natural images statistics and neural representation. *Annual Rev. Neurosci.*, **24**, 1193–1216.

Simpson, S.D., Meekan, M., Montgomery, J., McCauley, R., & Jeffs, A. (2005). Homeward sound. *Science*, **308**(5719), 221.

Sincich, L.C. & Horton, J.C. (2002). Divided by cytochrome oxidase: a map of the projections from V1 to V2 in macaques. *Science*, **295**(5560), 1734–1737.

Sincich, L.C. & Horton, J.C. (2005). Input to V2 thin stripes arises from V1 cytochrome oxidase patches. *J. Neurosci.*, **25**(44), 10087–10093.

Singh, S. (2003). A sense of wonder: interview with Upinder Bhala. *New Scientist*, **177**, 2384, 1 March.

Sininger, Y.S. & Cone-Wesson, B. (2004). Asymmetric cochlear processing mimics hemispheric specialization. *Science*, **305**(5690), 1581.

Slepian, D. & Pollak, H.O. (1961). Prolate spheroidal wave functions, Fourier Analysis and uncertainty, I. *Bell System Tech. J.*, **40**, 43–64.

Small, M. (1999). Nosing out a mate. *Scientific American*, **10**, 52–55.

Smith, D.V. & Margolskee, R.F. (2001). Making sense of smell. *Scientific American*, **284**, 26–33.

Snyder, A.W. (1973). The Stiles–Crawford effect explanation and consequences. *Vision Res.*, **13**, 1115–1137.

Snyder, A.W. & Barlow, H.B. (1988). Human vision: revealing the artist's touch. *Nature*, **331**(117–118).

Snyder, A.W. & Miller, W.H. (1978). Telephoto lens system of falconiform eyes. *Nature*, **275**, 127–129.

Snyder, A.W. & Mitchell, D.J. (1999). Is integer arithmetic fundamental to mental processing? The mind's secret arithmetic. *Proc. R. Soc. Lond. B*, **266**, 587–592.

Snyder, A.W. & Srinivasan, M.V. (1979). Human psychophysics: functional interpretation for contrast sensitivity versus spatial frequency curve. *Biol. Cybernetics*, **32**, 9–17.

Snyder, A.W., Laughlin, S.B., & Stavenga, D.G. (1977a). Information capacity of eyes. *Vision Res.*, **17**, 1163–1175.

Snyder, A.W., Stavenga, D.G., & Laughlin, S.B. (1977b). Spatial information capacity of compound eyes. *J. Comp. Physiol.*, **116**, 183–207.

Snyder, A.W., Bossomaier, T.R.J., & Hughes, A. (1986). Optical image quality and the retinal cone mosaic. *Science*, **231**, 499–501.

Snyder, A.W., Bossomaier, T.R.J., & Hughes, A. (1988). The theory of comparative eye design. Blakemore, C. (ed.), *Vision: Coding and Efficiency*. Oxford University Press, Oxford, UK, pp. 45–52.

Snyder, A.W., Bossomaier, T., & Mitchell, D.J. (2004). Concept formation: 'object' attributes dynamically inhibited from conscious awareness. *J. Integrative Neuroscience*, **3**(1), 31–46.

Sobel, N., Prabhakran, V., Hartley, C.A., Desmond, J.E., Glover, G.H., Sullivan, E.V., & Gabrielli, J.D.E. (1999). Blind smell: brain activation induced by an undetected airborne chemical. *Brain*, **122**, 209–217.

Sokolovski, A. (1974). Minimum audible field for the cat (monoaural) compared to the minimum audible field for man (binaural). *Audiology*, **13**, 423–436.

Solomon, J.H. & Hartmann, M.J. (2006). Robotic whiskers used to sense features. *Nature*, **443**, 525.

Sommer, M.A. & Wurtz, R.H. (2002). A pathway in primate brain for internal monitoring of movements. *Science*, **296**(5572), 1480–1482.

Spehr, M., Gisselmann, G., Poplawski, A., Riffell, J.A., Wetzel, C.H., Zimmer, R.K., & Hatt, H. (2003). Identification of a testicular odorant receptor mediating human sperm chemotaxis. *Science*, **299**(5615), 2054–2058.

Spherocam (n.d.). Spherocam: high dynamic range camera. http://www.spheron-usa.com/.

Sporns, O., Chialvo, D.R., Kaiser, M., & Hilgetag, C.C. (2004). Organisation, development and function of complex brain networks. *Trends in Cog. Sci.*, **8**, 418–425.

Srinivasan, M., Laughlin, S.B., & Dubs, A. (1982). Predictive coding: a fresh view of inhibition in the retina. *Proc. Roy. Soc. Lond. B*, **216**, 427–459.

Srinivasan, M.V., Chahl, J.C., Weber, K., Venkatesh, S., Nagle, M.G., Nagle, S.W., & Zhang, S. (1999). Robotic navigation inspired by principles of insect vision. *Robotics and Autonomous Systems*, **26**, 203–216.

Stakhovskaya, O., Sridhar, D., Bonham, B.H., & Leake, P.A. (2007). Frequency map for the human cochlear spiral ganglion: implications for cochlear implants. *J. Assoc. Res. Otolaryngol.*, **8**(2), 220–233.

Stern, R.M. & Trahiotis, C. (1995). Models of binaural interaction. Moore, B.C.J. (ed.), *Hearing*. Academic Press, San Diego, CA, pp. 297–345.

Steveninck, R.R. de van, Lewen, G.D., Strong, S.P., Koberle, R., & Bialek, W. (1997). Reproducibility and variability in neural spike trains. *Science*, **275**, 1805–1807.

Stevens, C.F. (2001). An evolutionary scaling law for the primate visual system and its basis in cortical function. *Nature*, **411**, 193–195.

Stevens, S.S. (1975). *Pschyophysics: Introduction to its Perceptual, Neural and Social Prospects*. Wiley, New York.

Stiles, W.S. & Crawford, B.H. (1933). The luminous efficiency of rays entering the eye pupil at different points. *Proc. R. Soc. Lond. B*, **112**, 428–450.

Stoddard, P.K. (1999). Predation enhances complexity in the evolution of electric fish signals. *Nature*, **400**, 254–256.

Stokstad, E. (2004). Songbirds check compass against sunset to stay on course. *Science*, **304**(5669), 373.

Strang, G. & Nguyen, T. (1996). *Wavelets and Filter Banks*. Wellesley-Cambridge Press, MA.

Strauss, E. (1999). Feeling the future. *Scientific American*, **10**(3), 44–47.

Strogatz, S. (2003). *Sync: The Emerging Science of Spontaneous Order*. Allen Lane, London.

Stryker, M.P. (2001). Drums keep pounding a rhythm in the brain. *Science*, **291**(5508), 1506–1507.

Stuart, G.W., Bossomaier, T.R.J., & Johnson, S. (1993). Preattentive processing of object size. *Perception*, **22**, 1175–1193.

Suga, N. (1990). Biosonar and neural computation in bats. *Scientific American*, **262**(6), 34–41.

Sugita, M. & Shiba, Y. (2005). Genetic tracing shows segregation of taste neuronal circuitries for bitter and sweet. *Science*, **309**(5735), 781–785.

Summerfield, C., Egner, T., Greene, M., Koechlin, E., Mangels, J., & Hirsch, J. (2006). Predictive codes for forthcoming perception in the frontal cortex. *Science*, **314**, 1311–1314.

Szmajda, B.A., Grünert, U., & Martin, P.R. (2008). Retinal ganglion cell inputs to the koniocellular pathway. *J. Compar. Neurol.*, **510**, 251–268.

Talavage, T.M., Martin I. Sereno, M.I., Melcher, J.R., Ledden, P.J., Rosen, B.R., & Dale, A.M. (2004). Tonotopic organization in human auditory cortex revealed by progressions of frequency sensitivity. *J. Neurophysiol.*, **91**, 1292–1296.

Taranda, J., Maison, S.F., Ballestero, J.A., Katz, E., Savino, J., Vetter, D.E., Boulter, J. *et al.* (2009). A point mutation in the hair cell nicotinic cholinergic receptor prolongs cochlear inhibition and enhances noise protection. *PLOS*, **7**(1), e1000018, doi: 10.1371.

Tatarchuk, N. (2005). Practical dynamic parallax occlusion mapping. *Siggraph*.

Terzopoulos, D., Tu, X., & Grzeszczuk, R. (1994). Artificial fishes with autonomous locomotion, perception, behavior, and learning in a simulated physical world. *Artificial Life*, **1**(4), 327–350.

Thiele, A., Henning, P., Kubischik, M., & Hoffman, K.-P. (2002). Neural mechanisms of saccadic suppression. *Science*, **295**, 2460–2462.

Tian, B., Reser, D., Durham, A., Kustov, A., & Rauschecker, J.P. (2001). Functional specialisation in rhesus monkey auditory cortex. *Science*, **292**(5515), 290–293.

Todd, N.P.M., Cody, F.W., & Banks, J. (2000). Is hearing all cochlear (revisited)? Frequency tuning and intensity thresholds in acoustically stimulated, myogenic vestibular-evoked potentials (MVEP) (A). *J. Acoust. Soc. Am.*, **107**, 2900.

Tomasello, M., Hare, B., Lehmann, H., & Call, J. (2007). Reliance on head versus eyes in the gaze following of great apes and human infants: the cooperative eyes hypothesis. *J. Human Evolution*, **52**, 314–320.

Tong, F. & Engel, S.A. (2001). Interocular rivalry revealed in the human cortical blind-spot representation. *Nature*, **411**, 195–199.

Tramo, M.J. (2001). Music of the hemispheres. *Science*, **291**(5501), 54–56.

Treede, R.-D., Apkarian, A.V., Bromm, B., Greenspan, J.D., & Lenz, F.A. (2000). Cortical representation of pain: the functional characterization of nocioreceptive areas near the lateral sulcus. *Pain*, **87**, 113–119.

Treisman, A. (1986). Features and objects in visual processing. *Scientific American*, **255**(5), 106–115.

Treloar, H.B., Feinstein, P., Mombaerts, P., & Greer, C.A. (2002). Specificity of glomerular targeting by olfactory sensory axons. *J. Neuroscience*, **22**(7), 2469–2477.

Troscianko, T., Benton, C.P., Lovell, G., Tolhurst, D.J., & Pizlo, Z. (2009). Camouflage and visual perception. *Phil. Trans. Roy. Soc. B*, **364**, 449–461.

Tsao, D.Y., Freiwald, W.A., Tootell, R.B.H., & Livingstone, M.S. (2006). A cortical region consisting entirely of face-selective cells. *Science*, **311**(5761), 670–674.

Tsuboi, A., Yoshihara, S., Yamazaki, N., Kasai, H., Asai-Tsuboi, H., Komatsu, M., Serizawa, S. *et al.* (1999). Olfactory neurons expressing closely linked and homologous ororant receptor genes tend to project their axons to neighbouring glomeruli on the olfactory bulb. *J. Neurosci.*, **19**, 8409–8418.

Van Boven, R.W. & Johnson, K.O. (1994). The limit of tactile spatial resolution in humans: grating orientation discrimination in the lip, tongue and finger. *Neurology*, **44**(12), 2361–2366.

Vanduffel, W., Fize, D., Peuskens, H., Denys, K., Sunaert, S., Todd, J.T., & Orban, G. (2002). Extracting 3D from motion: differences in human and monkey intraparietal cortex. *Science*, **298**(5592), 413–415.

Verfuss, U.K., Miller, L.A., & Schnitzler, H.-U. (2005). Spatial orientation in echolocating harbour porpoises (*Phocoena phocoena*). *J. Exp. Biol.*, **208**, 3385–3394.

Verrillo, R.T. & Bolanowski, J. (2003). Effects of temperature on the subjective magnitude of vibration. *Somatosensory and Motor Research*, **20**(2), 133–137.

Veselago, V.G. (1968). The electrodynamics of substances with simultaneously negative values of ϵ and μ. *Sov. Phys. Usp.*, **10**, 509–514.

Victor, J.D. (2005). Spike train metrics. *Curr. Opin. Neurobiol.*, **15**(5), 585–592.

Viitala, J., Korpimki, E., Palokangas, P., & Oivula, M. (1995). Attraction of kestrels to vole scent marks visible in ultraviolet light. *Nature*, **373**, 425–427.

Vincent, B.T. & Baddeley, R.J. (2003). Synaptic energy efficiency in retinal processing. *Vision Res.*, **43**, 1283–1290.

Vinje, W.E. & Gallant, J.L. (2000). Sparse coding and decorrelation in primary visual cortex during natural vision. *Science*, **287**(5456), 1273–1276.

Vodeley, L., Sineshchekov, O.A., Trivedi, V.D., Sasaki, J., Spudich, J.L., & Luecke, H. (2004). Anabaena sensory rhodopsin: a photochromic colour sensor at 2øa. *Science*, **306**(5700), 1390–1393.

Volgushev, M. & Eysel, U.T. (2000). Noise makes sense in neural computing. *Science*, **290**(5498), 1908–1909.

Vorobyev, M. (2003). Coloured oil droplets enhance colour discrimination. *Proc. Roy. Soc. Lond. B*, **270**, 1255–1261.

Wade, N.J. (1998). *A Natural History of Vision*. MIT Press, Cambridge, MA.

Walker, M.M., Diebel, C.E., Haugh, C.V., Pankhurst, P.M., Montgomery, J.C., & Green, C.R. (1997). Structure and function of the vertebrate magnetic sense. *Nature*, **390**, 371–376.

Walker, M.M., Dennis, T.E., & Kirschvink, J.L. (2002). The magnetic sense and its use in long-distance navigation by animals. *Curr. Opini. Neurobiol.*, **12**, 735–744.

Walker, R.G., Willingham, A.T., & Zuker, C.S. (2000). A *Drosophila* mechanosensory transduction channel. *Science*, **287**, 2229–2234.

Walsh, E.J., Wang, L.M., Armstrong, D.L., & McGee, J. (2003). Acoustic communication in *Panthera tigris*: a study of tiger vocalization and auditory receptivity. *Proc. of 145th Acoustical Society of America Meeting*. Architectural Engineering Faculty Publications, University of Nebraska, Lincoln, NE.

Ward, G. (1994). A contrast-based scale factor for luminance display. *Graphics Gems*, **IV**, 415–421.

Ward, G. (1998). Overcoming gamut and dynamic range limitations in digital images. *Proc. of IS&T 6th Color Imaging Conference*. Scottsdale, AZ, pp. 214–219.

Ward-Larson, G., Rushmeier, H., & Piatko, C. (1997). A visibility matching tone reproduction operator for high dynamic range scences. *IEEE Trans. Vis. Comp. Graphics*, **3**, 291–306.

Wässle, H. (1999). A patchwork of cones. *Nature*, **397**, 473–474.

Watts, D.J. (1999). *Small Worlds*. Princeton University Press, Princeton, NJ.

Weiss, G. (2001). Why is a soggy potato chip unappetizing? *Science*, **293**(5536), 1753–1754.

Westheimer, G. (1981). Visual hyperacuity. *Prog. Sensory Physiol*, **1**, 1–37.

Westheimer, G. (2008). Directional sensitivity of the retina: 75 years of Stiles–Crawford effect. *Proc. Roy. Soc. Lon. B*, **275**, 2777–2786.

Westheimer, G. & McKee, S.P. (1977). Perception of temporal order in adjacent visual stimuli. *Vision Res.*, **17**, 887–892.

Whalen, P.J., Kagan, J., Cook, R.G., Davis, F.C., Kim, H., Plis, S., McLaren, D.G. et al. (2004). Human amygdala responsivity to masked fearful eye whites. *Science*, **306**(5704), 2061.

Whitlock, K.E. (2004). Development of the nervus terminalis: origin and migration. *Microscopy Research and Technique*, **65**, 2–12.

Whitney, D., Goltz, H.C., Thomas, C.G., Gati, J.S., Menon, R.S., & Goodale, M.A. (2003). Flexible retinopy: motion-dependent position coding in the visual cortex. *Science*, **302**, 878–881.

Wickelgren, I. (2003). Tapping the mind. *Science*, **299**(5605), 496–499.

Wigner, E.P. (1932). On the quantum correlation for thermodynamic equilibrium. *Phys. Rev.*, **40**, 749–759.

Williams, C.M. & Kramer, E.M. (2010). The advantages of a tapered whisker. *PLOS One*, **5**(1), e8806.

Williams, D.R., Sekiguchi, N., Haake, W., Brainard, D., & Packer, O. (1991). The cost of trichromacy for spatial vision. Valberg, A. & Lee, B.B. (eds), *From Pigments to Perception*. Plenum Press, New York, pp. 11–22.

Williams, R. (2007). *Interview with V.S. Ramachandran*. ABC Science Show, 3 March.

Williams, W. (2001). Sound judgements. *Scientific American*, **285**(4), 12–13.

Willis, W.D., Jr. (1995). Neurobiology: cold, pain and the brain. *Nature*, **373**, 19–20.

Wilson, E.O. (1998). *Consilience*. Abacus, London.

Witten, I.H., Moffatt, A., & Bell, T.C. (1994). *Managing Gigabytes*. Van Nostrand Reinhold, New York.

Witten, H. & Frank, E. (2000). *Data Mining*. Morgan Kaufmann, San Francisco, CA.

Woodhouse, J.M. & Barlow, H.B. (1982). Spatial and temporal resolution and analysis. Barlow, H.B. & Mollon, J.D. (eds), *The Senses*. Cambridge University Press, Cambridge, UK, pp. 133–164.

Wysocki, L.E. & Ladich, F. (2005). Effects of noise exposure on click detection and the temporal resolution of the goldfish auditory system. *Hearing Res.*, **201**, 27–36.

Wysocki, L.E., Dittami, J.P., & Ladich, F. (2006). Ship noise and cortisol secretion in European freshwater fishes. *Biol. Convervation*, **128**, 501–508.

Yabuta, N.H., Sawatari, A., & Callaway, E.M. (2001). Two functional channels from primary visual cortex to dorsal visual cortical areas. *Science*, **292**(5515), 297–300.

Yakusheva, T.A., Shaikh, A.G., Green, A.M., Blazquez, P.M., Dickman, J.D., & Angelaki, D.E. (2007). Purkinje cells in posterior cerebellar vermis encode motion in an inertial reference frame. *Neuron*, **54**, 973–985.

Yantis, S. (2003). To see is to attend. *Science*, **299**(5603), 54–56.

Yoshida, A., Blanz, V., Myszkowski, K., & Seidel, H.-P. (2005). Perceptual evaluation of tone mapping operators with real-world scenes. *Proc. Spie.*, **5666**, 192–203.

Young, T. (1802). On the theory of light and colours. *Phil. Trans. R. Soc. Lond. B*, **92**, 20–71.

Yovel, Y., Franz, M.O., Stilz, P., & Schnitzler, H.-U. (2008). Plant classification from bat-like echolocation signals. *PLOS Comput. Biol.*, **4**, e1000032.

Zador, A.M. (2001). Synaptic connectivity and computation. *Nature Neuroscience*, **4**(12), 1157–1158.

Zandonella, C. (2001). Banish the blues. *New Scientist*, **2300**, 22.

Zanker, J.M. & Harris, J.P. (2002). On temporal hyperacuity in the human visual system. *Vision Res.*, **42**, 2499–2508.

Zatorre, R.J. & Krumhansl, C.L. (2002). Mental models and musical minds. *Science*, **298**(5601), 2138–2139.

Zeki, S. (1993). *A Vision of the Brain*. Blackwell Scientific Publications, Oxford, UK.

Zeki, S. (1999). *Inner Vision. An Exploration of Art and the Brain*. Oxford University Press.

Zeki, S. & Moutoussis, K. (1997). Temporal hierarchy of the visual perceptive systems in the mondrian world. *Proc. Roy. Soc. Lon. B*, **264**, 1415–1419.

Zhang, K. & Sejnowski, T.J. (2000). A universal scaling law between gray matter and white matter of cerebral cortex. *Proc. Nat. Acad. Sci.*, **97**, 105621–105626.

Zhao, G.Q., Zhang, Y., Hoon, M.A., Chandrashekar, J., Erlenbach, I., Ryba, N.J., & Zuker, C.S. (2003). The receptors for mammalian sweet and unami taste. *Cell*, **115**, 255–266.

Zhuo, Y., Zhou, T.G., Rao, H.Y., Wang, J.J., Meng, M., Chen, M., Zhou, C. *et al.* (2003). Contributions of the visual ventral pathway to long-range apparent motion. *Science*, **299**, 417–420.

Zou, D.-J., Feinstein, P., Rivers, A.L., Mathews, G.A., Kim, A., Greer, C.A., Mombaerts, P., & Firestein, S. (2004). Postnatal refinement of peripheral olfactory projections. *Science*, **304**(5679), 1976–1979.

Zrenner, E. (2002). Will retinal implants restore vision? *Science*, **295**(55577), 1022–1025.

Zwicker, E. & Fastl, H. (1990). *Psychoacoustics: Facts and Models*. Springer-Verlag, Berlin, Germany.

Index